D0583899

DIFFERENTIAL EQUATIONS, DYNAMICAL SYSTEMS, AND AN INTRODUCTION TO CHAOS

This is volume 60, 2ed in the PURE AND APPLIED MATHEMATICS Series
Founding Editors: Paul A. Smith and Samuel Eilenberg

DIFFERENTIAL EQUATIONS, DYNAMICAL SYSTEMS, AND AN INTRODUCTION TO CHAOS

Morris W. Hirsch

University of California, Berkeley

Stephen Smale

University of California, Berkeley

Robert L. Devaney

Boston University

ELSEVIER
ACADEMIC
PRESS

Amsterdam Boston Heidelberg London New York Oxford
Paris San Diego San Francisco Singapore Sydney Tokyo

Academic Press is an imprint of Elsevier

Senior Editor, Mathematics	Barbara Holland
Associate Editor	Tom Singer
Project Manager	Kyle Sarofeen
Marketing Manager	Linda Beattie
Production Services	Beth Callaway, Graphic World
Cover Design	Eric DeCicco
Copyediting	Graphic World
Composition	Cepha Imaging Pvt. Ltd.
Printer	Maple-Vail

This book is printed on acid-free paper. ⊚

Academic Press
An Imprint of Elsevier
525 B Street, Suite 1900, San Diego, California 92101-4495, USA
http://www.academicpress.com

Academic Press
An Imprint of Elsevier
84 Theobald's Road, London WC1X 8RR, UK
http://www.academicpress.com

Library of Congress Cataloging-in-Publication Data

Hirsch, Morris W., 1933-
  Differential equations, dynamical systems, and an introduction to chaos/Morris W. Hirsch, Stephen Smale, Robert L. Devaney.
    p. cm.
    Rev. ed. of: Differential equations, dynamical systems, and linear algebra/Morris W. Hirsch and Stephen Smale. 1974.
    Includes bibliographical references and index.
    ISBN-13: 978-0-12-349703-1   ISBN-10: 0-12-349703-5 (alk. paper)
    1. Differential equations. 2. Algebras, Linear. 3. Chaotic behavior in systems. I. Smale, Stephen, 1930- II. Devaney, Robert L., 1948- III. Hirsch, Morris W., 1933- Differential Equations, dynamical systems, and linear algebra. IV. Title.

  QA372.H67 2003
  515'.35--dc22

ISBN-13: 978-0-12-349703-1
ISBN-10: 0-12-349703-5   2003058255

PRINTED IN THE UNITED STATES OF AMERICA
06 07 08 9 8 7 6 5 4 3

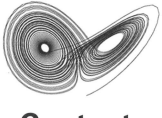

Contents

Preface

In the 30 years since the publication of the first edition of this book, much has changed in the field of mathematics known as dynamical systems. In the early 1970s, we had very little access to high-speed computers and computer graphics. The word *chaos* had never been used in a mathematical setting, and most of the interest in the theory of differential equations and dynamical systems was confined to a relatively small group of mathematicians.

Things have changed dramatically in the ensuing 3 decades. Computers are everywhere, and software packages that can be used to approximate solutions of differential equations and view the results graphically are widely available. As a consequence, the analysis of nonlinear systems of differential equations is much more accessible than it once was. The discovery of such complicated dynamical systems as the horseshoe map, homoclinic tangles, and the Lorenz system, and their mathematical analyses, convinced scientists that simple stable motions such as equilibria or periodic solutions were not always the most important behavior of solutions of differential equations. The beauty and relative accessibility of these chaotic phenomena motivated scientists and engineers in many disciplines to look more carefully at the important differential equations in their own fields. In many cases, they found chaotic behavior in these systems as well. Now dynamical systems phenomena appear in virtually every area of science, from the oscillating Belousov-Zhabotinsky reaction in chemistry to the chaotic Chua circuit in electrical engineering, from complicated motions in celestial mechanics to the bifurcations arising in ecological systems.

As a consequence, the audience for a text on differential equations and dynamical systems is considerably larger and more diverse than it was in the 1970s. We have accordingly made several major structural changes to this text, including the following:

1. The treatment of linear algebra has been scaled back. We have dispensed with the generalities involved with abstract vector spaces and normed linear spaces. We no longer include a complete proof of the reduction of all $n \times n$ matrices to canonical form. Rather we deal primarily with matrices no larger than 4×4.
2. We have included a detailed discussion of the chaotic behavior in the Lorenz attractor, the Shil'nikov system, and the double scroll attractor.
3. Many new applications are included; previous applications have been updated.
4. There are now several chapters dealing with discrete dynamical systems.
5. We deal primarily with systems that are C^∞, thereby simplifying many of the hypotheses of theorems.

The book consists of three main parts. The first part deals with linear systems of differential equations together with some first-order nonlinear equations. The second part of the book is the main part of the text: Here we concentrate on nonlinear systems, primarily two dimensional, as well as applications of these systems in a wide variety of fields. The third part deals with higher dimensional systems. Here we emphasize the types of chaotic behavior that do not occur in planar systems, as well as the principal means of studying such behavior, the reduction to a discrete dynamical system.

Writing a book for a diverse audience whose backgrounds vary greatly poses a significant challenge. We view this book as a text for a second course in differential equations that is aimed not only at mathematicians, but also at scientists and engineers who are seeking to develop sufficient mathematical skills to analyze the types of differential equations that arise in their disciplines. Many who come to this book will have strong backgrounds in linear algebra and real analysis, but others will have less exposure to these fields. To make this text accessible to both groups, we begin with a fairly gentle introduction to low-dimensional systems of differential equations. Much of this will be a review for readers with deeper backgrounds in differential equations, so we intersperse some new topics throughout the early part of the book for these readers.

For example, the first chapter deals with first-order equations. We begin this chapter with a discussion of linear differential equations and the logistic population model, topics that should be familiar to anyone who has a rudimentary acquaintance with differential equations. Beyond this review, we discuss the logistic model with harvesting, both constant and periodic. This allows us to introduce bifurcations at an early stage as well as to describe Poincaré maps and periodic solutions. These are topics that are not usually found in elementary differential equations courses, yet they are accessible to anyone with a background in multivariable calculus. Of course, readers with a limited background may wish to skip these specialized topics at first and concentrate on the more elementary material.

Chapters 2 through 6 deal with linear systems of differential equations. Again we begin slowly, with Chapters 2 and 3 dealing only with planar systems of differential equations and two-dimensional linear algebra. Chapters 5 and 6 introduce higher dimensional linear systems; however, our emphasis remains on three- and four-dimensional systems rather than completely general n-dimensional systems, though many of the techniques we describe extend easily to higher dimensions.

The core of the book lies in the second part. Here we turn our attention to nonlinear systems. Unlike linear systems, nonlinear systems present some serious theoretical difficulties such as existence and uniqueness of solutions, dependence of solutions on initial conditions and parameters, and the like. Rather than plunge immediately into these difficult theoretical questions, which require a solid background in real analysis, we simply state the important results in Chapter 7 and present a collection of examples that illustrate what these theorems say (and do not say). Proofs of all of these results are included in the final chapter of the book.

In the first few chapters in the nonlinear part of the book, we introduce such important techniques as linearization near equilibria, nullcline analysis, stability properties, limit sets, and bifurcation theory. In the latter half of this part, we apply these ideas to a variety of systems that arise in biology, electrical engineering, mechanics, and other fields.

Many of the chapters conclude with a section called "Exploration." These sections consist of a series of questions and numerical investigations dealing with a particular topic or application relevant to the preceding material. In each Exploration we give a brief introduction to the topic at hand and provide references for further reading about this subject. But we leave it to the reader to tackle the behavior of the resulting system using the material presented earlier. We often provide a series of introductory problems as well as hints as to how to proceed, but in many cases, a full analysis of the system could become a major research project. You will not find "answers in the back of the book" for these questions; in many cases nobody knows the complete answer. (Except, of course, you!)

The final part of the book is devoted to the complicated nonlinear behavior of higher dimensional systems known as chaotic behavior. We introduce these ideas via the famous Lorenz system of differential equations. As is often the case in dimensions three and higher, we reduce the problem of comprehending the complicated behavior of this differential equation to that of understanding the dynamics of a discrete dynamical system or iterated function. So we then take a detour into the world of discrete systems, discussing along the way how symbolic dynamics may be used to describe completely certain chaotic systems.

We then return to nonlinear differential equations to apply these techniques to other chaotic systems, including those that arise when homoclinic orbits are present.

We maintain a website at `math.bu.edu/hsd` devoted to issues regarding this text. Look here for errata, suggestions, and other topics of interest to teachers and students of differential equations. We welcome any contributions from readers at this site.

It is a special pleasure to thank Bard Ermentrout, John Guckenheimer, Tasso Kaper, Jerrold Marsden, and Gareth Roberts for their many fine comments about an earlier version of this edition. Thanks are especially due to Daniel Look and Richard Moeckel for a careful reading of the entire manuscript. Many of the phase plane drawings in this book were made using the excellent Mathematica package called DynPac: A Dynamical Systems Package for Mathematica written by Al Clark. See `www.me.rochester.edu/~clark/dynpac.html`. And, as always, Ki*ll*er Devaney digested the entire manuscript; all errors that remain are due to her.

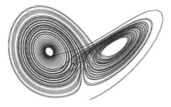

Acknowledgements

We would like to thank the following reviewers:

Bruce Peckham, University of Minnesota
Bard Ermentrout, University of Pittsburgh
Richard Moeckel, University of Minnesota
Jerry Marsden, CalTech
John Guckenheimer, Cornell University
Gareth Roberts, College of Holy Cross
Rick Moeckel, University of Minnesota
Hans Lindblad, University of California San Diego
Tom LoFaro, Gustavus Adolphus College
Daniel M. Look, Boston University

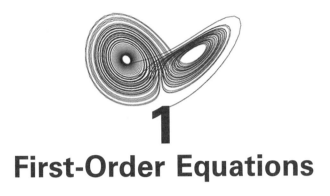

1
First-Order Equations

The purpose of this chapter is to develop some elementary yet important examples of first-order differential equations. These examples illustrate some of the basic ideas in the theory of ordinary differential equations in the simplest possible setting.

We anticipate that the first few examples in this chapter will be familiar to readers who have taken an introductory course in differential equations. Later examples, such as the logistic model with harvesting, are included to give the reader a taste of certain topics (bifurcations, periodic solutions, and Poincaré maps) that we will return to often throughout this book. In later chapters, our treatment of these topics will be much more systematic.

1.1 The Simplest Example

The differential equation familiar to all calculus students

$$\frac{dx}{dt} = ax$$

is the simplest differential equation. It is also one of the most important. First, what does it mean? Here $x = x(t)$ is an unknown real-valued function of a real variable t and dx/dt is its derivative (we will also use x' or $x'(t)$ for the derivative). Also, a is a parameter; for each value of a we have a

1

different differential equation. The equation tells us that for every value of t the relationship

$$x'(t) = ax(t)$$

is true.

The solutions of this equation are obtained from calculus: If k is any real number, then the function $x(t) = ke^{at}$ is a solution since

$$x'(t) = ake^{at} = ax(t).$$

Moreover, *there are no other solutions*. To see this, let $u(t)$ be any solution and compute the derivative of $u(t)e^{-at}$:

$$\frac{d}{dt}\left(u(t)e^{-at}\right) = u'(t)e^{-at} + u(t)(-ae^{-at})$$

$$= au(t)e^{-at} - au(t)e^{-at} = 0.$$

Therefore $u(t)e^{-at}$ is a constant k, so $u(t) = ke^{at}$. This proves our assertion. We have therefore found all possible solutions of this differential equation. We call the collection of all solutions of a differential equation the *general solution* of the equation.

The constant k appearing in this solution is completely determined if the value u_0 of a solution at a single point t_0 is specified. Suppose that a function $x(t)$ satisfying the differential equation is also required to satisfy $x(t_0) = u_0$. Then we must have $ke^{at_0} = u_0$, so that $k = u_0e^{-at_0}$. Thus we have determined k, and this equation therefore has a unique solution satisfying the specified *initial condition* $x(t_0) = u_0$. For simplicity, we often take $t_0 = 0$; then $k = u_0$. There is no loss of generality in taking $t_0 = 0$, for if $u(t)$ is a solution with $u(0) = u_0$, then the function $v(t) = u(t - t_0)$ is a solution with $v(t_0) = u_0$.

It is common to restate this in the form of an *initial value problem*:

$$x' = ax, \qquad x(0) = u_0.$$

A solution $x(t)$ of an initial value problem must not only solve the differential equation, but it must also take on the prescribed initial value u_0 at $t = 0$.

Note that there is a special solution of this differential equation when $k = 0$. This is the constant solution $x(t) \equiv 0$. A constant solution such as this is called an *equilibrium solution* or *equilibrium point* for the equation. Equilibria are often among the most important solutions of differential equations.

The constant a in the equation $x' = ax$ can be considered a parameter. If a changes, the equation changes and so do the solutions. Can we describe qualitatively the way the solutions change? The sign of a is crucial here:

1. If $a > 0$, $\lim_{t \to \infty} ke^{at}$ equals ∞ when $k > 0$, and equals $-\infty$ when $k < 0$;
2. If $a = 0$, $ke^{at} = $ constant;
3. If $a < 0$, $\lim_{t \to \infty} ke^{at} = 0$.

The qualitative behavior of solutions is vividly illustrated by sketching the graphs of solutions as in Figure 1.1. Note that the behavior of solutions is quite different when a is positive and negative. When $a > 0$, all nonzero solutions tend away from the equilibrium point at 0 as t increases, whereas when $a < 0$, solutions tend toward the equilibrium point. We say that the equilibrium point is a *source* when nearby solutions tend away from it. The equilibrium point is a *sink* when nearby solutions tend toward it.

We also describe solutions by drawing them on the *phase line*. Because the solution $x(t)$ is a function of time, we may view $x(t)$ as a particle moving along the real line. At the equilibrium point, the particle remains at rest (indicated

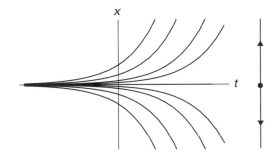

Figure 1.1 The solution graphs and phase line for $x' = ax$ for $a > 0$. Each graph represents a particular solution.

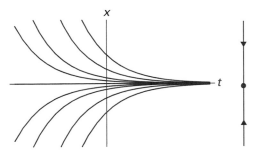

Figure 1.2 The solution graphs and phase line for $x' = ax$ for $a < 0$.

by a solid dot), while any other solution moves up or down the x-axis, as indicated by the arrows in Figure 1.1.

The equation $x' = ax$ is *stable* in a certain sense if $a \neq 0$. More precisely, if a is replaced by another constant b whose sign is the same as a, then the qualitative behavior of the solutions does not change. But if $a = 0$, the slightest change in a leads to a radical change in the behavior of solutions. We therefore say that we have a *bifurcation* at $a = 0$ in the one-parameter family of equations $x' = ax$.

1.2 The Logistic Population Model

The differential equation $x' = ax$ above can be considered a simplistic model of population growth when $a > 0$. The quantity $x(t)$ measures the population of some species at time t. The assumption that leads to the differential equation is that the rate of growth of the population (namely, dx/dt) is directly proportional to the size of the population. Of course, this naive assumption omits many circumstances that govern actual population growth, including, for example, the fact that actual populations cannot increase without bound.

To take this restriction into account, we can make the following further assumptions about the population model:

1. If the population is small, the growth rate is nearly directly proportional to the size of the population;
2. but if the population grows too large, the growth rate becomes negative.

One differential equation that satisfies these assumptions is the *logistic population growth model*. This differential equation is

$$x' = ax \left(1 - \frac{x}{N} \right).$$

Here a and N are positive parameters: a gives the rate of population growth when x is small, while N represents a sort of "ideal" population or "carrying capacity." Note that if x is small, the differential equation is essentially $x' = ax$ [since the term $1 - (x/N) \approx 1$], but if $x > N$, then $x' < 0$. Thus this simple equation satisfies the above assumptions. We should add here that there are many other differential equations that correspond to these assumptions; our choice is perhaps the simplest.

Without loss of generality we will assume that $N = 1$. That is, we will choose units so that the carrying capacity is exactly 1 unit of population, and $x(t)$ therefore represents the fraction of the ideal population present at time t.

Therefore the logistic equation reduces to

$$x' = f_a(x) = ax(1 - x).$$

This is an example of a first-order, autonomous, nonlinear differential equation. It is *first order* since only the first derivative of x appears in the equation. It is *autonomous* since the right-hand side of the equation depends on x alone, not on time t. And it is *nonlinear* since $f_a(x)$ is a nonlinear function of x. The previous example, $x' = ax$, is a first-order, autonomous, linear differential equation.

The solution of the logistic differential equation is easily found by the tried-and-true calculus method of separation and integration:

$$\int \frac{dx}{x(1 - x)} = \int a\, dt.$$

The method of partial fractions allows us to rewrite the left integral as

$$\int \left(\frac{1}{x} + \frac{1}{1 - x} \right) dx.$$

Integrating both sides and then solving for x yields

$$x(t) = \frac{Ke^{at}}{1 + Ke^{at}}$$

where K is the arbitrary constant that arises from integration. Evaluating this expression at $t = 0$ and solving for K gives

$$K = \frac{x(0)}{1 - x(0)}.$$

Using this, we may rewrite this solution as

$$\frac{x(0)e^{at}}{1 - x(0) + x(0)e^{at}}.$$

So this solution is valid for any initial population $x(0)$. When $x(0) = 1$, we have an equilibrium solution, since $x(t)$ reduces to $x(t) \equiv 1$. Similarly, $x(t) \equiv 0$ is also an equilibrium solution.

Thus we have "existence" of solutions for the logistic differential equation. We have no guarantee that these are all of the solutions to this equation at this stage; we will return to this issue when we discuss the existence and uniqueness problem for differential equations in Chapter 7.

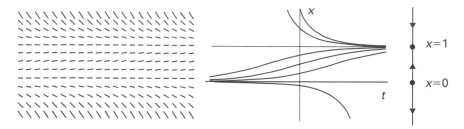

Figure 1.3   The slope field, solution graphs, and phase line for $x' = ax(1 - x)$.

To get a qualitative feeling for the behavior of solutions, we sketch the *slope field* for this equation. The right-hand side of the differential equation determines the slope of the graph of any solution at each time t. Hence we may plot little slope lines in the tx–plane as in Figure 1.3, with the slope of the line at (t, x) given by the quantity $ax(1 - x)$. Our solutions must therefore have graphs that are everywhere tangent to this slope field. From these graphs, we see immediately that, in agreement with our assumptions, all solutions for which $x(0) > 0$ tend to the ideal population $x(t) \equiv 1$. For $x(0) < 0$, solutions tend to $-\infty$, although these solutions are irrelevant in the context of a population model.

Note that we can also read this behavior from the graph of the function $f_a(x) = ax(1 - x)$. This graph, displayed in Figure 1.4, crosses the x-axis at the two points $x = 0$ and $x = 1$, so these represent our equilibrium points. When $0 < x < 1$, we have $f(x) > 0$. Hence slopes are positive at any (t, x) with $0 < x < 1$, and so solutions must increase in this region. When $x < 0$ or $x > 1$, we have $f(x) < 0$ and so solutions must decrease, as we see in both the solution graphs and the phase lines in Figure 1.3.

We may read off the fact that $x = 0$ is a source and $x = 1$ is a sink from the graph of f in similar fashion. Near 0, we have $f(x) > 0$ if $x > 0$, so slopes are positive and solutions increase, but if $x < 0$, then $f(x) < 0$, so slopes are negative and solutions decrease. Thus nearby solutions move away from 0 and so 0 is a source. Similarly, 1 is a sink.

We may also determine this information analytically. We have $f_a'(x) = a - 2ax$ so that $f_a'(0) = a > 0$ and $f_a'(1) = -a < 0$. Since $f_a'(0) > 0$, slopes must increase through the value 0 as x passes through 0. That is, slopes are negative below $x = 0$ and positive above $x = 0$. Hence solutions must tend away from $x = 0$. In similar fashion, $f_a'(1) < 0$ forces solutions to tend toward $x = 1$, making this equilibrium point a sink. We will encounter many such "derivative tests" like this that predict the qualitative behavior near equilibria in subsequent chapters.

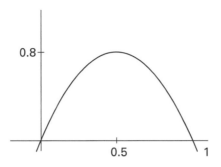

Figure 1.4 The graph of the function $f(x) = ax(1 - x)$ with $a = 3.2$.

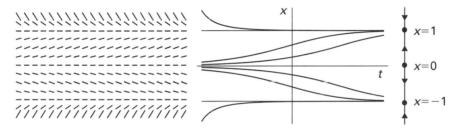

Figure 1.5 The slope field, solution graphs, and phase line for $x' = x - x^3$.

Example. As a further illustration of these qualitative ideas, consider the differential equation

$$x' = g(x) = x - x^3.$$

There are three equilibrium points, at $x = 0, \pm 1$. Since $g'(x) = 1 - 3x^2$, we have $g'(0) = 1$, so the equilibrium point 0 is a source. Also, $g'(\pm 1) = -2$, so the equilibrium points at ± 1 are both sinks. Between these equilibria, the sign of the slope field of this equation is nonzero. From this information we can immediately display the phase line, which is shown in Figure 1.5. ■

1.3 Constant Harvesting and Bifurcations

Now let's modify the logistic model to take into account harvesting of the population. Suppose that the population obeys the logistic assumptions with the

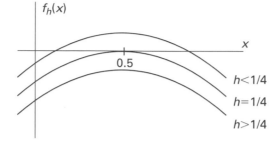

Figure 1.6   The graphs of the function
$f_h(x) = x(1 - x) - h$.

parameter $a = 1$, but is also harvested at the constant rate h. The differential equation becomes

$$x' = x(1 - x) - h$$

where $h \geq 0$ is a new parameter.

Rather than solving this equation explicitly (which can be done — see Exercise 6 at the end of this chapter), we use the graphs of the functions

$$f_h(x) = x(1 - x) - h$$

to "read off" the qualitative behavior of solutions. In Figure 1.6 we display the graph of f_h in three different cases: $0 < h < 1/4$, $h = 1/4$, and $h > 1/4$. It is straightforward to check that f_h has two roots when $0 \leq h < 1/4$, one root when $h = 1/4$, and no roots if $h > 1/4$, as illustrated in the graphs. As a consequence, the differential equation has two equilibrium points x_ℓ and x_r with $0 \leq x_\ell < x_r$ when $0 < h < 1/4$. It is also easy to check that $f'_h(x_\ell) > 0$, so that x_ℓ is a source, and $f'_h(x_r) < 0$ so that x_r is a sink.

As h passes through $h = 1/4$, we encounter another example of a bifurcation. The two equilibria x_ℓ and x_r coalesce as h increases through $1/4$ and then disappear when $h > 1/4$. Moreover, when $h > 1/4$, we have $f_h(x) < 0$ for all x. Mathematically, this means that all solutions of the differential equation decrease to $-\infty$ as time goes on.

We record this visually in the *bifurcation diagram*. In this diagram we plot the parameter h horizontally. Over each h-value we plot the corresponding phase line. The curve in this picture represents the equilibrium points for each value of h. This gives another view of the sink and source merging into a single equilibrium point and then disappearing as h passes through $1/4$ (see Figure 1.7).

Ecologically, this bifurcation corresponds to a disaster for the species under study. For rates of harvesting $1/4$ or lower, the population persists, provided

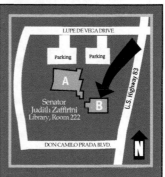

Deadlines

- ☐ Apply between **Jan. 1st and March 1st** and receive the MAXIMUM amount of financial aid.
- ☐ Apply between **March 2nd and May 1st** and receive financial aid for fall classes.
- ☐ Apply between **May 2nd and Aug. 15th** – the priority deadline for using financial aid to pay for Fall has passed. You will need to make payment arrangements to pay for your fall tuition and fees.

Student Information

- ☐ Full Legal Name
- ☐ FSA ID (If you don't have one, apply online at: www.fsaid.ed.gov)
- ☐ Date of Birth
- ☐ Social Security number
- ☐ Texas Driver's License number
- ☐ E-Mail Address
- ☐ Permanent Resident Alien Registration # (if applicable)

Parent Information

- ☐ Full Legal Name (Parent 1 and/or Parent 2)
- ☐ FSA ID (Parent 1 or Parent 2) (If Parent does not have one, apply online at: www. fsaid.ed.gov)
- ☐ Parent 1/Parent 2 Date of Birth
- ☐ Parent 1/Parent 2 Social Security number
- ☐ Parent 1/Parent 2 Date of Birth
- ☐ Parent 1/Parent 2 Social Security number
- ☐ Month and Year parents were married, separated, divorced, or widowed

Family Finance information for 2015

- ☐ W2 Forms (for students and both parents if applicable)
- ☐ 1040, 1040A, or 1040EZ Income Tax Forms (for student and both parents if applicable)
- ☐ Household size (# of people living with you; include yourself, parents, siblings, etc.)
- ☐ Total Social Security Benefits received (all family members)
- ☐ Total Disability received
- ☐ Total Welfare benefits received (TANF)
- ☐ Total Child Support received
- ☐ Total Child Support paid
- ☐ Total untaxed income or benefits not reported elsewhere (e.g., workers' compensation, untaxed income, etc.)

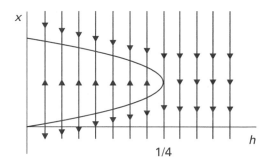

Figure 1.7   The bifurcation diagram for
$f_h(x) = x(1 - x) - h$.

the initial population is sufficiently large $(x(0) \geq x_\ell)$. But a very small change in the rate of harvesting when $h = 1/4$ leads to a major change in the fate of the population: At any rate of harvesting $h > 1/4$, the species becomes extinct.

This phenomenon highlights the importance of detecting bifurcations in families of differential equations, a procedure that we will encounter many times in later chapters. We should also mention that, despite the simplicity of this population model, the prediction that small changes in harvesting rates can lead to disastrous changes in population has been observed many times in real situations on earth.

Example. As another example of a bifurcation, consider the family of differential equations

$$x' = g_a(x) = x^2 - ax = x(x - a)$$

which depends on a parameter a. The equilibrium points are given by $x = 0$ and $x = a$. We compute $g_a'(0) = -a$, so 0 is a sink if $a > 0$ and a source if $a < 0$. Similarly, $g_a'(a) = a$, so $x = a$ is a sink if $a < 0$ and a source if $a > 0$. We have a bifurcation at $a = 0$ since there is only one equilibrium point when $a = 0$. Moreover, the equilibrium point at 0 changes from a source to a sink as a increases through 0. Similarly, the equilibrium at $x = a$ changes from a sink to a source as a passes through 0. The bifurcation diagram for this family is depicted in Figure 1.8. ∎

1.4 Periodic Harvesting and Periodic Solutions

Now let's change our assumptions about the logistic model to reflect the fact that harvesting does not always occur at a constant rate. For example,

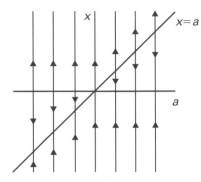

Figure 1.8   The bifurcation
diagram for $x' = x^2 - ax$.

populations of many species of fish are harvested at a higher rate in warmer
seasons than in colder months. So we assume that the population is harvested
at a periodic rate. One such model is then

$$x' = f(t, x) = ax(1 - x) - h(1 + \sin(2\pi t))$$

where again a and h are positive parameters. Thus the harvesting reaches a
maximum rate $-2h$ at time $t = \frac{1}{4} + n$ where n is an integer (representing the
year), and the harvesting reaches its minimum value 0 when $t = \frac{3}{4} + n$, exactly
one-half year later. Note that this differential equation now depends explicitly
on time; this is an example of a *nonautonomous* differential equation. As in
the autonomous case, a solution $x(t)$ of this equation must satisfy $x'(t) = f(t, x(t))$ for all t. Also, this differential equation is no longer separable, so we
cannot generate an analytic formula for its solution using the usual methods
from calculus. Thus we are forced to take a more qualitative approach.

To describe the fate of the population in this case, we first note that the right-
hand side of the differential equation is periodic with period 1 in the time
variable. That is, $f(t + 1, x) = f(t, x)$. This fact simplifies somewhat the
problem of finding solutions. Suppose that we know the solution of all initial
value problems, not for all times, but only for $0 \leq t \leq 1$. Then in fact we
know the solutions *for all time*. For example, suppose $x_1(t)$ is the solution that
is defined for $0 \leq t \leq 1$ and satisfies $x_1(0) = x_0$. Suppose that $x_2(t)$ is the
solution that satisfies $x_2(0) = x_1(1)$. Then we may extend the solution x_1 by
defining $x_1(t + 1) = x_2(t)$ for $0 \leq t \leq 1$. The extended function is a solution
since we have

$$x_1'(t + 1) = x_2'(t) = f(t, x_2(t))$$
$$= f(t + 1, x_1(t + 1)).$$

Figure 1.9 The slope field for $f(x) = x(1 - x) - h(1 + \sin(2\pi t))$.

Thus if we know the behavior of all solutions in the interval $0 \le t \le 1$, then we can extrapolate in similar fashion to all time intervals and thereby know the behavior of solutions for all time.

Secondly, suppose that we know the value at time $t = 1$ of the solution satisfying any initial condition $x(0) = x_0$. Then, to each such initial condition x_0, we can associate the value $x(1)$ of the solution $x(t)$ that satisfies $x(0) = x_0$. This gives us a function $p(x_0) = x(1)$. If we compose this function with itself, we derive the value of the solution through x_0 at time 2; that is, $p(p(x_0)) = x(2)$. If we compose this function with itself n times, then we can compute the value of the solution curve at time n and hence we know the fate of the solution curve.

The function p is called a *Poincaré map* for this differential equation. Having such a function allows us to move from the realm of continuous dynamical systems (differential equations) to the often easier-to-understand realm of discrete dynamical systems (iterated functions). For example, suppose that we know that $p(x_0) = x_0$ for some initial condition x_0. That is, x_0 is a *fixed point* for the function p. Then from our previous observations, we know that $x(n) = x_0$ for each integer n. Moreover, for each time t with $0 < t < 1$, we also have $x(t) = x(t + 1)$ and hence $x(t + n) = x(t)$ for each integer n. That is, the solution satisfying the initial condition $x(0) = x_0$ is a periodic function of t with period 1. Such solutions are called *periodic solutions* of the differential equation. In Figure 1.10, we have displayed several solutions of the logistic equation with periodic harvesting. Note that the solution satisfying the initial condition $x(0) = x_0$ is a periodic solution, and we have $x_0 = p(x_0) = p(p(x_0))\ldots$. Similarly, the solution satisfying the initial condition $x(0) = \hat{x}_0$ also appears to be a periodic solution, so we should have $p(\hat{x}_0) = \hat{x}_0$.

Unfortunately, it is usually the case that computing a Poincaré map for a differential equation is impossible, but for the logistic equation with periodic harvesting we get lucky.

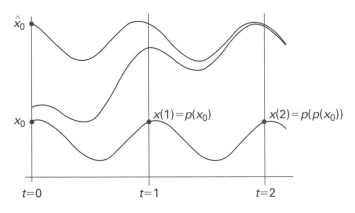

Figure 1.10   The Poincaré map for $x' = 5x(1 - x) -$
$0.8(1 + \sin(2\pi t))$.

1.5 Computing the Poincaré Map

Before computing the Poincaré map for this equation, we introduce some important terminology. To emphasize the dependence of a solution on the initial value x_0, we will denote the corresponding solution by $\phi(t, x_0)$. This function $\phi : \mathbb{R} \times \mathbb{R} \to \mathbb{R}$ is called the *flow* associated to the differential equation. If we hold the variable x_0 fixed, then the function

$$t \to \phi(t, x_0)$$

is just an alternative expression for the solution of the differential equation satisfying the initial condition x_0. Sometimes we write this function as $\phi_t(x_0)$.

Example. For our first example, $x' = ax$, the flow is given by

$$\phi(t, x_0) = x_0 e^{at}.$$

For the logistic equation (without harvesting), the flow is

$$\phi(t, x_0) = \frac{x(0)e^{at}}{1 - x(0) + x(0)e^{at}}.$$

Now we return to the logistic differential equation with periodic harvesting

$$x' = f(t, x) = ax(1 - x) - h(1 + \sin(2\pi t)).$$

The solution satisfying the initial condition $x(0) = x_0$ is given by $t \to \phi(t, x_0)$. While we do not have a formula for this expression, we do know that, by the fundamental theorem of calculus, this solution satisfies

$$\phi(t, x_0) = x_0 + \int_0^t f(s, \phi(s, x_0)) \, ds$$

since

$$\frac{\partial \phi}{\partial t}(t, x_0) = f(t, \phi(t, x_0))$$

and $\phi(0, x_0) = x_0$.

If we differentiate this solution with respect to x_0, we obtain, using the chain rule:

$$\frac{\partial \phi}{\partial x_0}(t, x_0) = 1 + \int_0^t \frac{\partial f}{\partial \phi}(s, \phi(s, x_0)) \cdot \frac{\partial \phi}{\partial x_0}(s, x_0) \, ds.$$

Now let

$$z(t) = \frac{\partial \phi}{\partial x_0}(t, x_0).$$

Note that

$$z(0) = \frac{\partial \phi}{\partial x_0}(0, x_0) = 1.$$

Differentiating z with respect to t, we find

$$z'(t) = \frac{\partial f}{\partial \phi}(t, \phi(t, x_0)) \cdot \frac{\partial \phi}{\partial x_0}(t, x_0)$$

$$= \frac{\partial f}{\partial \phi}(t, \phi(t, x_0)) \cdot z(t).$$

Again, we do not know $\phi(t, x_0)$ explicitly, but this equation does tell us that $z(t)$ solves the differential equation

$$z'(t) = \frac{\partial f}{\partial \phi}(t, \phi(t, x_0)) z(t)$$

with $z(0) = 1$. Consequently, via separation of variables, we may compute that the solution of this equation is

$$z(t) = \exp \int_0^t \frac{\partial f}{\partial \phi}(s, \phi(s, x_0)) \, ds$$

and so we find

$$\frac{\partial \phi}{\partial x_0}(1, x_0) = \exp \int_0^1 \frac{\partial f}{\partial x_0}(s, \phi(s, x_0)) \, ds.$$

Since $p(x_0) = \phi(1, x_0)$, we have determined the derivative $p'(x_0)$ of the Poincaré map. Note that $p'(x_0) > 0$. Therefore p is an increasing function.

Differentiating once more, we find

$$p''(x_0) = p'(x_0) \left(\int_0^1 \frac{\partial^2 f}{\partial x_0 \partial x_0}(s, \phi(s, x_0)) \cdot \exp \left(\int_0^s \frac{\partial f}{\partial x_0}(u, \phi(u, x_0)) \, du \right) ds \right),$$

which looks pretty intimidating. However, since

$$f(t, x_0) = a x_0 (1 - x_0) - h(1 + \sin(2\pi t)),$$

we have

$$\frac{\partial^2 f}{\partial x_0 \partial x_0} \equiv -2a.$$

Thus we know in addition that $p''(x_0) < 0$. Consequently, the graph of the Poincaré map is concave down. This implies that the graph of p can cross the diagonal line $y = x$ at most two times. That is, there can be at most two values of x for which $p(x) = x$. Therefore the Poincaré map has at most two fixed points. These fixed points yield periodic solutions of the original differential equation. These are solutions that satisfy $x(t+1) = x(t)$ for all t. Another way to say this is that the flow $\phi(t, x_0)$ is a periodic function in t with period 1 when the initial condition x_0 is one of the fixed points. We saw these two solutions in the particular case when $h = 0.8$ in Figure 1.10. In Figure 1.11, we again see two solutions that appear to be periodic. Note that one of these solutions appears to attract all nearby solutions, while the other appears to repel them. We will return to these concepts often and make them more precise later in the book.

Recall that the differential equation also depends on the harvesting parameter h. For small values of h there will be two fixed points such as shown in Figure 1.11. Differentiating f with respect to h, we find

$$\frac{\partial f}{\partial h}(t, x_0) = -(1 + \sin 2\pi t)$$

Hence $\partial f / \partial h < 0$ (except when $t = 3/4$). This implies that the slopes of the slope field lines at each point (t, x_0) decrease as h increases. As a consequence, the values of the Poincaré map also decrease as h increases. Hence there is a unique value h_* for which the Poincaré map has exactly one fixed point. For $h > h_*$, there are no fixed points for p and so $p(x_0) < x_0$ for all initial values. It then follows that the population again dies out. ∎

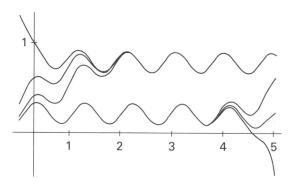

Figure 1.11 Several solutions of $x' =$ $5x(1 - x) - 0.8(1 + \sin(2\pi t))$.

1.6 Exploration: A Two-Parameter Family

Consider the family of differential equations

$$x' = f_{a,b}(x) = ax - x^3 - b$$

which depends on two parameters, a and b. The goal of this exploration is to combine all of the ideas in this chapter to put together a complete picture of the two-dimensional parameter plane (the ab–plane) for this differential equation. Feel free to use a computer to experiment with this differential equation at first, but then try to verify your observations rigorously.

1. First fix $a = 1$. Use the graph of $f_{1,b}$ to construct the bifurcation diagram for this family of differential equations depending on b.
2. Repeat the previous step for $a = 0$ and then for $a = -1$.
3. What does the bifurcation diagram look like for other values of a?
4. Now fix b and use the graph to construct the bifurcation diagram for this family, which this time depends on a.
5. In the ab–plane, sketch the regions where the corresponding differential equation has different numbers of equilibrium points, including a sketch of the boundary between these regions.
6. Describe, using phase lines and the graph of $f_{a,b}(x)$, the bifurcations that occur as the parameters pass through this boundary.
7. Describe in detail the bifurcations that occur at $a = b = 0$ as a and/or b vary.

8. Consider the differential equation $x' = x - x^3 - b\sin(2\pi t)$ where $|b|$ is small. What can you say about solutions of this equation? Are there any periodic solutions?

9. Experimentally, what happens as $|b|$ increases? Do you observe any bifurcations? Explain what you observe.

EXERCISES

1. Find the general solution of the differential equation $x' = ax + 3$ where a is a parameter. What are the equilibrium points for this equation? For which values of a are the equilibria sinks? For which are they sources?

2. For each of the following differential equations, find all equilibrium solutions and determine if they are sinks, sources, or neither. Also, sketch the phase line.

   (a) $x' = x^3 - 3x$

   (b) $x' = x^4 - x^2$

   (c) $x' = \cos x$

   (d) $x' = \sin^2 x$

   (e) $x' = |1 - x^2|$

3. Each of the following families of differential equations depends on a parameter a. Sketch the corresponding bifurcation diagrams.

   (a) $x' = x^2 - ax$

   (b) $x' = x^3 - ax$

   (c) $x' = x^3 - x + a$

4. Consider the function $f(x)$ whose graph is displayed in Figure 1.12.

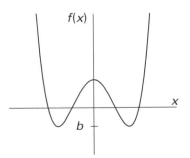

Figure 1.12   The graph of the function f.

(a) Sketch the phase line corresponding to the differential equation $x' = f(x)$.

(b) Let $g_a(x) = f(x) + a$. Sketch the bifurcation diagram corresponding to the family of differential equations $x' = g_a(x)$.

(c) Describe the different bifurcations that occur in this family.

5. Consider the family of differential equations

$$x' = ax + \sin x$$

where a is a parameter.

(a) Sketch the phase line when $a = 0$.

(b) Use the graphs of ax and $\sin x$ to determine the qualitative behavior of all of the bifurcations that occur as a increases from -1 to 1.

(c) Sketch the bifurcation diagram for this family of differential equations.

6. Find the general solution of the logistic differential equation with constant harvesting

$$x' = x(1 - x) - h$$

for all values of the parameter $h > 0$.

7. Consider the nonautonomous differential equation

$$x' = \begin{cases} x - 4 & \text{if } t < 5 \\ 2 - x & \text{if } t \geq 5 \end{cases}.$$

(a) Find a solution of this equation satisfying $x(0) = 4$. Describe the qualitative behavior of this solution.

(b) Find a solution of this equation satisfying $x(0) = 3$. Describe the qualitative behavior of this solution.

(c) Describe the qualitative behavior of any solution of this system as $t \to \infty$.

8. Consider a first-order linear equation of the form $x' = ax + f(t)$ where $a \in \mathbb{R}$. Let $y(t)$ be any solution of this equation. Prove that the general solution is $y(t) + c \exp(at)$ where $c \in \mathbb{R}$ is arbitrary.

9. Consider a first-order, linear, nonautonomous equation of the form $x'(t) = a(t)x$.

(a) Find a formula involving integrals for the solution of this system.

(b) Prove that your formula gives the general solution of this system.

10. Consider the differential equation $x' = x + \cos t$.

   (a) Find the general solution of this equation.

   (b) Prove that there is a unique periodic solution for this equation.

   (c) Compute the Poincaré map $p: \{t = 0\} \to \{t = 2\pi\}$ for this equation and use this to verify again that there is a unique periodic solution.

11. First-order differential equations need not have solutions that are defined for all times.

   (a) Find the general solution of the equation $x' = x^2$.

   (b) Discuss the domains over which each solution is defined.

   (c) Give an example of a differential equation for which the solution satisfying $x(0) = 0$ is defined only for $-1 < t < 1$.

12. First-order differential equations need not have unique solutions satisfying a given initial condition.

   (a) Prove that there are infinitely many different solutions of the differential equations $x' = x^{1/3}$ satisfying $x(0) = 0$.

   (b) Discuss the corresponding situation that occurs for $x' = x/t$, $x(0) = x_0$.

   (c) Discuss the situation that occurs for $x' = x/t^2$, $x(0) = 0$.

13. Let $x' = f(x)$ be an autonomous first-order differential equation with an equilibrium point at x_0.

   (a) Suppose $f'(x_0) = 0$. What can you say about the behavior of solutions near x_0? Give examples.

   (b) Suppose $f'(x_0) = 0$ and $f''(x_0) \neq 0$. What can you now say?

   (c) Suppose $f'(x_0) = f''(x_0) = 0$ but $f'''(x_0) \neq 0$. What can you now say?

14. Consider the first-order nonautonomous equation $x' = p(t)x$ where $p(t)$ is differentiable and periodic with period T. Prove that all solutions of this equation are periodic with period T if and only if

$$\int_0^T p(s)\, ds = 0.$$

15. Consider the differential equation $x' = f(t, x)$ where $f(t, x)$ is continuously differentiable in t and x. Suppose that

$$f(t + T, x) = f(t, x)$$

for all t. Suppose there are constants p, q such that

$$f(t, p) > 0, \ f(t, q) < 0$$

for all t. Prove that there is a periodic solution $x(t)$ for this equation with $p < x(0) < q$.

16. Consider the differential equation $x' = x^2 - 1 - \cos(t)$. What can be said about the existence of periodic solutions for this equation?

2

Planar Linear Systems

In this chapter we begin the study of *systems of differential equations*. A system of differential equations is a collection of n interrelated differential equations of the form

$$x_1' = f_1(t, x_1, x_2, \ldots, x_n)$$
$$x_2' = f_2(t, x_1, x_2, \ldots, x_n)$$
$$\vdots$$
$$x_n' = f_n(t, x_1, x_2, \ldots, x_n).$$

Here the functions f_j are real-valued functions of the $n + 1$ variables $x_1, x_2, \ldots,$ x_n, and t. Unless otherwise specified, we will always assume that the f_j are C^∞ functions. This means that the partial derivatives of all orders of the f_j exist and are continuous.

To simplify notation, we will use vector notation:

$$X = \begin{pmatrix} x_1 \\ \vdots \\ x_n \end{pmatrix}.$$

We often write the vector X as $(x_1, \ldots, x_n)$ to save space.

Our system may then be written more concisely as

$$X' = F(t, X)$$

where

$$F(t, X) = \begin{pmatrix} f_1(t, x_1, \ldots, x_n) \\ \vdots \\ f_n(t, x_1, \ldots, x_n) \end{pmatrix}.$$

A solution of this system is then a function of the form $X(t) = (x_1(t), \ldots, x_n(t))$ that satisfies the equation, so that

$$X'(t) = F(t, X(t))$$

where $X'(t) = (x'_1(t), \ldots, x'_n(t))$. Of course, at this stage, we have no guarantee that there is such a solution, but we will begin to discuss this complicated question in Section 2.7.

The system of equations is called *autonomous* if none of the f_j depends on t, so the system becomes $X' = F(X)$. For most of the rest of this book we will be concerned with autonomous systems.

In analogy with first-order differential equations, a vector X_0 for which $F(X_0) = 0$ is called an *equilibrium point* for the system. An equilibrium point corresponds to a constant solution $X(t) \equiv X_0$ of the system as before.

Just to set some notation once and for all, we will always denote real variables by lowercase letters such as x, y, x_1, x_2, t, and so forth. Real-valued functions will also be written in lowercase such as $f(x, y)$ or $f_1(x_1, \ldots, x_n, t)$. We will reserve capital letters for vectors such as $X = (x_1, \ldots, x_n)$, or for vector-valued functions such as

$$F(x, y) = (f(x, y), g(x, y))$$

or

$$H(x_1, \ldots, x_n) = \begin{pmatrix} h_1(x_1, \ldots, x_n) \\ \vdots \\ h_n(x_1, \ldots, x_n) \end{pmatrix}.$$

We will denote n-dimensional Euclidean space by $\mathbb{R}^n$, so that $\mathbb{R}^n$ consists of all vectors of the form $X = (x_1, \ldots, x_n)$.

2.1 Second-Order Differential Equations

Many of the most important differential equations encountered in science and engineering are second-order differential equations. These are differential equations of the form

$$x'' = f(t, x, x').$$

Important examples of second-order equations include Newton's equation

$$mx'' = f(x),$$

the equation for an RLC circuit in electrical engineering

$$LCx'' + RCx' + x = v(t),$$

and the mainstay of most elementary differential equations courses, the forced harmonic oscillator

$$mx'' + bx' + kx = f(t).$$

We will discuss these and more complicated relatives of these equations at length as we go along. First, however, we note that these equations are a special subclass of two-dimensional systems of differential equations that are defined by simply introducing a second variable $y = x'$.

For example, consider a second-order constant coefficient equation of the form

$$x'' + ax' + bx = 0.$$

If we let $y = x'$, then we may rewrite this equation as a system of first-order equations

$$x' = y$$
$$y' = -bx - ay.$$

Any second-order equation can be handled in a similar manner. Thus, for the remainder of this book, we will deal primarily with systems of equations.

2.2 Planar Systems

For the remainder of this chapter we will deal with autonomous systems in $\mathbb{R}^2$, which we will write in the form

$$x' = f(x, y)$$
$$y' = g(x, y)$$

thus eliminating the annoying subscripts on the functions and variables. As above, we often use the abbreviated notation $X' = F(X)$ where $X = (x, y)$ and $F(X) = F(x, y) = (f(x, y), g(x, y))$.

In analogy with the slope fields of Chapter 1, we regard the right-hand side of this equation as defining a *vector field* on $\mathbb{R}^2$. That is, we think of $F(x, y)$ as representing a vector whose x- and y-components are $f(x, y)$ and $g(x, y)$, respectively. We visualize this vector as being based at the point (x, y). For example, the vector field associated to the system

$$x' = y$$
$$y' = -x$$

is displayed in Figure 2.1. Note that, in this case, many of the vectors overlap, making the pattern difficult to visualize. For this reason, we always draw a *direction field* instead, which consists of scaled versions of the vectors.

A solution of this system should now be thought of as a parameterized curve in the plane of the form $(x(t), y(t))$ such that, for each t, the tangent vector at the point $(x(t), y(t))$ is $F(x(t), y(t))$. That is, the solution curve $(x(t), y(t))$ winds its way through the plane always tangent to the given vector $F(x(t), y(t))$ based at $(x(t), y(t))$.

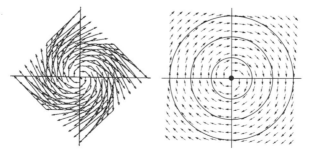

Figure 2.1 The vector field, direction field, and several solutions for the system $x' = y$, $y' = -x$.

Example. The curve

$$\begin{pmatrix} x(t) \\ y(t) \end{pmatrix} = \begin{pmatrix} a \sin t \\ a \cos t \end{pmatrix}$$

for any $a \in \mathbb{R}$ is a solution of the system

$$x' = y$$
$$y' = -x$$

since

$$x'(t) = a \cos t = y(t)$$
$$y'(t) = -a \sin t = -x(t)$$

as required by the differential equation. These curves define circles of radius $|a|$ in the plane that are traversed in the clockwise direction as t increases. When $a = 0$, the solutions are the constant functions $x(t) \equiv 0 \equiv y(t)$. ■

Note that this example is equivalent to the second-order differential equation $x'' = -x$ by simply introducing the second variable $y = x'$. This is an example of a *linear* second-order differential equation, which, in more general form, can be written

$$a(t)x'' + b(t)x' + c(t)x = f(t).$$

An important special case of this is the linear, *constant coefficient* equation

$$ax'' + bx' + cx = f(t),$$

which we write as a system as

$$x' = y$$
$$y' = -\frac{c}{a}x - \frac{b}{a}y + \frac{f(t)}{a}.$$

An even more special case is the *homogeneous* equation in which $f(t) \equiv 0$.

Example. One of the simplest yet most important second-order, linear, constant coefficient differential equations is the equation for a *harmonic oscillator*. This equation models the motion of a mass attached to a spring. The spring is attached to a vertical wall and the mass is allowed to slide along a horizontal track. We let x denote the displacement of the mass from its natural resting place (with $x > 0$ if the spring is stretched and $x < 0$ if the spring is compressed). Therefore the velocity of the moving mass is $x'(t)$ and the acceleration is $x''(t)$. The spring exerts a restorative force proportional to $x(t)$. In addition there is a frictional force proportional to $x'(t)$ in the direction opposite to that of the motion. There are three parameters for this system: m denotes the mass of the oscillator, $b \geq 0$ is the *damping constant*, and $k > 0$ is the *spring constant*. Newton's law states that the force acting on the oscillator is equal to mass times acceleration. Therefore the differential equation for the damped harmonic oscillator is

$$mx'' + bx' + kx = 0.$$

If $b = 0$, the oscillator is said to be *undamped*; otherwise, we have a *damped* harmonic oscillator. This is an example of a second-order, linear, constant coefficient, homogeneous differential equation. As a system, the harmonic oscillator equation becomes

$$x' = y$$
$$y' = -\frac{k}{m}x - \frac{b}{m}y.$$

More generally, the motion of the mass-spring system can be subjected to an external force (such as moving the vertical wall back and forth periodically). Such an external force usually depends only on time, not position, so we have a more general forced harmonic oscillator system

$$mx'' + bx' + kx = f(t)$$

where $f(t)$ represents the external force. This is now a nonautonomous, second-order, linear equation.   ∎

2.3 Preliminaries from Algebra

Before proceeding further with systems of differential equations, we need to recall some elementary facts regarding systems of algebraic equations. We will often encounter simultaneous equations of the form

$$ax + by = \alpha$$
$$cx + dy = \beta$$

where the values of a, b, c, and d as well as α and β are given. In matrix form, we may write this equation as

$$\begin{pmatrix} a & b \\ c & d \end{pmatrix} \begin{pmatrix} x \\ y \end{pmatrix} = \begin{pmatrix} \alpha \\ \beta \end{pmatrix}.$$

We denote by A the 2×2 coefficient matrix

$$A = \begin{pmatrix} a & b \\ c & d \end{pmatrix}.$$

This system of equations is easy to solve, assuming that there is a solution. There is a unique solution of these equations if and only if the *determinant* of A is nonzero. Recall that this determinant is the quantity given by

$$\det A = ad - bc.$$

If $\det A = 0$, we may or may not have solutions, but if there is a solution, then in fact there must be infinitely many solutions.

In the special case where $\alpha = \beta = 0$, we always have infinitely many solutions of

$$A \begin{pmatrix} x \\ y \end{pmatrix} = \begin{pmatrix} 0 \\ 0 \end{pmatrix}$$

when $\det A = 0$. Indeed, if the coefficient a of A is nonzero, we have $x = -(b/a)y$ and so

$$-c \left(\frac{b}{a} \right) y + dy = 0.$$

Thus $(ad - bc)y = 0$. Since $\det A = 0$, the solutions of the equation assume the form $(-(b/a)y, y)$ where y is arbitrary. This says that every solution lies on a straight line through the origin in the plane. A similar line of solutions occurs as long as at least one of the entries of A is nonzero. We will not worry too much about the case where all entries of A are 0; in fact, we will completely ignore it.

Let V and W be vectors in the plane. We say that V and W are *linearly independent* if V and W do not lie along the same straight line through the origin. The vectors V and W are *linearly dependent* if either V or W is the zero vector or if both lie on the same line through the origin.

A geometric criterion for two vectors in the plane to be linearly independent is that they do not point in the same or opposite directions. That is, two nonzero vectors V and W are linearly independent if and only if $V \neq \lambda W$ for any real number λ. An equivalent algebraic criterion for linear independence is as follows:

Proposition. *Suppose $V = (v_1, v_2)$ and $W = (w_1, w_2)$. Then V and W are linearly independent if and only if*

$$\det \begin{pmatrix} v_1 & w_1 \\ v_2 & w_2 \end{pmatrix} \neq 0.$$

For a proof, see Exercise 11 at the end of this chapter. ∎

Whenever we have a pair of linearly independent vectors V and W, we may always write any vector $Z \in \mathbb{R}^2$ in a unique way as a *linear combination* of V and W. That is, we may always find a pair of real numbers α and β such that

$$Z = \alpha V + \beta W.$$

Moreover, α and β are unique. To see this, suppose $Z = (z_1, z_2)$. Then we must solve the equations

$$z_1 = \alpha v_1 + \beta w_1$$
$$z_2 = \alpha v_2 + \beta w_2$$

where the v_i, w_i, and z_i are known. But this system has a unique solution (α, β) since

$$\det \begin{pmatrix} v_1 & w_1 \\ v_2 & w_2 \end{pmatrix} \neq 0.$$

The linearly independent vectors V and W are said to define a *basis* for $\mathbb{R}^2$. Any vector Z has unique "coordinates" relative to V and W. These coordinates are the pair (α, β) for which $Z = \alpha V + \beta W$.

Example. The unit vectors $E_1 = (1, 0)$ and $E_2 = (0, 1)$ obviously form a basis called the *standard basis* of $\mathbb{R}^2$. The coordinates of Z in this basis are just the "usual" Cartesian coordinates (x, y) of Z. ∎

Example. The vectors $V_1 = (1, 1)$ and $V_2 = (-1, 1)$ also form a basis of $\mathbb{R}^2$. Relative to this basis, the coordinates of E_1 are $(1/2, -1/2)$ and those of E_2 are

$(1/2, 1/2)$, because

$$\begin{pmatrix} 1 \\ 0 \end{pmatrix} = \frac{1}{2}\begin{pmatrix} 1 \\ 1 \end{pmatrix} - \frac{1}{2}\begin{pmatrix} -1 \\ 1 \end{pmatrix}$$

$$\begin{pmatrix} 0 \\ 1 \end{pmatrix} = \frac{1}{2}\begin{pmatrix} 1 \\ 1 \end{pmatrix} + \frac{1}{2}\begin{pmatrix} -1 \\ 1 \end{pmatrix}$$

These "changes of coordinates" will become important later. ∎

Example. The vectors $V_1 = (1, 1)$ and $V_2 = (-1, -1)$ do not form a basis of $\mathbb{R}^2$ since these vectors are collinear. Any linear combination of these vectors is of the form

$$\alpha V_1 + \beta V_2 = \begin{pmatrix} \alpha - \beta \\ \alpha - \beta \end{pmatrix},$$

which yields only vectors on the straight line through the origin, V_1, and V_2. ∎

2.4 Planar Linear Systems

We now further restrict our attention to the most important class of planar systems of differential equations, namely, linear systems. In the autonomous case, these systems assume the simple form

$$x' = ax + by$$
$$y' = cx + dy$$

where a, b, c, and d are constants. We may abbreviate this system by using the *coefficient matrix A* where

$$A = \begin{pmatrix} a & b \\ c & d \end{pmatrix}.$$

Then the linear system may be written as

$$X' = AX.$$

Note that the origin is always an equilibrium point for a linear system. To find other equilibria, we must solve the linear system of algebraic equations

$$ax + by = 0$$
$$cx + dy = 0.$$

This system has a nonzero solution if and only if $\det A = 0$. As we saw previously, if $\det A = 0$, then there is a straight line through the origin on which each point is an equilibrium. Thus we have:

Proposition.   *The planar linear system $X' = AX$ has*

1. *A unique equilibrium point $(0, 0)$ if $\det A \neq 0$.*
2. *A straight line of equilibrium points if $\det A = 0$ (and A is not the 0 matrix).*   ∎

2.5 Eigenvalues and Eigenvectors

Now we turn to the question of finding nonequilibrium solutions of the linear system $X' = AX$. The key observation here is this: Suppose V_0 is a nonzero vector for which we have $AV_0 = \lambda V_0$ where $\lambda \in \mathbb{R}$. Then the function

$$X(t) = e^{\lambda t} V_0$$

is a solution of the system. To see this, we compute

$$X'(t) = \lambda e^{\lambda t} V_0$$
$$= e^{\lambda t}(\lambda V_0)$$
$$= e^{\lambda t}(AV_0)$$
$$= A(e^{\lambda t} V_0)$$
$$= AX(t)$$

so $X(t)$ does indeed solve the system of equations. Such a vector V_0 and its associated scalar have names:

Definition

A nonzero vector V_0 is called an *eigenvector* of A if $AV_0 = \lambda V_0$ for some λ. The constant λ is called an *eigenvalue* of A.

As we observed, there is an important relationship between eigenvalues, eigenvectors, and solutions of systems of differential equations:

Theorem.   *Suppose that V_0 is an eigenvector for the matrix A with associated eigenvalue λ. Then the function $X(t) = e^{\lambda t} V_0$ is a solution of the system $X' = AX$.*   ∎

Note that if V_0 is an eigenvector for A with eigenvalue λ, then any nonzero scalar multiple of V_0 is also an eigenvector for A with eigenvalue λ. Indeed, if $AV_0 = \lambda V_0$, then

$$A(\alpha V_0) = \alpha AV_0 = \lambda(\alpha V_0)$$

for any nonzero constant α.

Example. Consider

$$A = \begin{pmatrix} 1 & 3 \\ 1 & -1 \end{pmatrix}.$$

Then A has an eigenvector $V_0 = (3, 1)$ with associated eigenvalue $\lambda = 2$ since

$$\begin{pmatrix} 1 & 3 \\ 1 & -1 \end{pmatrix} \begin{pmatrix} 3 \\ 1 \end{pmatrix} = \begin{pmatrix} 6 \\ 2 \end{pmatrix} = 2 \begin{pmatrix} 3 \\ 1 \end{pmatrix}. \qquad ■$$

Similarly, $V_1 = (1, -1)$ is an eigenvector with associated eigenvalue $\lambda = -2$. Thus, for the system

$$X' = \begin{pmatrix} 1 & 3 \\ 1 & -1 \end{pmatrix} X$$

we now know three solutions: the equilibrium solution at the origin together with

$$X_1(t) = e^{2t} \begin{pmatrix} 3 \\ 1 \end{pmatrix} \quad \text{and} \quad X_2(t) = e^{-2t} \begin{pmatrix} 1 \\ -1 \end{pmatrix}.$$

We will see that we can use these solutions to generate *all* solutions of this system in a moment, but first we address the question of how to find eigenvectors and eigenvalues.

To produce an eigenvector $V = (x, y)$, we must find a nonzero solution (x, y) of the equation

$$A \begin{pmatrix} x \\ y \end{pmatrix} = \lambda \begin{pmatrix} x \\ y \end{pmatrix}.$$

Note that there are three unknowns in this system of equations: the two components of V as well as λ. Let I denote the 2×2 identity matrix

$$I = \begin{pmatrix} 1 & 0 \\ 0 & 1 \end{pmatrix}.$$

Then we may rewrite the equation in the form

$$(A - \lambda I)V = 0,$$

where 0 denotes the vector $(0, 0)$.

Now $A - \lambda I$ is just a 2×2 matrix (having entries involving the variable λ), so this linear system of equations has nonzero solutions if and only if $\det (A - \lambda I) = 0$, as we saw previously. But this equation is just a quadratic equation in λ, whose roots are therefore easy to find. This equation will appear over and over in the sequel; it is called the *characteristic equation*. As a function of λ, we call $\det(A - \lambda I)$ the *characteristic polynomial*. Thus the strategy to generate eigenvectors is first to find the roots of the characteristic equation. This yields the eigenvalues. Then we use each of these eigenvalues to generate in turn an associated eigenvector.

Example. We return to the matrix

$$A = \begin{pmatrix} 1 & 3 \\ 1 & -1 \end{pmatrix}.$$

We have

$$A - \lambda I = \begin{pmatrix} 1 - \lambda & 3 \\ 1 & -1 - \lambda \end{pmatrix}.$$

So the characteristic equation is

$$\det(A - \lambda I) = (1 - \lambda)(-1 - \lambda) - 3 = 0.$$

Simplifying, we find

$$\lambda^2 - 4 = 0,$$

which yields the two eigenvalues $\lambda = \pm 2$. Then, for $\lambda = 2$, we next solve the equation

$$(A - 2I) \begin{pmatrix} x \\ y \end{pmatrix} = \begin{pmatrix} 0 \\ 0 \end{pmatrix}.$$

In component form, this reduces to the system of equations

$$(1 - 2)x + 3y = 0$$
$$x + (-1 - 2)y = 0$$

or $-x + 3y = 0$, because these equations are redundant. Thus any vector of the form $(3y, y)$ with $y \neq 0$ is an eigenvector associated to $\lambda = 2$. In similar fashion, any vector of the form $(y, -y)$ with $y \neq 0$ is an eigenvector associated to $\lambda = -2$. ∎

Of course, the astute reader will notice that there is more to the story of eigenvalues, eigenvectors, and solutions of differential equations than what we have described previously. For example, the roots of the characteristic equation may be complex, or they may be repeated real numbers. We will handle all of these cases shortly, but first we return to the problem of solving linear systems.

2.6 Solving Linear Systems

As we saw in the example in the previous section, if we find two real roots λ_1 and λ_2 (with $\lambda_1 \neq \lambda_2$) of the characteristic equation, then we may generate a pair of solutions of the system of differential equations of the form $X_i(t) = e^{\lambda_i t} V_i$ where V_i is the eigenvector associated to λ_i. Note that each of these solutions is a *straight-line solution*. Indeed, we have $X_i(0) = V_i$, which is a nonzero point in the plane. For each t, $e^{\lambda_i t} V_i$ is a scalar multiple of V_i and so lies on the straight ray emanating from the origin and passing through V_i. Note that, if $\lambda_i > 0$, then

$$\lim_{t \to \infty} |X_i(t)| = \infty$$

and

$$\lim_{t \to -\infty} X_i(t) = (0, 0).$$

The magnitude of the solution $X_i(t)$ increases monotonically to ∞ along the ray through V_i as t increases, and $X_i(t)$ tends to the origin along this ray in backward time. The exact opposite situation occurs if $\lambda_i < 0$, whereas, if $\lambda_i = 0$, the solution $X_i(t)$ is the constant solution $X_i(t) = V_i$ for all t.

So how do we find all solutions of the system given this pair of special solutions? The answer is now easy and important. Suppose we have two distinct real eigenvalues λ_1 and λ_2 with eigenvectors V_1 and V_2. Then V_1 and V_2 are linearly independent, as is easily checked (see Exercise 14). Thus V_1 and V_2 form a basis of $\mathbb{R}^2$, so, given any point $Z_0 \in \mathbb{R}^2$, we may find a unique pair of real numbers α and β for which

$$\alpha V_1 + \beta V_2 = Z_0.$$

Now consider the function $Z(t) = \alpha X_1(t) + \beta X_2(t)$ where the $X_i(t)$ are the straight-line solutions previously. We claim that $Z(t)$ is a solution of $X' = AX$. To see this we compute

$$Z'(t) = \alpha X_1'(t) + \beta X_2'(t)$$
$$= \alpha AX_1(t) + \beta AX_2(t)$$
$$= A(\alpha X_1(t) + \beta X_2(t)).$$
$$= AZ(t)$$

This last step follows from the linearity of matrix multiplication (see Exercise 13). Hence we have shown that $Z'(t) = AZ(t)$, so $Z(t)$ is a solution. Moreover, $Z(t)$ is a solution that satisfies $Z(0) = Z_0$. Finally, we claim that $Z(t)$ is the unique solution of $X' = AX$ that satisfies $Z(0) = Z_0$. Just as in Chapter 1, we suppose that $Y(t)$ is another such solution with $Y(0) = Z_0$. Then we may write

$$Y(t) = \zeta(t)V_1 + \mu(t)V_2$$

with $\zeta(0) = \alpha, \mu(0) = \beta$. Hence

$$AY(t) = Y'(t) = \zeta'(t)V_1 + \mu'(t)V_2.$$

But

$$AY(t) = \zeta(t)AV_1 + \mu(t)AV_2$$
$$= \lambda_1 \zeta(t)V_1 + \lambda_2 \mu(t)V_2.$$

Therefore we have

$$\zeta'(t) = \lambda_1 \zeta(t)$$
$$\mu'(t) = \lambda_2 \mu(t)$$

with $\zeta(0) = \alpha, \mu(0) = \beta$. As we saw in Chapter 1, it follows that

$$\zeta(t) = \alpha e^{\lambda_1 t}, \mu(t) = \beta e^{\lambda_2 t}$$

so that $Y(t)$ is indeed equal to $Z(t)$.

As a consequence, we have now found the unique solution to the system $X' = AX$ that satisfies $X(0) = Z_0$ for any $Z_0 \in \mathbb{R}^2$. The collection of all such solutions is called the *general solution* of $X' = AX$. That is, the general solution is the collection of solutions of $X' = AX$ that features a unique solution of the initial value problem $X(0) = Z_0$ for each $Z_0 \in \mathbb{R}^2$.

We therefore have shown the following:

Theorem. *Suppose A has a pair of real eigenvalues $\lambda_1 \neq \lambda_2$ and associated eigenvectors V_1 and V_2. Then the general solution of the linear system $X' = AX$ is given by*

$$X(t) = \alpha e^{\lambda_1 t} V_1 + \beta e^{\lambda_2 t} V_2.$$ ∎

Example. Consider the second-order differential equation

$$x'' + 3x' + 2x = 0.$$

This is a specific case of the damped harmonic oscillator discussed earlier, where the mass is 1, the spring constant is 2, and the damping constant is 3. As a system, this equation may be rewritten:

$$X' = \begin{pmatrix} 0 & 1 \\ -2 & -3 \end{pmatrix} X = AX.$$

The characteristic equation is

$$\lambda^2 + 3\lambda + 2 = (\lambda + 2)(\lambda + 1) = 0,$$

so the system has eigenvalues -1 and -2. The eigenvector corresponding to the eigenvalue -1 is given by solving the equation

$$(A + I) \begin{pmatrix} x \\ y \end{pmatrix} = \begin{pmatrix} 0 \\ 0 \end{pmatrix}.$$

In component form this equation becomes

$$x + y = 0$$
$$-2x - 2y = 0.$$

Hence, one eigenvector associated to the eigenvalue -1 is $(1, -1)$. In similar fashion we compute that an eigenvector associated to the eigenvalue -2 is $(1, -2)$. Note that these two eigenvectors are linearly independent. Therefore, by the previous theorem, the general solution of this system is

$$X(t) = \alpha e^{-t} \begin{pmatrix} 1 \\ -1 \end{pmatrix} + \beta e^{-2t} \begin{pmatrix} 1 \\ -2 \end{pmatrix}.$$

That is, the position of the mass is given by the first component of the solution

$$x(t) = \alpha e^{-t} + \beta e^{-2t}$$

and the velocity is given by the second component

$$y(t) = x'(t) = -\alpha e^{-t} - 2\beta e^{-2t}.$$ ∎

2.7 The Linearity Principle

The theorem discussed in the previous section is a very special case of the fundamental theorem for n-dimensional linear systems, which we shall prove in Section 6.1 of Chapter 6. For the two-dimensional version of this result, note that if $X' = AX$ is a planar linear system for which $Y_1(t)$ and $Y_2(t)$ are both solutions, then just as before, the function $\alpha Y_1(t) + \beta Y_2(t)$ is also a solution of this system. We do not need real and distinct eigenvalues to prove this. This fact is known as the *linearity principle*. More importantly, if the initial conditions $Y_1(0)$ and $Y_2(0)$ are linearly independent vectors, then these vectors form a basis of $\mathbb{R}^2$. Hence, given any vector $X_0 \in \mathbb{R}^2$, we may determine constants α and β such that $X_0 = \alpha Y_1(0) + \beta Y_2(0)$. Then the linearity principle tells us that the solution $X(t)$ satisfying the initial condition $X(0) = X_0$ is given by $X(t) = \alpha Y_1(t) + \beta Y_2(t)$. Hence we have produced a solution of the system that solves any given initial value problem. The existence and uniqueness theorem for linear systems in Chapter 6 will show that this solution is also unique. This important result may then be summarized:

Theorem. *Let $X' = AX$ be a planar system. Suppose that $Y_1(t)$ and $Y_2(t)$ are solutions of this system, and that the vectors $Y_1(0)$ and $Y_2(0)$ are linearly independent. Then*

$$X(t) = \alpha Y_1(t) + \beta Y_2(t)$$

is the unique solution of this system that satisfies $X(0) = \alpha Y_1(0) + \beta Y_2(0)$. ∎

EXERCISES

1. Find the eigenvalues and eigenvectors of each of the following 2×2 matrices:

(a) $\begin{pmatrix} 3 & 1 \\ 1 & 3 \end{pmatrix}$ (b) $\begin{pmatrix} 2 & 1 \\ 1 & 1 \end{pmatrix}$

(c) $\begin{pmatrix} a & b \\ 0 & c \end{pmatrix}$ (d) $\begin{pmatrix} 1 & 3 \\ \sqrt{2} & 3\sqrt{2} \end{pmatrix}$

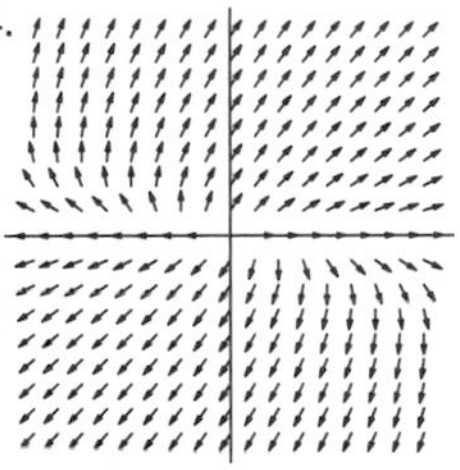

Figure 2.2   Match these direction fields with
the systems in Exercise 2.

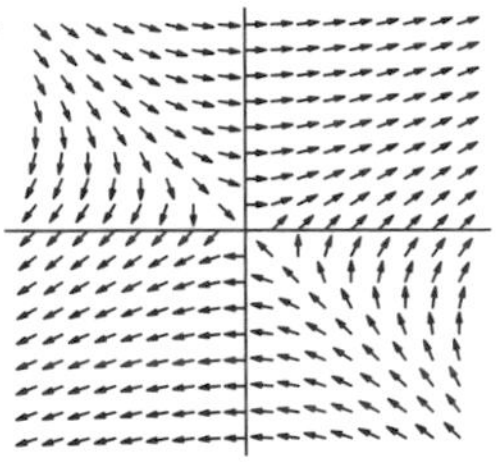

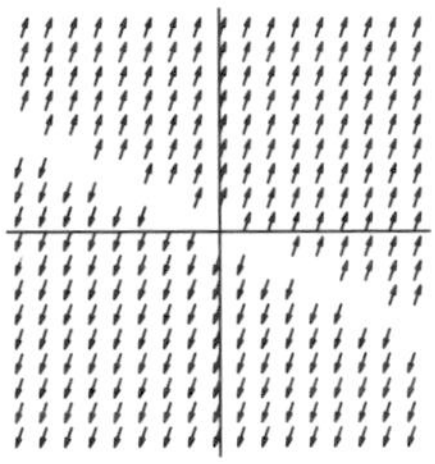

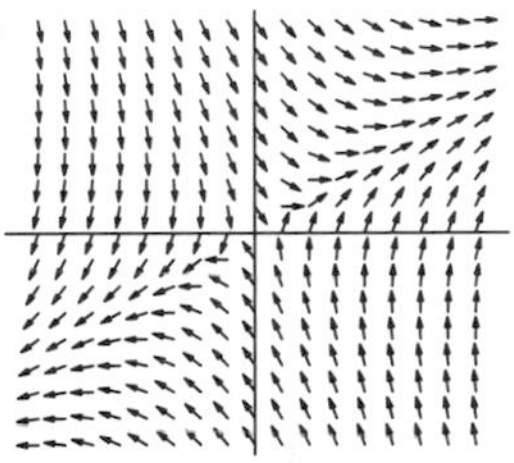

2. Find the general solution of each of the following linear systems:

$$\text{(a) } X' = \begin{pmatrix} 1 & 2 \\ 0 & 3 \end{pmatrix} X \quad \text{(b) } X' = \begin{pmatrix} 1 & 2 \\ 3 & 6 \end{pmatrix} X$$

$$\text{(c) } X' = \begin{pmatrix} 1 & 2 \\ 1 & 0 \end{pmatrix} X \quad \text{(d) } X' = \begin{pmatrix} 1 & 2 \\ 3 & -3 \end{pmatrix} X$$

3. In Figure 2.2, you see four direction fields. Match each of these direction
fields with one of the systems in the previous question.

4. Find the general solution of the system

$$X' = \begin{pmatrix} a & b \\ c & a \end{pmatrix} X$$

where $bc > 0$.

5. Find the general solution of the system

$$X' = \begin{pmatrix} 0 & 0 \\ 0 & 0 \end{pmatrix} X.$$

6. For the harmonic oscillator system $x'' + bx' + kx = 0$, find all values
of b and k for which this system has real, distinct eigenvalues. Find the

general solution of this system in these cases. Find the solution of the system that satisfies the initial condition $(0, 1)$. Describe the motion of the mass in this particular case.

7. Consider the 2×2 matrix

$$A = \begin{pmatrix} a & 1 \\ 0 & 1 \end{pmatrix}.$$

Find the value a_0 of the parameter a for which A has repeated real eigenvalues. What happens to the eigenvectors of this matrix as a approaches a_0?

8. Describe all possible 2×2 matrices whose eigenvalues are 0 and 1.

9. Give an example of a linear system for which (e^{-t}, α) is a solution for every constant α. Sketch the direction field for this system. What is the general solution of this system.

10. Give an example of a system of differential equations for which $(t, 1)$ is a solution. Sketch the direction field for this system. What is the general solution of this system?

11. Prove that two vectors $V = (v_1, v_2)$ and $W = (w_1, w_2)$ are linearly independent if and only if

$$\det \begin{pmatrix} v_1 & w_1 \\ v_2 & w_2 \end{pmatrix} \neq 0.$$

12. Prove that if λ, μ are real eigenvalues of a 2×2 matrix, then any nonzero column of the matrix $A - \lambda I$ is an eigenvector for μ.

13. Let A be a 2×2 matrix and V_1 and V_2 vectors in $\mathbb{R}^2$. Prove that $A(\alpha V_1 + \beta V_2) = \alpha A V_1 + \beta A V_2$.

14. Prove that the eigenvectors of a 2×2 matrix corresponding to distinct real eigenvalues are always linearly independent.

3

Phase Portraits for Planar Systems

Given the linearity principle from the previous chapter, we may now compute the general solution of any planar system. There is a seemingly endless number of distinct cases, but we will see that these represent in the simplest possible form nearly all of the types of solutions we will encounter in the higher dimensional case.

3.1 Real Distinct Eigenvalues

Consider $X' = AX$ and suppose that A has two real eigenvalues $\lambda_1 < \lambda_2$. Assuming for the moment that $\lambda_i \neq 0$, there are three cases to consider:

1. $\lambda_1 < 0 < \lambda_2$;
2. $\lambda_1 < \lambda_2 < 0$;
3. $0 < \lambda_1 < \lambda_2$.

We give a specific example of each case; any system that falls into any one of these three categories may be handled in a similar manner, as we show later.

Example. (Saddle) First consider the simple system $X' = AX$ where

$$A = \begin{pmatrix} \lambda_1 & 0 \\ 0 & \lambda_2 \end{pmatrix}$$

with $\lambda_1 < 0 < \lambda_2$. This can be solved immediately since the system decouples into two unrelated first-order equations:

$$x' = \lambda_1 x$$
$$y' = \lambda_2 y.$$

We already know how to solve these equations, but, having in mind what comes later, let's find the eigenvalues and eigenvectors. The characteristic equation is

$$(\lambda - \lambda_1)(\lambda - \lambda_2) = 0$$

so λ_1 and λ_2 are the eigenvalues. An eigenvector corresponding to λ_1 is $(1, 0)$ and to λ_2 is $(0, 1)$. Hence we find the general solution

$$X(t) = \alpha e^{\lambda_1 t} \begin{pmatrix} 1 \\ 0 \end{pmatrix} + \beta e^{\lambda_2 t} \begin{pmatrix} 0 \\ 1 \end{pmatrix}.$$

Since $\lambda_1 < 0$, the straight-line solutions of the form $\alpha e^{\lambda_1 t}(1, 0)$ lie on the x-axis and tend to $(0, 0)$ as $t \to \infty$. This axis is called the *stable line*. Since $\lambda_2 > 0$, the solutions $\beta e^{\lambda_2 t}(0, 1)$ lie on the y-axis and tend away from $(0, 0)$ as $t \to \infty$; this axis is the *unstable line*. All other solutions (with $\alpha, \beta \neq 0$) tend to ∞ in the direction of the unstable line, as $t \to \infty$, since $X(t)$ comes closer and closer to $(0, \beta e^{\lambda_2 t})$ as t increases. In backward time, these solutions tend to ∞ in the direction of the stable line. ■

In Figure 3.1 we have plotted the *phase portrait* of this system. The phase portrait is a picture of a collection of representative solution curves of the

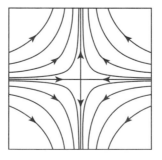

Figure 3.1 Saddle phase portrait for $x' = -x$, $y' = y$.

system in $\mathbb{R}^2$, which we call the *phase plane*. The equilibrium point of a system of this type (eigenvalues satisfying $\lambda_1 < 0 < \lambda_2$) is called a *saddle*.

For a slightly more complicated example of this type, consider $X' = AX$ where

$$A = \begin{pmatrix} 1 & 3 \\ 1 & -1 \end{pmatrix}.$$

As we saw in Chapter 2, the eigenvalues of A are ± 2. The eigenvector associated to $\lambda = 2$ is the vector $(3, 1)$; the eigenvector associated to $\lambda = -2$ is $(1, -1)$. Hence we have an unstable line that contains straight-line solutions of the form

$$X_1(t) = \alpha e^{2t} \begin{pmatrix} 3 \\ 1 \end{pmatrix},$$

each of which tends away from the origin as $t \to \infty$. The stable line contains the straight-line solutions

$$X_2(t) = \beta e^{-2t} \begin{pmatrix} 1 \\ -1 \end{pmatrix},$$

which tend toward the origin as $t \to \infty$. By the linearity principle, any other solution assumes the form

$$X(t) = \alpha e^{2t} \begin{pmatrix} 3 \\ 1 \end{pmatrix} + \beta e^{-2t} \begin{pmatrix} 1 \\ -1 \end{pmatrix}$$

for some α, β. Note that, if $\alpha \neq 0$, as $t \to \infty$, we have

$$X(t) \sim \alpha e^{2t} \begin{pmatrix} 3 \\ 1 \end{pmatrix} = X_1(t)$$

whereas, if $\beta \neq 0$, as $t \to -\infty$,

$$X(t) \sim \beta e^{-2t} \begin{pmatrix} 1 \\ -1 \end{pmatrix} = X_2(t).$$

Thus, as time increases, the typical solution approaches $X_1(t)$ while, as time decreases, this solution tends toward $X_2(t)$, just as in the previous case. Figure 3.2 displays this phase portrait.

In the general case where A has a positive and negative eigenvalue, we always find a similar stable and unstable line on which solutions tend toward or away

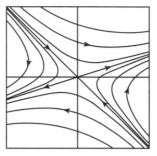

Figure 3.2   Saddle phase
portrait for $x' = x + 3y,$
$y' = x - y.$

from the origin. All other solutions approach the unstable line as $t \to \infty$, and
tend toward the stable line as $t \to -\infty$.

Example. (Sink) Now consider the case $X' = AX$ where

$$A = \begin{pmatrix} \lambda_1 & 0 \\ 0 & \lambda_2 \end{pmatrix}$$

but $\lambda_1 < \lambda_2 < 0$. As above we find two straight-line solutions and then the
general solution:

$$X(t) = \alpha e^{\lambda_1 t} \begin{pmatrix} 1 \\ 0 \end{pmatrix} + \beta e^{\lambda_2 t} \begin{pmatrix} 0 \\ 1 \end{pmatrix}.$$

Unlike the saddle case, now all solutions tend to $(0, 0)$ as $t \to \infty$. The question
is: How do they approach the origin? To answer this, we compute the slope
dy/dx of a solution with $\beta \neq 0$. We write

$$x(t) = \alpha e^{\lambda_1 t}$$

$$y(t) = \beta e^{\lambda_2 t}$$

and compute

$$\frac{dy}{dx} = \frac{dy/dt}{dx/dt} = \frac{\lambda_2 \beta e^{\lambda_2 t}}{\lambda_1 \alpha e^{\lambda_1 t}} = \frac{\lambda_2 \beta}{\lambda_1 \alpha} e^{(\lambda_2 - \lambda_1)t}.$$

Since $\lambda_2 - \lambda_1 > 0$, it follows that these slopes approach $\pm \infty$ (provided $\beta \neq 0$).
Thus these solutions tend to the origin tangentially to the y-axis.   ∎

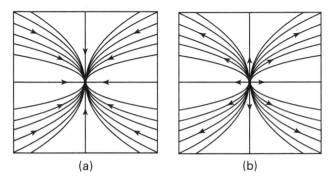

(a)   (b)

Figure 3.3   Phase portraits for (a) a sink and
(b) a source.

Since $\lambda_1 < \lambda_2 < 0$, we call λ_1 the stronger eigenvalue and λ_2 the weaker eigenvalue. The reason for this in this particular case is that the x-coordinates of solutions tend to 0 much more quickly than the y-coordinates. This accounts for why solutions (except those on the line corresponding to the λ_1 eigenvector) tend to "hug" the straight-line solution corresponding to the weaker eigenvalue as they approach the origin.

The phase portrait for this system is displayed in Figure 3.3a. In this case the equilibrium point is called a *sink*.

More generally, if the system has eigenvalues $\lambda_1 < \lambda_2 < 0$ with eigenvectors (u_1, u_2) and (v_1, v_2), respectively, then the general solution is

$$\alpha e^{\lambda_1 t}\begin{pmatrix} u_1 \\ u_2 \end{pmatrix} + \beta e^{\lambda_2 t}\begin{pmatrix} v_1 \\ v_2 \end{pmatrix}.$$

The slope of this solution is given by

$$\frac{dy}{dx} = \frac{\lambda_1 \alpha e^{\lambda_1 t} u_2 + \lambda_2 \beta e^{\lambda_2 t} v_2}{\lambda_1 \alpha e^{\lambda_1 t} u_1 + \lambda_2 \beta e^{\lambda_2 t} v_1}$$

$$= \left(\frac{\lambda_1 \alpha e^{\lambda_1 t} u_2 + \lambda_2 \beta e^{\lambda_2 t} v_2}{\lambda_1 \alpha e^{\lambda_1 t} u_1 + \lambda_2 \beta e^{\lambda_2 t} v_1} \right) \frac{e^{-\lambda_2 t}}{e^{-\lambda_2 t}}$$

$$= \frac{\lambda_1 \alpha e^{(\lambda_1 - \lambda_2)t} u_2 + \lambda_2 \beta v_2}{\lambda_1 \alpha e^{(\lambda_1 - \lambda_2)t} u_1 + \lambda_2 \beta v_1},$$

which tends to the slope v_2/v_1 of the λ_2 eigenvector, unless we have $\beta = 0$. If $\beta = 0$, our solution is the straight-line solution corresponding to the eigenvalue λ_1. Hence all solutions (except those on the straight line corresponding to the stronger eigenvalue) tend to the origin tangentially to the straight-line solution corresponding to the weaker eigenvalue in this case as well.

Example. (Source) When the matrix

$$A = \begin{pmatrix} \lambda_1 & 0 \\ 0 & \lambda_2 \end{pmatrix}$$

satisfies $0 < \lambda_2 < \lambda_1$, our vector field may be regarded as the negative of the previous example. The general solution and phase portrait remain the same, except that all solutions now tend away from $(0, 0)$ along the same paths. See Figure 3.3b.   ∎

Now one may argue that we are presenting examples here that are much too simple. While this is true, we will soon see that any system of differential equations whose matrix has real distinct eigenvalues can be manipulated into the above special forms by changing coordinates.

Finally, a special case occurs if one of the eigenvalues is equal to 0. As we have seen, there is a straight-line of equilibrium points in this case. If the other eigenvalue λ is nonzero, then the sign of λ determines whether the other solutions tend toward or away from these equilibria (see Exercises 10 and 11 at the end of this chapter).

3.2 Complex Eigenvalues

It may happen that the roots of the characteristic polynomial are complex numbers. In analogy with the real case, we call these roots *complex eigenvalues*. When the matrix A has complex eigenvalues, we no longer have straight line solutions. However, we can still derive the general solution as before by using a few tricks involving complex numbers and functions. The following examples indicate the general procedure.

Example. (Center) Consider $X' = AX$ with

$$A = \begin{pmatrix} 0 & \beta \\ -\beta & 0 \end{pmatrix}$$

and $\beta \neq 0$. The characteristic polynomial is $\lambda^2 + \beta^2 = 0$, so the eigenvalues are now the imaginary numbers $\pm i\beta$. Without worrying about the resulting complex vectors, we react just as before to find the eigenvector corresponding to $\lambda = i\beta$. We therefore solve

$$\begin{pmatrix} -i\beta & \beta \\ -\beta & -i\beta \end{pmatrix} \begin{pmatrix} x \\ y \end{pmatrix} = \begin{pmatrix} 0 \\ 0 \end{pmatrix}$$

or $i\beta x = \beta y$, since the second equation is redundant. Thus we find a complex eigenvector $(1, i)$, and so the function

$$X(t) = e^{i\beta t} \begin{pmatrix} 1 \\ i \end{pmatrix}$$

is a complex solution of $X' = AX$.

Now in general it is not polite to hand someone a complex solution to a real system of differential equations, but we can remedy this with the help of Euler's formula

$$e^{i\beta t} = \cos \beta t + i \sin \beta t.$$

Using this fact, we rewrite the solution as

$$X(t) = \begin{pmatrix} \cos \beta t + i \sin \beta t \\ i(\cos \beta t + i \sin \beta t) \end{pmatrix} = \begin{pmatrix} \cos \beta t + i \sin \beta t \\ -\sin \beta t + i \cos \beta t \end{pmatrix}.$$

Better yet, by breaking $X(t)$ into its real and imaginary parts, we have

$$X(t) = X_{\text{Re}}(t) + iX_{\text{Im}}(t)$$

where

$$X_{\text{Re}}(t) = \begin{pmatrix} \cos \beta t \\ -\sin \beta t \end{pmatrix}, \quad X_{\text{Im}}(t) = \begin{pmatrix} \sin \beta t \\ \cos \beta t \end{pmatrix}.$$

But now we see that both $X_{\text{Re}}(t)$ and $X_{\text{Im}}(t)$ are (real!) solutions of the original system. To see this, we simply check

$$X'_{\text{Re}}(t) + iX'_{\text{Im}}(t) = X'(t)$$
$$= AX(t)$$
$$= A(X_{\text{Re}}(t) + iX_{\text{Im}}(t))$$
$$= AX_{\text{Re}} + iAX_{\text{Im}}(t).$$

Equating the real and imaginary parts of this equation yields $X'_{\text{Re}} = AX_{\text{Re}}$ and $X'_{\text{Im}} = AX_{\text{Im}}$, which shows that both are indeed solutions. Moreover, since

$$X_{\text{Re}}(0) = \begin{pmatrix} 1 \\ 0 \end{pmatrix}, \quad X_{\text{Im}}(0) = \begin{pmatrix} 0 \\ 1 \end{pmatrix},$$

the linear combination of these solutions

$$X(t) = c_1 X_{\text{Re}}(t) + c_2 X_{\text{Im}}(t)$$

where c_1 and c_2 are arbitrary constants provides a solution to any initial value problem.

We claim that this is the general solution of this equation. To prove this, we need to show that these are the only solutions of this equation. Suppose that this is not the case. Let

$$Y(t) = \begin{pmatrix} u(t) \\ v(t) \end{pmatrix}$$

be another solution. Consider the complex function $f(t) = (u(t) + iv(t))e^{i\beta t}$. Differentiating this expression and using the fact that $Y(t)$ is a solution of the equation yields $f'(t) = 0$. Hence $u(t) + iv(t)$ is a complex constant times $e^{-i\beta t}$. From this it follows directly that $Y(t)$ is a linear combination of $X_{\text{Re}}(t)$ and $X_{\text{Im}}(t)$.

Note that each of these solutions is a periodic function with period $2\pi/\beta$. Indeed, the phase portrait shows that all solutions lie on circles centered at the origin. These circles are traversed in the clockwise direction if $\beta > 0$, counterclockwise if $\beta < 0$. See Figure 3.4. This type of system is called a *center*. ∎

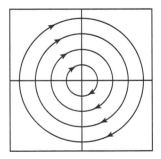

Figure 3.4 Phase portrait for a center.

Example. (Spiral Sink, Spiral Source) More generally, consider $X' = AX$ where

$$A = \begin{pmatrix} \alpha & \beta \\ -\beta & \alpha \end{pmatrix}$$

and $\alpha, \beta \neq 0$. The characteristic equation is now $\lambda^2 - 2\alpha\lambda + \alpha^2 + \beta^2$, so the eigenvalues are $\lambda = \alpha \pm i\beta$. An eigenvector associated to $\alpha + i\beta$ is determined by the equation

$$(\alpha - (\alpha + i\beta))x + \beta y = 0.$$

Thus $(1, i)$ is again an eigenvector. Hence we have complex solutions of the form

$$
\begin{aligned}
X(t) &= e^{(\alpha+i\beta)t} \begin{pmatrix} 1 \\ i \end{pmatrix} \\
&= e^{\alpha t} \begin{pmatrix} \cos\beta t \\ -\sin\beta t \end{pmatrix} + ie^{\alpha t} \begin{pmatrix} \sin\beta t \\ \cos\beta t \end{pmatrix} \\
&= X_{\text{Re}}(t) + iX_{\text{Im}}(t).
\end{aligned}
$$

As above, both $X_{\text{Re}}(t)$ and $X_{\text{Im}}(t)$ yield real solutions of the system whose initial conditions are linearly independent. Thus we find the general solution

$$X(t) = c_1 e^{\alpha t} \begin{pmatrix} \cos\beta t \\ -\sin\beta t \end{pmatrix} + c_2 e^{\alpha t} \begin{pmatrix} \sin\beta t \\ \cos\beta t \end{pmatrix}.$$

Without the term $e^{\alpha t}$, these solutions would wind periodically around circles centered at the origin. The $e^{\alpha t}$ term converts solutions into spirals that either spiral into the origin (when $\alpha < 0$) or away from the origin ($\alpha > 0$). In these cases the equilibrium point is called a *spiral sink* or *spiral source*, respectively. See Figure 3.5.   ■

3.3 Repeated Eigenvalues

The only remaining cases occur when A has repeated real eigenvalues. One simple case occurs when A is a diagonal matrix of the form

$$A = \begin{pmatrix} \lambda & 0 \\ 0 & \lambda \end{pmatrix}.$$

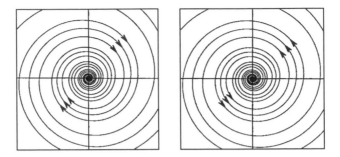

Figure 3.5   Phase portraits for a spiral sink and a spiral source.

The eigenvalues of A are both equal to λ. In this case every nonzero vector is an eigenvector since

$$AV = \lambda V$$

for any $V \in \mathbb{R}^2$. Hence solutions are of the form

$$X(t) = \alpha e^{\lambda t} V.$$

Each such solution lies on a straight line through $(0, 0)$ and either tends to $(0, 0)$ (if $\lambda < 0$) or away from $(0, 0)$ (if $\lambda > 0$). So this is an easy case.

A more interesting case occurs when

$$A = \begin{pmatrix} \lambda & 1 \\ 0 & \lambda \end{pmatrix}.$$

Again both eigenvalues are equal to λ, but now there is only one linearly independent eigenvector given by $(1, 0)$. Hence we have one straight-line solution

$$X_1(t) = \alpha e^{\lambda t} \begin{pmatrix} 1 \\ 0 \end{pmatrix}.$$

To find other solutions, note that the system can be written

$$x' = \lambda x + y$$
$$y' = \lambda y.$$

Thus, if $y \neq 0$, we must have

$$y(t) = \beta e^{\lambda t}.$$

Therefore the differential equation for $x(t)$ reads

$$x' = \lambda x + \beta e^{\lambda t}.$$

This is a nonautonomous, first-order differential equation for $x(t)$. One might first expect solutions of the form $e^{\lambda t}$, but the nonautonomous term is also in this form. As you perhaps saw in calculus, the best option is to guess a solution of the form

$$x(t) = \alpha e^{\lambda t} + \mu t e^{\lambda t}$$

for some constants α and μ. This technique is often called "the method of undetermined coefficients." Inserting this guess into the differential equation shows that $\mu = \beta$ while α is arbitrary. Hence the solution of the system may be written

$$\alpha e^{\lambda t} \begin{pmatrix} 1 \\ 0 \end{pmatrix} + \beta e^{\lambda t} \begin{pmatrix} t \\ 1 \end{pmatrix}.$$

This is in fact the general solution (see Exercise 12).

Note that, if $\lambda < 0$, each term in this solution tends to 0 as $t \to \infty$. This is clear for the $\alpha e^{\lambda t}$ and $\beta e^{\lambda t}$ terms. For the term $\beta t e^{\lambda t}$ this is an immediate consequence of l'Hôpital's rule. Hence all solutions tend to $(0,0)$ as $t \to \infty$. When $\lambda > 0$, all solutions tend away from $(0,0)$. See Figure 3.6. In fact, solutions tend toward or away from the origin in a direction tangent to the eigenvector $(1,0)$ (see Exercise 7).

3.4 Changing Coordinates

Despite differences in the associated phase portraits, we really have dealt with only three types of matrices in these past three sections:

$$\begin{pmatrix} \lambda & 0 \\ 0 & \mu \end{pmatrix}, \quad \begin{pmatrix} \alpha & \beta \\ -\beta & \alpha \end{pmatrix}, \quad \begin{pmatrix} \lambda & 1 \\ 0 & \lambda \end{pmatrix},$$

where λ may equal μ in the first case.

Any 2×2 matrix that is in one of these three forms is said to be in *canonical form*. Systems in this form may seem rather special, but they are not. Given

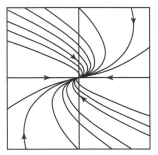

Figure 3.6   Phase
portrait for a system with
repeated negative
eigenvalues.

any linear system $X' = AX$, we can always "change coordinates" so that the new system's coefficient matrix is in canonical form and hence easily solved. Here is how to do this.

A *linear map* (or *linear transformation*) on $\mathbb{R}^2$ is a function $T : \mathbb{R}^2 \to \mathbb{R}^2$ of the form

$$T\begin{pmatrix} x \\ y \end{pmatrix} = \begin{pmatrix} ax + by \\ cx + dy \end{pmatrix}.$$

That is, T simply multiplies any vector by the 2×2 matrix

$$\begin{pmatrix} a & b \\ c & d \end{pmatrix}.$$

We will thus think of the linear map and its matrix as being interchangeable, so that we also write

$$T = \begin{pmatrix} a & b \\ c & d \end{pmatrix}.$$

We hope no confusion will result from this slight imprecision.

Now suppose that T is *invertible*. This means that the matrix T has an *inverse matrix* S that satisfies $TS = ST = I$ where I is the 2×2 identity matrix. It is traditional to denote the inverse of a matrix T by T^{-1}. As is easily checked, the matrix

$$S = \frac{1}{\det T} \begin{pmatrix} d & -b \\ -c & a \end{pmatrix}$$

serves as T^{-1} if $\det T \neq 0$. If $\det T = 0$, then we know from Chapter 2 that there are infinitely many vectors (x, y) for which

$$T \begin{pmatrix} x \\ y \end{pmatrix} = \begin{pmatrix} 0 \\ 0 \end{pmatrix}.$$

Hence there is no inverse matrix in this case, for we would need

$$\begin{pmatrix} x \\ y \end{pmatrix} = T^{-1} T \begin{pmatrix} x \\ y \end{pmatrix} = T^{-1} \begin{pmatrix} 0 \\ 0 \end{pmatrix}$$

for each such vector. We have shown:

Proposition. *The 2×2 matrix T is invertible if and only if* $\det T \neq 0$. ∎

Now, instead of considering a linear system $X' = AX$, suppose we consider a different system

$$Y' = (T^{-1} A T) Y$$

for some invertible linear map T. Note that if $Y(t)$ is a solution of this new system, then $X(t) = TY(t)$ solves $X' = AX$. Indeed, we have

$$(TY(t))' = TY'(t)$$
$$= T(T^{-1} A T) Y(t)$$
$$= A(TY(t))$$

as required. That is, the linear map T converts solutions of $Y' = (T^{-1} A T) Y$ to solutions of $X' = AX$. Alternatively, T^{-1} takes solutions of $X' = AX$ to solutions of $Y' = (T^{-1} A T) Y$.

We therefore think of T as a change of coordinates that converts a given linear system into one whose coefficient matrix is different. What we hope to be able to do is find a linear map T that converts the given system into a system of the form $Y' = (T^{-1} A T) Y$ that is easily solved. And, as you may have guessed, we can always do this by finding a linear map that converts a given linear system to one in canonical form.

Example. (Real Eigenvalues) Suppose the matrix A has two real, distinct eigenvalues λ_1 and λ_2 with associated eigenvectors V_1 and V_2. Let T be the matrix whose columns are V_1 and V_2. Thus $TE_j = V_j$ for $j = 1, 2$ where the

E_j form the standard basis of $\mathbb{R}^2$. Also, $T^{-1}V_j = E_j$. Therefore we have

$$(T^{-1}AT)E_j = T^{-1}AV_j = T^{-1}(\lambda_j V_j)$$
$$= \lambda_j T^{-1} V_j$$
$$= \lambda_j E_j.$$

Thus the matrix $T^{-1}AT$ assumes the canonical form

$$T^{-1}AT = \begin{pmatrix} \lambda_1 & 0 \\ 0 & \lambda_2 \end{pmatrix}$$

and the corresponding system is easy to solve.   ∎

Example. As a further specific example, suppose

$$A = \begin{pmatrix} -1 & 0 \\ 1 & -2 \end{pmatrix}$$

The characteristic equation is $\lambda^2 + 3\lambda + 2$, which yields eigenvalues $\lambda = -1$ and $\lambda = -2$. An eigenvector corresponding to $\lambda = -1$ is given by solving

$$(A + I)\begin{pmatrix} x \\ y \end{pmatrix} = \begin{pmatrix} 0 & 0 \\ 1 & -1 \end{pmatrix}\begin{pmatrix} x \\ y \end{pmatrix} = \begin{pmatrix} 0 \\ 0 \end{pmatrix}$$

which yields an eigenvector $(1, 1)$. Similarly an eigenvector associated to $\lambda = -2$ is given by $(0, 1)$.

We therefore have a pair of straight-line solutions, each tending to the origin as $t \to \infty$. The straight-line solution corresponding to the weaker eigenvalue lies along the line $y = x$; the straight-line solution corresponding to the stronger eigenvalue lies on the y-axis. All other solutions tend to the origin tangentially to the line $y = x$.

To put this system in canonical form, we choose T to be the matrix whose columns are these eigenvectors:

$$T = \begin{pmatrix} 1 & 0 \\ 1 & 1 \end{pmatrix}$$

so that

$$T^{-1} = \begin{pmatrix} 1 & 0 \\ -1 & 1 \end{pmatrix}.$$

Finally, we compute

$$T^{-1}AT = \begin{pmatrix} -1 & 0 \\ 0 & -2 \end{pmatrix},$$

so $T^{-1}AT$ is in canonical form. The general solution of the system $Y' = (T^{-1}AT)Y$ is

$$Y(t) = \alpha e^{-t} \begin{pmatrix} 1 \\ 0 \end{pmatrix} + \beta e^{-2t} \begin{pmatrix} 0 \\ 1 \end{pmatrix}$$

so the general solution of $X' = AX$ is

$$TY(t) = \begin{pmatrix} 1 & 0 \\ 1 & 1 \end{pmatrix} \left(\alpha e^{-t} \begin{pmatrix} 1 \\ 0 \end{pmatrix} + \beta e^{-2t} \begin{pmatrix} 0 \\ 1 \end{pmatrix} \right)$$

$$= \alpha e^{-t} \begin{pmatrix} 1 \\ 1 \end{pmatrix} + \beta e^{-2t} \begin{pmatrix} 0 \\ 1 \end{pmatrix}.$$

Thus the linear map T converts the phase portrait for the system

$$Y' = \begin{pmatrix} -1 & 0 \\ 0 & -2 \end{pmatrix} Y$$

to that of $X' = AX$ as shown in Figure 3.7. ∎

Note that we really do not have to go through the step of converting a specific system to one in canonical form; once we have the eigenvalues and eigenvectors, we can simply write down the general solution. We take this extra step because, when we attempt to classify all possible linear systems, the canonical form of the system will greatly simplify this process.

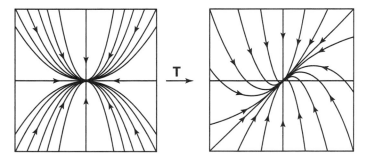

Figure 3.7   The change of variables T in the case of a (real) sink.

Example. (Complex Eigenvalues) Now suppose that the matrix A has complex eigenvalues $\alpha \pm i\beta$ with $\beta \neq 0$. Then we may find a complex eigenvector $V_1 + iV_2$ corresponding to $\alpha + i\beta$, where both V_1 and V_2 are real vectors. We claim that V_1 and V_2 are linearly independent vectors in $\mathbb{R}^2$. If this were not the case, then we would have $V_1 = cV_2$ for some $c \in \mathbb{R}$. But then we have

$$A(V_1 + iV_2) = (\alpha + i\beta)(V_1 + iV_2) = (\alpha + i\beta)(c + i)V_2.$$

But we also have

$$A(V_1 + iV_2) = (c + i)AV_2.$$

So we conclude that $AV_2 = (\alpha + i\beta)V_2$. This is a contradiction since the left-hand side is a real vector while the right is complex.

Since $V_1 + iV_2$ is an eigenvector associated to $\alpha + i\beta$, we have

$$A(V_1 + iV_2) = (\alpha + i\beta)(V_1 + iV_2).$$

Equating the real and imaginary components of this vector equation, we find

$$AV_1 = \alpha V_1 - \beta V_2$$
$$AV_2 = \beta V_1 + \alpha V_2.$$

Let T be the matrix whose columns are V_1 and V_2. Hence $TE_j = V_j$ for $j = 1, 2$. Now consider $T^{-1}AT$. We have

$$(T^{-1}AT)E_1 = T^{-1}(\alpha V_1 - \beta V_2)$$
$$= \alpha E_1 - \beta E_2$$

and similarly

$$(T^{-1}AT)E_2 = \beta E_1 + \alpha E_2.$$

Thus the matrix $T^{-1}AT$ is in the canonical form

$$T^{-1}AT = \begin{pmatrix} \alpha & \beta \\ -\beta & \alpha \end{pmatrix}.$$

We saw that the system $Y' = (T^{-1}AT)Y$ has phase portrait corresponding to a spiral sink, center, or spiral source depending on whether $\alpha < 0$, $\alpha = 0$, or

$\alpha > 0$. Therefore the phase portrait of $X' = AX$ is equivalent to one of these after changing coordinates using T. ∎

Example. (Another Harmonic Oscillator) Consider the second-order equation

$$x'' + 4x = 0.$$

This corresponds to an undamped harmonic oscillator with mass 1 and spring constant 4. As a system, we have

$$X' = \begin{pmatrix} 0 & 1 \\ -4 & 0 \end{pmatrix} X = AX.$$

The characteristic equation is

$$\lambda^2 + 4 = 0$$

so that the eigenvalues are $\pm 2i$. A complex eigenvector associated to $\lambda = 2i$ is a solution of the system

$$-2ix + y = 0$$
$$-4x - 2iy = 0.$$

One such solution is the vector $(1, 2i)$. So we have a complex solution of the form

$$e^{2it} \begin{pmatrix} 1 \\ 2i \end{pmatrix}.$$

Breaking this solution into its real and imaginary parts, we find the general solution

$$X(t) = c_1 \begin{pmatrix} \cos 2t \\ -2 \sin 2t \end{pmatrix} + c_2 \begin{pmatrix} \sin 2t \\ 2 \cos 2t \end{pmatrix}.$$

Thus the position of this oscillator is given by

$$x(t) = c_1 \cos 2t + c_2 \sin 2t,$$

which is a periodic function of period π.

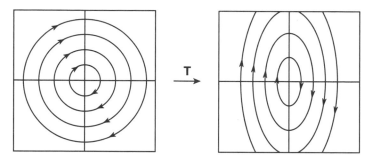

Figure 3.8   The change of variables T in the case of a center.

Now, let T be the matrix whose columns are the real and imaginary parts of the eigenvector $(1, 2i)$. That is

$$T = \begin{pmatrix} 1 & 0 \\ 0 & 2 \end{pmatrix}.$$

Then, we compute easily that

$$T^{-1}AT = \begin{pmatrix} 0 & 2 \\ -2 & 0 \end{pmatrix},$$

which is in canonical form. The phase portraits of these systems are shown in Figure 3.8. Note that T maps the circular solutions of the system $Y' = (T^{-1}AT)Y$ to elliptic solutions of $X' = AX$. ∎

Example. (Repeated Eigenvalues) Suppose A has a single real eigenvalue λ. If there exist a pair of linearly independent eigenvectors, then in fact A must be in the form

$$A = \begin{pmatrix} \lambda & 0 \\ 0 & \lambda \end{pmatrix},$$

so the system $X' = AX$ is easily solved (see Exercise 15).

For the more complicated case, let's assume that V is an eigenvector and that every other eigenvector is a multiple of V. Let W be any vector for which V and W are linearly independent. Then we have

$$AW = \mu V + \nu W$$

for some constants $\mu, \nu \in \mathbb{R}$. Note that $\mu \neq 0$, for otherwise we would have a second linearly independent eigenvector W with eigenvalue ν. We claim that $\nu = \lambda$. If $\nu - \lambda \neq 0$, a computation shows that

$$A\left(W + \left(\frac{\mu}{\nu - \lambda}\right)V\right) = \nu\left(W + \left(\frac{\mu}{\nu - \lambda}\right)V\right).$$

This says that ν is a second eigenvalue different from λ. Hence we must have $\nu = \lambda$.

Finally, let $U = (1/\mu)W$. Then

$$AU = V + \frac{\lambda}{\mu}W = V + \lambda U.$$

Thus if we define $TE_1 = V$, $TE_2 = U$, we get

$$T^{-1}AT = \begin{pmatrix} \lambda & 1 \\ 0 & \lambda \end{pmatrix}$$

as required. Thus $X' = AX$ is again in canonical form after this change of coordinates. ∎

EXERCISES

1. In Figure 3.9 on page 58, you see six phase portraits. Match each of these phase portraits with one of the following linear systems:

(a) $\begin{pmatrix} 3 & 5 \\ -2 & -2 \end{pmatrix}$ (b) $\begin{pmatrix} -3 & -2 \\ 5 & 2 \end{pmatrix}$ (c) $\begin{pmatrix} 3 & -2 \\ 5 & -2 \end{pmatrix}$

(d) $\begin{pmatrix} -3 & 5 \\ -2 & 3 \end{pmatrix}$ (e) $\begin{pmatrix} 3 & 5 \\ -2 & -3 \end{pmatrix}$ (f) $\begin{pmatrix} -3 & 5 \\ -2 & 2 \end{pmatrix}$

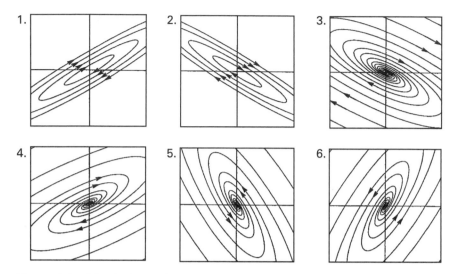

Figure 3.9   Match these phase portraits with the systems in Exercise 1.

2. For each of the following systems of the form $X' = AX$

(a)   Find the eigenvalues and eigenvectors of A.

(b)   Find the matrix T that puts A in canonical form.

(c)   Find the general solution of both $X' = AX$ and $Y' = (T^{-1}AT)Y$.

(d)   Sketch the phase portraits of both systems.

$$\text{(i) } A = \begin{pmatrix} 0 & 1 \\ 1 & 0 \end{pmatrix} \qquad \text{(ii) } A = \begin{pmatrix} 1 & 1 \\ 1 & 0 \end{pmatrix}$$

$$\text{(iii) } A = \begin{pmatrix} 1 & 1 \\ -1 & 0 \end{pmatrix} \qquad \text{(iv) } A = \begin{pmatrix} 1 & 1 \\ -1 & 3 \end{pmatrix}$$

$$\text{(v) } A = \begin{pmatrix} 1 & 1 \\ -1 & -3 \end{pmatrix} \qquad \text{(vi) } A = \begin{pmatrix} 1 & 1 \\ 1 & -1 \end{pmatrix}$$

3. Find the general solution of the following harmonic oscillator equations:

(a) $x'' + x' + x = 0$

(b) $x'' + 2x' + x = 0$

4. Consider the harmonic oscillator system

$$X' = \begin{pmatrix} 0 & 1 \\ -k & -b \end{pmatrix} X$$

where $b \geq 0, k > 0$ and the mass $m = 1$.

(a) For which values of k, b does this system have complex eigenvalues? Repeated eigenvalues? Real and distinct eigenvalues?

(b) Find the general solution of this system in each case.

(c) Describe the motion of the mass when the mass is released from the initial position $x = 1$ with zero velocity in each of the cases in part (a).

5. Sketch the phase portrait of $X' = AX$ where

$$A = \begin{pmatrix} a & 1 \\ 2a & 2 \end{pmatrix}.$$

For which values of a do you find a bifurcation? Describe the phase portrait for a-values above and below the bifurcation point.

6. Consider the system

$$X' = \begin{pmatrix} 2a & b \\ b & 0 \end{pmatrix} X.$$

Sketch the regions in the ab–plane where this system has different types of canonical forms.

7. Consider the system

$$X' = \begin{pmatrix} \lambda & 1 \\ 0 & \lambda \end{pmatrix} X$$

with $\lambda \neq 0$. Show that all solutions tend to (respectively, away from) the origin tangentially to the eigenvector $(1, 0)$ when $\lambda < 0$ (respectively, $\lambda > 0$).

8. Find all 2×2 matrices that have pure imaginary eigenvalues. That is, determine conditions on the entries of a matrix that guarantee that the matrix has pure imaginary eigenvalues.

9. Determine a computable condition that guarantees that, if a matrix A has complex eigenvalues with nonzero imaginary part, then solutions of $X' = AX$ travel around the origin in the counterclockwise direction.

10. Consider the system

$$X' = \begin{pmatrix} a & b \\ c & d \end{pmatrix} X$$

where $a + d \neq 0$ but $ad - bc = 0$. Find the general solution of this system and sketch the phase portrait.

11. Find the general solution and describe completely the phase portrait for

$$X' = \begin{pmatrix} 0 & 1 \\ 0 & 0 \end{pmatrix} X.$$

12. Prove that

$$\alpha e^{\lambda t} \begin{pmatrix} 1 \\ 0 \end{pmatrix} + \beta e^{\lambda t} \begin{pmatrix} t \\ 1 \end{pmatrix}$$

is the general solution of

$$X' = \begin{pmatrix} \lambda & 1 \\ 0 & \lambda \end{pmatrix} X.$$

13. Prove that a 2×2 matrix A always satisfies its own characteristic equation. That is, if $\lambda^2 + \alpha\lambda + \beta = 0$ is the characteristic equation associated to A, then the matrix $A^2 + \alpha A + \beta I$ is the 0 matrix.

14. Suppose the 2×2 matrix A has repeated eigenvalues λ. Let $V \in \mathbb{R}^2$. Using the previous problem, show that either V is an eigenvector for A or else $(A - \lambda I)V$ is an eigenvector for A.

15. Suppose the matrix A has repeated real eigenvalues λ and there exists a pair of linearly independent eigenvectors associated to A. Prove that

$$A = \begin{pmatrix} \lambda & 0 \\ 0 & \lambda \end{pmatrix}.$$

16. Consider the (nonlinear) system

$$x' = |y|$$
$$y' = -x.$$

Use the methods of this chapter to describe the phase portrait.

4
Classification of Planar Systems

In this chapter, we summarize what we have accomplished so far using a dynamical systems point of view. Among other things, this means that we would like to have a complete "dictionary" of all possible behaviors of 2×2 autonomous linear systems. One of the dictionaries we present here is geometric: the trace-determinant plane. The other dictionary is more dynamic: this involves the notion of conjugate systems.

4.1 The Trace-Determinant Plane

For a matrix

$$A = \begin{pmatrix} a & b \\ c & d \end{pmatrix}$$

we know that the eigenvalues are the roots of the characteristic equation, which can be written

$$\lambda^2 - (a + d)\lambda + (ad - bc) = 0.$$

The constant term in this equation is $\det A$. The coefficient of λ also has a name: The quantity $a + d$ is called the *trace* of A and is denoted by $\text{tr} \, A$.

Thus the eigenvalues satisfy

$$\lambda^2 - (\operatorname{tr} A)\lambda + \det A = 0$$

and are given by

$$\lambda_\pm = \frac{1}{2}\left(\operatorname{tr} A \pm \sqrt{(\operatorname{tr} A)^2 - 4\det A}\right).$$

Note that $\lambda_+ + \lambda_- = \operatorname{tr} A$ and $\lambda_+\lambda_- = \det A$, so the trace is the sum of the eigenvalues of A while the determinant is the product of the eigenvalues of A. We will also write $T = \operatorname{tr} A$ and $D = \det A$. Knowing T and D tells us the eigenvalues of A and therefore virtually everything about the geometry of solutions of $X' = AX$. For example, the values of T and D tell us whether solutions spiral into or away from the origin, whether we have a center, and so forth.

We may display this classification visually by painting a picture in the *trace-determinant plane*. In this picture a matrix with trace T and determinant D corresponds to the point with coordinates (T, D). The location of this point in the TD–plane then determines the geometry of the phase portrait as above. For example, the sign of $T^2 - 4D$ tells us that the eigenvalues are:

1. Complex with nonzero imaginary part if $T^2 - 4D < 0$;
2. Real and distinct if $T^2 - 4D > 0$;
3. Real and repeated if $T^2 - 4D = 0$.

Thus the location of (T, D) relative to the parabola $T^2 - 4D = 0$ in the TD–plane tells us all we need to know about the eigenvalues of A from an algebraic point of view.

In terms of phase portraits, however, we can say more. If $T^2 - 4D < 0$, then the real part of the eigenvalues is $T/2$, and so we have a

1. Spiral sink if $T < 0$;
2. Spiral source if $T > 0$;
3. Center if $T = 0$.

If $T^2 - 4D > 0$ we have a similar breakdown into cases. In this region, both eigenvalues are real. If $D < 0$, then we have a saddle. This follows since D is the product of the eigenvalues, one of which must be positive, the other negative. Equivalently, if $D < 0$, we compute

$$T^2 < T^2 - 4D$$

so that

$$\pm T < \sqrt{T^2 - 4D}.$$

Thus we have

$$T + \sqrt{T^2 - 4D} > 0$$

$$T - \sqrt{T^2 - 4D} < 0$$

so the eigenvalues are real and have different signs. If $D > 0$ and $T < 0$ then both

$$T \pm \sqrt{T^2 - 4D} < 0,$$

so we have a (real) sink. Similarly, $T > 0$ and $D > 0$ leads to a (real) source.

When $D = 0$ and $T \neq 0$, we have one zero eigenvalue, while both eigenvalues vanish if $D = T = 0$.

Plotting all of this verbal information in the TD–plane gives us a visual summary of all of the different types of linear systems. The equations above partition the TD–plane into various regions in which systems of a particular type reside. See Figure 4.1. This yields a geometric classification of 2×2 linear systems.

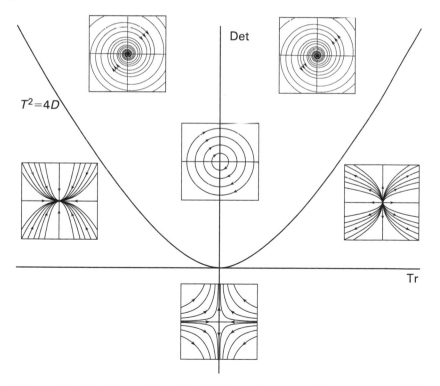

Figure 4.1   The trace-determinant plane. Any resemblance to any of the authors' faces is purely coincidental.

A couple of remarks are in order. First, the trace-determinant plane is a two-dimensional representation of what is really a four-dimensional space, since 2×2 matrices are determined by four parameters, the entries of the matrix. Thus there are infinitely many different matrices corresponding to each point in the TD-plane. While all of these matrices share the same eigenvalue configuration, there may be subtle differences in the phase portraits, such as the direction of rotation for centers and spiral sinks and sources, or the possibility of one or two independent eigenvectors in the repeated eigenvalue case.

We also think of the trace-determinant plane as the analog of the bifurcation diagram for planar linear systems. A one-parameter family of linear systems corresponds to a curve in the TD-plane. When this curve crosses the T-axis, the positive D-axis, or the parabola $T^2 - 4D = 0$, the phase portrait of the linear system undergoes a bifurcation: A major change occurs in the geometry of the phase portrait.

Finally, note that we may obtain quite a bit of information about the system from D and T without ever computing the eigenvalues. For example, if $D < 0$, we know that we have a saddle at the origin. Similarly, if both D and T are positive, then we have a source at the origin.

4.2 Dynamical Classification

In this section we give a different, more dynamical classification of planar linear systems. From a dynamical systems point of view, we are usually interested primarily in the long-term behavior of solutions of differential equations. Thus two systems are equivalent if their solutions share the same fate. To make this precise we recall some terminology introduced in Section 1.5 of Chapter 1.

To emphasize the dependence of solutions on both time and the initial conditions X_0, we let $\phi_t(X_0)$ denote the solution that satisfies the initial condition X_0. That is, $\phi_0(X_0) = X_0$. The function $\phi(t, X_0) = \phi_t(X_0)$ is called the *flow* of the differential equation, whereas ϕ_t is called the *time t map* of the flow.

For example, let

$$X' = \begin{pmatrix} 2 & 0 \\ 0 & 3 \end{pmatrix} X.$$

Then the time t map is given by

$$\phi_t(x_0, y_0) = \left(x_0 e^{2t}, y_0 e^{3t} \right).$$

Thus the flow is a function that depends on both time and the initial values.

We will consider two systems to be dynamically equivalent if there is a function h that takes one flow to the other. We require that this function be a *homeomorphism*, that is, h is a one-to-one, onto, and continuous function whose inverse is also continuous.

Definition

Suppose $X' = AX$ and $X' = BX$ have flows ϕ^A and ϕ^B. These two systems are (topologically) *conjugate* if there exists a homeomorphism $h : \mathbb{R}^2 \to \mathbb{R}^2$ that satisfies

$$\phi^B(t, h(X_0)) = h(\phi^A(t, X_0)).$$

The homeomorphism h is called a *conjugacy*. Thus a conjugacy takes the solution curves of $X' = AX$ to those of $X' = BX$.

Example. For the one-dimensional linear differential equations

$$x' = \lambda_1 x \quad \text{and} \quad x' - \lambda_2 x$$

we have the flows

$$\phi^j(t, x_0) = x_0 e^{\lambda_j t}$$

for $j = 1, 2$. Suppose that λ_1 and λ_2 are nonzero and have the same sign. Then let

$$h(x) = \begin{cases} x^{\lambda_2/\lambda_1} & \text{if } x \geq 0 \\[2mm] -|x|^{\lambda_2/\lambda_1} & \text{if } x < 0 \end{cases}$$

where we recall that

$$x^{\lambda_2/\lambda_1} = \exp\left(\frac{\lambda_2}{\lambda_1} \log(x)\right).$$

Note that h is a homeomorphism of the real line. We claim that h is a conjugacy between $x' = \lambda_1 x$ and $x' = \lambda_2 x$. To see this, we check that when $x_0 > 0$

$$h(\phi^1(t, x_0)) = \left(x_0 e^{\lambda_1 t}\right)^{\lambda_2/\lambda_1}$$

$$= x_0^{\lambda_2/\lambda_1} e^{\lambda_2 t}$$
$$= \phi^2(t, h(x_0))$$

as required. A similar computation works when $x_0 < 0$.   ■

There are several things to note here. First, λ_1 and λ_2 must have the same sign, because otherwise we would have $|h(0)| = \infty$, in which case h is not a homeomorphism. This agrees with our notion of dynamical equivalence: If λ_1 and λ_2 have the same sign, then their solutions behave similarly as either both tend to the origin or both tend away from the origin. Also, note that if $\lambda_2 < \lambda_1$, then h is not differentiable at the origin, whereas if $\lambda_2 > \lambda_1$ then $h^{-1}(x) = x^{\lambda_1/\lambda_2}$ is not differentiable at the origin. This is the reason why we require h to be only a homeomorphism and not a *diffeomorphism* (a differentiable homeomorphism with differentiable inverse): If we assume differentiability, then we must have $\lambda_1 = \lambda_2$, which does not yield a very interesting notion of "equivalence."

This gives a classification of (autonomous) linear first-order differential equations, which agrees with our qualitative observations in Chapter 1. There are three conjugacy "classes": the sinks, the sources, and the special "in-between" case, $x' = 0$, where all solutions are constants.

Now we move to the planar version of this scenario. We first note that we only need to decide on conjugacies among systems whose matrices are in canonical form. For, as we saw in Chapter 3, if the linear map $T : \mathbb{R}^2 \to \mathbb{R}^2$ puts A in canonical form, then T takes the time t map of the flow of $Y' = (T^{-1}AT)Y$ to the time t map for $X' = AX$.

Our classification of planar linear systems now proceeds just as in the one-dimensional case. We will stay away from the case where the system has eigenvalues with real part equal to 0, but you will tackle this case in the exercises.

Definition
A matrix A is *hyperbolic* if none of its eigenvalues has real part 0. We also say that the system $X' = AX$ is *hyperbolic*.

Theorem. *Suppose that the 2×2 matrices A_1 and A_2 are hyperbolic. Then the linear systems $X' = A_i X$ are conjugate if and only if each matrix has the same number of eigenvalues with negative real part.*   ■

Thus two hyperbolic matrices yield conjugate linear systems if both sets of eigenvalues fall into the same category below:

1. One eigenvalue is positive and the other is negative;

2. Both eigenvalues have negative real parts;
3. Both eigenvalues have positive real parts.

Before proving this, note that this theorem implies that a system with a spiral sink is conjugate to a system with a (real) sink. Of course! Even though their phase portraits look very different, it is nevertheless the case that all solutions of both systems share the same fate: They tend to the origin as $t \to \infty$.

Proof: Recall from the previous discussion that we may assume all systems are in canonical form. Then the proof divides into three distinct cases.

Case 1

Suppose we have two linear systems $X' = A_i X$ for $i = 1, 2$ such that each A_i has eigenvalues $\lambda_i < 0 < \mu_i$. Thus each system has a saddle at the origin. This is the easy case. As we saw previously, the real differential equations $x' = \lambda_i x$ have conjugate flows via the homeomorphism

$$h_1(x) = \begin{cases} x^{\lambda_2/\lambda_1} & \text{if } x \geq 0 \\ -|x|^{\lambda_2/\lambda_1} & \text{if } x < 0 \end{cases}.$$

Similarly, the equations $y' = \mu_i y$ also have conjugate flows via an analogous function h_2. Now define

$$H(x, y) = (h_1(x), h_2(y)).$$

Then one checks immediately that H provides a conjugacy between these two systems.

Case 2

Consider the system $X' = AX$ where A is in canonical form with eigenvalues that have negative real parts. We further assume that the matrix A is not in the form

$$\begin{pmatrix} \lambda & 1 \\ 0 & \lambda \end{pmatrix}$$

with $\lambda < 0$. Thus, in canonical form, A assumes one of the following two forms:

$$\text{(a)} \begin{pmatrix} \alpha & \beta \\ -\beta & \alpha \end{pmatrix} \qquad \text{(b)} \begin{pmatrix} \lambda & 0 \\ 0 & \mu \end{pmatrix}$$

with $\alpha, \lambda, \mu < 0$. We will show that, in either case, the system is conjugate to $X' = BX$ where

$$B = \begin{pmatrix} -1 & 0 \\ 0 & -1 \end{pmatrix}.$$

It then follows that any two systems of this form are conjugate.

Consider the unit circle in the plane parameterized by the curve $X(\theta) = (\cos\theta, \sin\theta)$, $0 \leq \theta \leq 2\pi$. We denote this circle by S^1. We first claim that the vector field determined by a matrix in the above form must point inside S^1. In case (a), we have that the vector field on S^1 is given by

$$AX(\theta) = \begin{pmatrix} \alpha\cos\theta + \beta\sin\theta \\ -\beta\cos\theta + \alpha\sin\theta \end{pmatrix}.$$

The outward pointing normal vector to S^1 at $X(\theta)$ is

$$N(\theta) = \begin{pmatrix} \cos\theta \\ \sin\theta \end{pmatrix}.$$

The dot product of these two vectors satisfies

$$AX(\theta) \cdot N(\theta) = \alpha(\cos^2\theta + \sin^2\theta) < 0$$

since $\alpha < 0$. This shows that $AX(\theta)$ does indeed point inside S^1. Case (b) is even easier.

As a consequence, each nonzero solution of $X' = AX$ crosses S^1 exactly once. Let ϕ_t^A denote the time t map for this system, and let $\tau = \tau(x, y)$ denote the time at which $\phi_t^A(x, y)$ meets S^1. Thus

$$\left| \phi_{\tau(x,y)}^A(x, y) \right| = 1.$$

Let ϕ_t^B denote the time t map for the system $X' = BX$. Clearly,

$$\phi_t^B(x, y) = (e^{-t}x, e^{-t}y).$$

We now define a conjugacy H between these two systems. If $(x, y) \neq (0, 0)$, let

$$H(x, y) = \phi_{-\tau(x,y)}^B \phi_{\tau(x,y)}^A(x, y)$$

and set $H(0, 0) = (0, 0)$. Geometrically, the value of $H(x, y)$ is given by following the solution curve of $X' = AX$ exactly $\tau(x, y)$ time units (forward

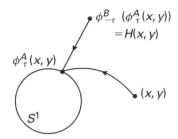

Figure 4.2   The definition of $\tau\,(x, y)$.

or backward) until the solution reaches S^1, and then following the solution of $X' = BX$ starting at that point on S^1 and proceeding in the opposite time direction exactly τ time units. See Figure 4.2.

To see that H gives a conjugacy, note first that

$$\tau\left(\phi_s^A(x, y)\right) = \tau(x, y) - s$$

since

$$\phi_{\tau-s}^A \phi_s^A(x, y) = \phi_\tau^A(x, y) \in S^1.$$

Therefore we have

$$
\begin{aligned}
H\left(\phi_s^A(x, y)\right) &= \phi_{-\tau+s}^B \phi_{\tau-s}^A \left(\phi_s^A(x, y)\right) \\
&= \phi_s^B \phi_{-\tau}^B \phi_\tau^A(x, y) \\
&= \phi_s^B \left(H(x, y)\right).
\end{aligned}
$$

So H is a conjugacy.

Now we show that H is a homeomorphism. We can construct an inverse for H by simply reversing the process defining H. That is, let

$$G(x, y) = \phi_{-\tau_1(x,y)}^A \phi_{\tau_1(x,y)}^B(x, y)$$

and set $G(0, 0) = (0, 0)$. Here $\tau_1(x, y)$ is the time for the solution of $X' = BX$ through (x, y) to reach S^1. An easy computation shows that $\tau_1(x, y) = \log r$ where $r^2 = x^2 + y^2$. Clearly, $G = H^{-1}$ so H is one to one and onto. Also, G is continuous at $(x, y) \neq (0, 0)$ since G may be written

$$G(x, y) = \phi_{-\log r}^A \left(\frac{x}{r}, \frac{y}{r}\right),$$

which is a composition of continuous functions. For continuity of G at the origin, suppose that (x, y) is close to the origin, so that r is small. Observe that as $r \to 0$, $-\log r \to \infty$. Now $(x/r, y/r)$ is a point on S^1 and for r sufficiently small, $\phi^A_{-\log r}$ maps the unit circle very close to $(0, 0)$. This shows that G is continuous at $(0, 0)$.

We thus need only show continuity of H. For this, we need to show that $\tau(x, y)$ is continuous. But τ is determined by the equation

$$\left| \phi^A_t(x, y) \right| = 1.$$

We write $\phi^A_t(x, y) = (x(t), y(t))$. Taking the partial derivative of $|\phi^A_t(x, y)|$ with respect to t, we find

$$\frac{\partial}{\partial t} \left| \phi^A_t(x, y) \right| = \frac{\partial}{\partial t} \sqrt{(x(t))^2 + (y(t))^2}$$

$$= \frac{1}{\sqrt{(x(t))^2 + (y(t))^2}} \left(x(t)x'(t) + y(t)y'(t) \right)$$

$$= \frac{1}{\left| \phi^A_t(x, y) \right|} \left(\begin{pmatrix} x(t) \\ y(t) \end{pmatrix} \cdot \begin{pmatrix} x'(t) \\ y'(t) \end{pmatrix} \right).$$

But the latter dot product is nonzero when $t = \tau(x, y)$ since the vector field given by $(x'(t), y'(t))$ points inside S^1. Hence

$$\frac{\partial}{\partial t} \left| \phi^A_t(x, y) \right| \neq 0$$

at $(\tau(x, y), x, y)$. Thus we may apply the implicit function theorem to show that τ is differentiable at (x, y) and hence continuous. Continuity of H at the origin follows as in the case of $G = H^{-1}$. Thus H is a homeomorphism and we have a conjugacy between $X' = AX$ and $X' = BX$.

Note that this proof works equally well if the eigenvalues have positive real parts.

Case 3

Finally, suppose that

$$A = \begin{pmatrix} \lambda & 1 \\ 0 & \lambda \end{pmatrix}$$

with $\lambda < 0$. The associated vector field need not point inside the unit circle in this case. However, if we let

$$T = \begin{pmatrix} 1 & 0 \\ 0 & \epsilon \end{pmatrix},$$

then the vector field given by

$$Y' = (T^{-1}AT)Y$$

now does have this property, provided $\epsilon > 0$ is sufficiently small. Indeed

$$T^{-1}AT = \begin{pmatrix} \lambda & \epsilon \\ 0 & \lambda \end{pmatrix}$$

so that

$$\left(T^{-1}AT \begin{pmatrix} \cos\theta \\ \sin\theta \end{pmatrix} \right) \cdot \begin{pmatrix} \cos\theta \\ \sin\theta \end{pmatrix} = \lambda + \epsilon \sin\theta \cos\theta.$$

Thus if we choose $\epsilon < -\lambda$, this dot product is negative. Therefore the change of variables T puts us into the situation where the same proof as in Case 2 applies. This proves the "if" part of the Theorem. The "only if" part follows from our earlier observation that, if the eigenvalues are negative, then the conjugacy takes curves tending to the equilibrium point to other such curves. So if there is only one negative eigenvalue, then we only get a single curve of such solutions, but if there are two eigenvalues with negative real parts, we get a whole open set of such solutions and so there is no conjugacy. This completes the proof. ∎

4.3 Exploration: A 3D Parameter Space

Consider the three-parameter family of linear systems given by

$$X' = \begin{pmatrix} a & b \\ c & 0 \end{pmatrix} X$$

where a, b, and c are parameters.

1. First, fix $a > 0$. Describe the analog of the trace-determinant plane in the bc–plane. That is, identify the bc–values in this plane where the corresponding system has saddles, centers, spiral sinks, etc. Sketch these regions in the bc–plane.

2. Repeat the previous question when $a < 0$ and when $a = 0$.
3. Describe the bifurcations that occur as a changes from positive to negative.
4. Now put all of the previous information together and give a description of the full three-dimensional parameter space for this system. You could build a 3D model of this space, create a flip-book animation of the changes as, say, a varies, or use a computer model to visualize this image. In any event, your model should accurately capture all of the distinct regions in this space.

EXERCISES

1. Consider the one-parameter family of linear systems given by

$$X' = \begin{pmatrix} a & \sqrt{2} + (a/2) \\ \sqrt{2} - (a/2) & 0 \end{pmatrix} X.$$

(a) Sketch the path traced out by this family of linear systems in the trace-determinant plane as a varies.

(b) Discuss any bifurcations that occur along this path and compute the corresponding values of a.

2. Sketch the analog of the trace-determinant plane for the two-parameter family of systems

$$X' = \begin{pmatrix} a & b \\ b & a \end{pmatrix} X$$

in the ab–plane. That is, identify the regions in the ab–plane where this system has similar phase portraits.

3. Consider the harmonic oscillator equation (with $m = 1$)

$$x'' + bx' + kx = 0$$

where $b \geq 0$ and $k > 0$. Identify the regions in the relevant portion of the bk–plane where the corresponding system has similar phase portraits.

4. Prove that $H(x, y) = (x, -y)$ provides a conjugacy between

$$X' = \begin{pmatrix} 1 & 1 \\ -1 & 1 \end{pmatrix} X \quad \text{and} \quad Y' = \begin{pmatrix} 1 & -1 \\ 1 & 1 \end{pmatrix} Y.$$

5. For each of the following systems, find an explicit conjugacy between their flows.

(a) $X' = \begin{pmatrix} -1 & 1 \\ 0 & 2 \end{pmatrix} X$ and $Y' = \begin{pmatrix} 1 & 0 \\ 1 & -2 \end{pmatrix} Y$

(b) $X' = \begin{pmatrix} 0 & 1 \\ -4 & 0 \end{pmatrix} X$ and $Y' = \begin{pmatrix} 0 & 2 \\ -2 & 0 \end{pmatrix} Y.$

6. Prove that any two linear systems with the same eigenvalues $\pm i\beta$, $\beta \neq 0$ are conjugate. What happens if the systems have eigenvalues $\pm i\beta$ and $\pm i\gamma$ with $\beta \neq \gamma$? What if $\gamma = -\beta$?

7. Consider all linear systems with exactly one eigenvalue equal to 0. Which of these systems are conjugate? Prove this.

8. Consider all linear systems with two zero eigenvalues. Which of these systems are conjugate? Prove this.

9. Provide a complete description of the conjugacy classes for 2×2 systems in the nonhyperbolic case.

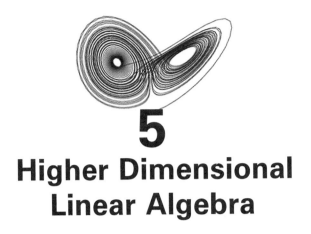

5

Higher Dimensional Linear Algebra

As in Chapter 2, we need to make another detour into the world of linear algebra before proceeding to the solution of higher dimensional linear systems of differential equations. There are many different canonical forms for matrices in higher dimensions, but most of the algebraic ideas involved in changing coordinates to put matrices into these forms are already present in the 2×2 case. In particular, the case of matrices with distinct (real or complex) eigenvalues can be handled with minimal additional algebraic complications, so we deal with this case first. This is the "generic case," as we show in Section 5.6. Matrices with repeated eigenvalues demand more sophisticated concepts from linear algebra; we provide this background in Section 5.4. We assume throughout this chapter that the reader is familiar with solving systems of linear algebraic equations by putting the associated matrix in (reduced) row echelon form.

5.1 Preliminaries from Linear Algebra

In this section we generalize many of the algebraic notions of Section 2.3 to higher dimensions. We denote a vector $X \in \mathbb{R}^n$ in coordinate form as

$$X = \begin{pmatrix} x_1 \\ \vdots \\ x_n \end{pmatrix}.$$

In the plane, we called a pair of vectors V and W linearly independent if they were not collinear. Equivalently, V and W were linearly independent if there were no (nonzero) real numbers α and β such that $\alpha V + \beta W$ was the zero vector.

More generally, in $\mathbb{R}^n$, a collection of vectors $V_1, \ldots, V_k$ in $\mathbb{R}^n$ is said to be *linearly independent* if, whenever

$$\alpha_1 V_1 + \cdots + \alpha_k V_k = 0$$

with $\alpha_j \in \mathbb{R}$, it follows that each $\alpha_j = 0$. If we can find such $\alpha_1, \ldots, \alpha_k$, not all of which are 0, then the vectors are *linearly dependent*. Note that if $V_1, \ldots, V_k$ are linearly independent and W is the linear combination

$$W = \beta_1 V_1 + \cdots + \beta_k V_k,$$

then the β_j are unique. This follows since, if we could also write

$$W = \gamma_1 V_1 + \cdots + \gamma_k V_k,$$

then we would have

$$0 = W - W = (\beta_1 - \gamma_1) V_1 + \cdots (\beta_k - \gamma_k) V_k,$$

which forces $\beta_j = \gamma_j$ for each j, by linear independence of the V_j.

Example. The vectors $(1, 0, 0)$, $(0, 1, 0)$, and $(0, 0, 1)$ are clearly linearly independent in $\mathbb{R}^3$. More generally, let E_j be the vector in $\mathbb{R}^n$ whose jth component is 1 and all other components are 0. Then the vectors $E_1, \ldots, E_n$ are linearly independent in $\mathbb{R}^n$. The collection of vectors $E_1, \ldots, E_n$ is called the *standard basis* of $\mathbb{R}^n$. We will discuss the concept of a basis in Section 5.4. ∎

Example. The vectors $(1, 0, 0)$, $(1, 1, 0)$, and $(1, 1, 1)$ in $\mathbb{R}^3$ are also linearly independent, because if we have

$$\alpha_1 \begin{pmatrix} 1 \\ 0 \\ 0 \end{pmatrix} + \alpha_2 \begin{pmatrix} 1 \\ 1 \\ 0 \end{pmatrix} + \alpha_3 \begin{pmatrix} 1 \\ 1 \\ 1 \end{pmatrix} = \begin{pmatrix} \alpha_1 + \alpha_2 + \alpha_3 \\ \alpha_2 + \alpha_3 \\ \alpha_3 \end{pmatrix} = \begin{pmatrix} 0 \\ 0 \\ 0 \end{pmatrix},$$

then the third component says that $\alpha_3 = 0$. The fact that $\alpha_3 = 0$ in the second component then says that $\alpha_2 = 0$, and finally the first component similarly tells us that $\alpha_1 = 0$. On the other hand, the vectors $(1, 1, 1)$, $(1, 2, 3)$, and $(2, 3, 4)$ are linearly dependent, for we have

$$1 \begin{pmatrix} 1 \\ 1 \\ 1 \end{pmatrix} + 1 \begin{pmatrix} 1 \\ 2 \\ 3 \end{pmatrix} - 1 \begin{pmatrix} 2 \\ 3 \\ 4 \end{pmatrix} = \begin{pmatrix} 0 \\ 0 \\ 0 \end{pmatrix}.$$

∎

When solving linear systems of differential equations, we will often encounter special subsets of $\mathbb{R}^n$ called *subspaces*. A *subspace* of $\mathbb{R}^n$ is a collection of all possible linear combinations of a given set of vectors. More precisely, given $V_1, \ldots, V_k \in \mathbb{R}^n$, the set

$$S = \{\alpha_1 V_1 + \cdots + \alpha_k V_k \mid \alpha_j \in \mathbb{R}\}$$

is a subspace of $\mathbb{R}^n$. In this case we say that S is *spanned* by $V_1, \ldots, V_k$. Equivalently, it can be shown (see Exercise 12 at the end of this chapter) that a subspace S is a nonempty subset of $\mathbb{R}^n$ having the following two properties:

1. If $X, Y \in S$, then $X + Y \in S$;
2. If $X \in S$ and $\alpha \in \mathbb{R}$, then $\alpha X \in S$.

Note that the zero vector lies in every subspace of $\mathbb{R}^n$ and that any linear combination of vectors in a subspace S also lies in S.

Example. Any straight line through the origin in $\mathbb{R}^n$ is a subspace of $\mathbb{R}^n$, since this line may be written as $\{tV \mid t \in \mathbb{R}\}$ for some nonzero $V \in \mathbb{R}^n$. The single vector V spans this subspace. The plane $\mathcal{P}$ defined by $x + y + z = 0$ in $\mathbb{R}^3$ is a subspace of $\mathbb{R}^3$. Indeed, any vector V in $\mathcal{P}$ may be written in the form $(x, y, -x - y)$ or

$$V = x \begin{pmatrix} 1 \\ 0 \\ -1 \end{pmatrix} + y \begin{pmatrix} 0 \\ 1 \\ -1 \end{pmatrix},$$

which shows that the vectors $(1, 0, -1)$ and $(0, 1, -1)$ span $\mathcal{P}$. ■

In linear algebra, one often encounters rectangular $n \times m$ matrices, but in differential equations, most often these matrices are square ($n \times n$). Consequently we will assume that all matrices in this chapter are $n \times n$. We write such a matrix

$$A = \begin{pmatrix} a_{11} & a_{12} & \cdots & a_{1n} \\ a_{21} & a_{22} & \cdots & a_{2n} \\ & & \vdots & \\ a_{n1} & a_{n2} & \cdots & a_{nn} \end{pmatrix}$$

more compactly as $A = [a_{ij}]$.

For $X = (x_1, \ldots, x_n) \in \mathbb{R}^n$, we define the product AX to be the vector

$$AX = \begin{pmatrix} \sum_{j=1}^{n} a_{1j} x_j \\ \vdots \\ \sum_{j=1}^{n} a_{nj} x_j \end{pmatrix},$$

so that the ith entry in this vector is the dot product of the ith row of A with the vector X.

Matrix sums are defined in the obvious way. If $A = [a_{ij}]$ and $B = [b_{ij}]$ are $n \times n$ matrices, then we define $A + B = C$ where $C = [a_{ij} + b_{ij}]$. Matrix arithmetic has some obvious linearity properties:

1. $A(k_1 X_1 + k_2 X_2) = k_1 A X_1 + k_2 A X_2$ where $k_j \in \mathbb{R}$, $X_j \in \mathbb{R}^n$;
2. $A + B = B + A$;
3. $(A + B) + C = A + (B + C)$.

The product of the $n \times n$ matrices A and B is defined to be the $n \times n$ matrix $AB = [c_{ij}]$ where

$$c_{ij} = \sum_{k=1}^{n} a_{ik} b_{kj},$$

so that c_{ij} is the dot product of the ith row of A with the jth column of B. We can easily check that, if A, B, and C are $n \times n$ matrices, then

1. $(AB)C = A(BC)$;
2. $A(B + C) = AB + AC$;
3. $(A + B)C = AC + BC$;
4. $k(AB) = (kA)B = A(kB)$ for any $k \in \mathbb{R}$.

All of the above properties of matrix arithmetic are easily checked by writing out the ij entries of the corresponding matrices. It is important to remember that matrix multiplication is not commutative, so that $AB \neq BA$ in general. For example

$$\begin{pmatrix} 1 & 0 \\ 1 & 1 \end{pmatrix} \begin{pmatrix} 1 & 1 \\ 0 & 1 \end{pmatrix} = \begin{pmatrix} 1 & 1 \\ 1 & 2 \end{pmatrix}$$

whereas

$$\begin{pmatrix} 1 & 1 \\ 0 & 1 \end{pmatrix} \begin{pmatrix} 1 & 0 \\ 1 & 1 \end{pmatrix} = \begin{pmatrix} 2 & 1 \\ 1 & 1 \end{pmatrix}.$$

Also, matrix cancellation is usually forbidden; if $AB = AC$, then we do not necessarily have $B = C$ as in

$$\begin{pmatrix} 1 & 1 \\ 1 & 1 \end{pmatrix} \begin{pmatrix} 1 & 0 \\ 0 & 0 \end{pmatrix} = \begin{pmatrix} 1 & 0 \\ 1 & 0 \end{pmatrix} = \begin{pmatrix} 1 & 1 \\ 1 & 1 \end{pmatrix} \begin{pmatrix} 1/2 & 1/2 \\ 1/2 & -1/2 \end{pmatrix}.$$

In particular, if AB is the zero matrix, it does not follow that one of A or B is also the zero matrix.

The $n \times n$ matrix A is *invertible* if there exists an $n \times n$ matrix C for which $AC = CA = I$ where I is the $n \times n$ identity matrix that has 1s along the diagonal and 0s elsewhere. The matrix C is called the *inverse* of A. Note that if A has an inverse, then this inverse is unique. For if $AB = BA = I$ as well, then

$$C = CI = C(AB) = (CA)B = IB = B.$$

The inverse of A is denoted by A^{-1}.

If A is invertible, then the vector equation $AX = V$ has a unique solution for any $V \in \mathbb{R}^n$. Indeed, $A^{-1}V$ is one solution. Moreover, it is the only one, for if Y is another solution, then we have

$$Y = (A^{-1}A)Y = A^{-1}(AY) = A^{-1}V.$$

For the converse of this statement, recall that the equation $AX = V$ has unique solutions if and only if the *reduced row echelon form* of the matrix A is the identity matrix. The reduced row echelon form of A is obtained by applying to A a sequence of *elementary row operations* of the form

1. Add k times row i of A to row j;
2. Interchange row i and j;
3. Multiply row i by $k \neq 0$.

Note that these elementary row operations correspond exactly to the operations that are used to solve linear systems of algebraic equations:

1. Add k times equation i to equation j;
2. Interchange equations i and j;
3. Multiply equation i by $k \neq 0$.

Each of these elementary row operations may be represented by multiplying A by an *elementary* matrix. For example, if $L = [\ell_{ij}]$ is the matrix that has 1's along the diagonal, $\ell_{ji} = k$ for some choice of i and j, $i \neq j$, and all other entries 0, then LA is the matrix that is obtained by performing row operation 1 on A. Similarly, if L has 1's along the diagonal with the exception that $\ell_{ii} = \ell_{jj} = 0$, but $\ell_{ij} = \ell_{ji} = 1$, and all other entries are 0, then LA is the matrix that results after performing row operation 2 on A. Finally, if L is the identity matrix with a k instead of 1 in the ii position, then LA is the matrix obtained by performing row operation 3. A matrix L in one of these three forms is called an elementary matrix.

Each elementary matrix is invertible, since its inverse is given by the matrix that simply "undoes" the corresponding row operation. As a consequence, any product of elementary matrices is invertible. Therefore, if $L_1, \ldots, L_n$ are the elementary matrices that correspond to the row operations that put

A into the reduced row echelon form, which is the identity matrix, then $(L_n \cdots L_1) = A^{-1}$. That is, if the vector equation $AX = V$ has unique solutions for any $V \in \mathbb{R}^n$, then *A* is invertible. Thus we have our first important result.

Proposition.   *Let A be an n × n matrix. Then the system of algebraic equations $AX = V$ has a unique solution for any $V \in \mathbb{R}^n$ if and only if A is invertible.* ■

Thus the natural question now is: How do we tell if *A* is invertible? One answer is provided by the following result.

Proposition.   *The matrix A is invertible if and only if the columns of A form a linearly independent set of vectors.*

Proof: Suppose first that *A* is invertible and has columns $V_1, \ldots, V_n$. We have $AE_j = V_j$ where the E_j form the standard basis of $\mathbb{R}^n$. If the V_j are not linearly independent, we may find real numbers $\alpha_1, \ldots, \alpha_n$, not all zero, such that $\sum_j \alpha_j V_j = 0$. But then

$$0 = \sum_{j=1}^{n} \alpha_j AE_j = A \left(\sum_{j=1}^{n} \alpha_j E_j \right).$$

Hence the equation $AX = 0$ has two solutions, the nonzero vector $(\alpha_1, \ldots, \alpha_n)$ and the 0 vector. This contradicts the previous proposition.

Conversely, suppose that the V_j are linearly independent. If *A* is not invertible, then we may find a pair of vectors X_1 and X_2 with $X_1 \neq X_2$ and $AX_1 = AX_2$. Therefore the nonzero vector $Z = X_1 - X_2$ satisfies $AZ = 0$. Let $Z = (\alpha_1, \ldots, \alpha_n)$. Then we have

$$0 = AZ = \sum_{j=1}^{n} \alpha_j V_j,$$

so that the V_j are not linearly independent. This contradiction establishes the result. ■

A more computable criterion for determining whether or not a matrix is invertible, as in the 2×2 case, is given by the determinant of *A*. Given the $n \times n$ matrix *A*, we will denote by A_{ij} the $(n-1) \times (n-1)$ matrix obtained by deleting the *i*th row and *j*th column of *A*.

Definition
The *determinant* of $A = [a_{ij}]$ is defined inductively by

$$\det A = \sum_{k=1}^{n} (-1)^{1+k} a_{1k} \det A_{1k}.$$

Note that we know the determinant of a 2×2 matrix, so this induction makes sense for $k > 2$.

Example. From the definition we compute

$$\det \begin{pmatrix} 1 & 2 & 3 \\ 4 & 5 & 6 \\ 7 & 8 & 9 \end{pmatrix} = 1 \det \begin{pmatrix} 5 & 6 \\ 8 & 9 \end{pmatrix} - 2 \det \begin{pmatrix} 4 & 6 \\ 7 & 9 \end{pmatrix} + 3 \det \begin{pmatrix} 4 & 5 \\ 7 & 8 \end{pmatrix}$$

$$= -3 + 12 - 9 = 0.$$

We remark that the definition of $\det A$ given above involves "expanding along the first row" of A. One can equally well expand along the jth row so that

$$\det A = \sum_{k=1}^{n} (-1)^{j+k} a_{jk} \det A_{jk}.$$

We will not prove this fact; the proof is an entirely straightforward though tedious calculation. Similarly, $\det A$ can be calculated by expanding along a given column (see Exercise 1). ∎

Example. Expanding the matrix in the previous example along the second and third rows yields the same result:

$$\det \begin{pmatrix} 1 & 2 & 3 \\ 4 & 5 & 6 \\ 7 & 8 & 9 \end{pmatrix} = -4 \det \begin{pmatrix} 2 & 3 \\ 8 & 9 \end{pmatrix} + 5 \det \begin{pmatrix} 1 & 3 \\ 7 & 9 \end{pmatrix} - 6 \det \begin{pmatrix} 1 & 2 \\ 7 & 8 \end{pmatrix}$$

$$= 24 - 60 + 36 = 0$$

$$= 7 \det \begin{pmatrix} 2 & 3 \\ 5 & 6 \end{pmatrix} - 8 \det \begin{pmatrix} 1 & 3 \\ 4 & 6 \end{pmatrix} + 9 \det \begin{pmatrix} 1 & 2 \\ 4 & 5 \end{pmatrix}$$

$$= -21 + 48 - 27 = 0.$$

Incidentally, note that this matrix is not invertible, since

$$\begin{pmatrix} 1 & 2 & 3 \\ 4 & 5 & 6 \\ 7 & 8 & 9 \end{pmatrix} \begin{pmatrix} 1 \\ -2 \\ 1 \end{pmatrix} = \begin{pmatrix} 0 \\ 0 \\ 0 \end{pmatrix}.$$ ∎

The determinant of certain types of matrices is easy to compute. A matrix $[a_{ij}]$ is called upper triangular if all entries below the main diagonal are 0. That is, $a_{ij} = 0$ if $i > j$. Lower triangular matrices are defined similarly. We have

Proposition.   *If A is an upper or lower triangular $n \times n$ matrix, then* det *A is the product of the entries along the diagonal. That is,* $\det[a_{ij}] = a_{11} \ldots a_{nn}$. ∎

The proof is a straightforward application of induction. The following proposition describes the effects that elementary row operations have on the determinant of a matrix.

Proposition.   *Let A and B be $n \times n$ matrices.*

1. *Suppose the matrix B is obtained by adding a multiple of one row of A to another row of A. Then* det $B =$ det *A.*
2. *Suppose B is obtained by interchanging two rows of A. Then* det $B = -$ det *A.*
3. *Suppose B is obtained by multiplying each element of a row of A by k. Then* det $B = k$ det *A.*

Proof: The proof of the proposition is straightforward when A is a 2×2 matrix, so we use induction. Suppose A is $k \times k$ with $k > 2$. To compute det B, we expand along a row that is left untouched by the row operation. By induction on k, we see that det B is a sum of determinants of size $(k - 1) \times (k - 1)$. Each of these subdeterminants has precisely the same row operation performed on it as in the case of the full matrix. By induction, it follows that each of these subdeterminants is multiplied by $1, -1$, or k in cases 1, 2, and 3, respectively. Hence det B has the same property. ∎

In particular, we note that if L is an elementary matrix, then

$$\det(LA) = (\det L)(\det A).$$

Indeed, det $L = 1, -1$, or k in cases 1–3 above (see Exercise 7). The preceding proposition now yields a criterion for A to be invertible:

Corollary.   **(Invertibility Criterion)**   *The matrix A is invertible if and only if* det $A \neq 0$.

Proof: By elementary row operations, we can manipulate any matrix A into an upper triangular matrix. Then A is invertible if and only if all diagonal entries of this row reduced matrix are nonzero. In particular, the determinant of this matrix is nonzero. Now, by the previous observation, row operations multiply det A by nonzero numbers, so we see that all of the diagonal entries are nonzero if and only if det A is also nonzero. This concludes the proof. ∎

We conclude this section with a further important property of determinants.

Proposition. $\det(AB) = (\det A)(\det B)$.

Proof: If either A or B is noninvertible, then AB is also noninvertible (see Exercise 11). Hence the proposition is true since both sides of the equation are zero. If A is invertible, then we can write

$$A = L_1 \ldots L_n \cdot I$$

where each L_j is an elementary matrix. Hence

$$
\begin{aligned}
\det(AB) &= \det(L_1 \ldots L_n B) \\
&= \det(L_1) \det(L_2 \ldots L_n B) \\
&= \det(L_1)(\det L_2) \ldots (\det L_n)(\det B) \\
&= \det(L_1 \ldots L_n) \det(B) \\
&= \det(A) \det(B).
\end{aligned}
$$
∎

5.2 Eigenvalues and Eigenvectors

As we saw in Chapter 3, eigenvalues and eigenvectors play a central role in the process of solving linear systems of differential equations.

Definition
A vector V is an *eigenvector* of an $n \times n$ matrix A if V is a nonzero solution to the system of linear equations $(A - \lambda I)V = 0$. The quantity λ is called an *eigenvalue* of A, and V is an eigenvector associated to λ.

Just as in Chapter 2, the eigenvalues of a matrix A may be real or complex and the associated eigenvectors may have complex entries.

By the invertibility criterion of the previous section, it follows that λ is an eigenvalue of A if and only if λ is a root of the *characteristic equation*

$$\det(A - \lambda I) = 0.$$

Since A is $n \times n$, this is a polynomial equation of degree n, which therefore has exactly n roots (counted with multiplicity).

As we saw in $\mathbb{R}^2$, there are many different types of solutions of systems of differential equations, and these types depend on the configuration of the eigenvalues of A and the resulting canonical forms. There are many, many more types of canonical forms in higher dimensions. We will describe these types in this and the following sections, but we will relegate some of the more specialized proofs of these facts to the exercises.

Suppose first that $\lambda_1, \ldots, \lambda_\ell$ are real and distinct eigenvalues of A with associated eigenvectors $V_1, \ldots, V_\ell$. Here "distinct" means that no two of the eigenvalues are equal. Thus $AV_k = \lambda_k V_k$ for each k. We claim that the V_k are linearly independent. If not, we may choose a maximal subset of the V_i that are linearly independent, say, $V_1, \ldots, V_j$. Then any other eigenvector may be written in a unique way as a linear combination of $V_1, \ldots, V_j$. Say V_{j+1} is one such eigenvector. Then we may find α_i, not all 0, such that

$$V_{j+1} = \alpha_1 V_1 + \cdots + \alpha_j V_j.$$

Multiplying both sides of this equation by A, we find

$$\lambda_{j+1} V_{j+1} = \alpha_1 AV_1 + \cdots + \alpha_j AV_j$$
$$= \alpha_1 \lambda_1 V_1 + \cdots + \alpha_j \lambda_j V_j.$$

Now $\lambda_{j+1} \neq 0$ for otherwise we would have

$$\alpha_1 \lambda_1 V_1 + \cdots + \alpha_j \lambda_j V_j = 0,$$

with each $\lambda_i \neq 0$. This contradicts the fact that $V_1, \ldots, V_j$ are linearly independent. Hence we have

$$V_{j+1} = \alpha_1 \frac{\lambda_1}{\lambda_{j+1}} V_1 + \cdots + \alpha_j \frac{\lambda_j}{\lambda_{j+1}} V_j.$$

Since the λ_i are distinct, we have now written V_{j+1} in two different ways as a linear combination of $V_1, \ldots, V_j$. This contradicts the fact that this set of vectors is linearly independent. We have proved:

Proposition. *Suppose $\lambda_1, \ldots, \lambda_\ell$ are real and distinct eigenvalues for A with associated eigenvectors $V_1, \ldots, V_\ell$. Then the V_j are linearly independent.* ■

Of primary importance when we return to differential equations is the

Corollary. *Suppose A is an $n \times n$ matrix with real, distinct eigenvalues. Then there is a matrix T such that*

$$T^{-1}AT = \begin{pmatrix} \lambda_1 & & \\ & \ddots & \\ & & \lambda_n \end{pmatrix}$$

where all of the entries off the diagonal are 0.

Proof: Let V_j be an eigenvector associated to λ_j. Consider the linear map T for which $TE_j = V_j$, where the E_j form the standard basis of $\mathbb{R}^n$. That is, T is the matrix whose columns are $V_1, \ldots, V_n$. Since the V_j are linearly independent, T is invertible and we have

$$(T^{-1}AT)E_j = T^{-1}AV_j$$
$$= \lambda_j T^{-1}V_j$$
$$= \lambda_j E_j.$$

That is, the jth column of $T^{-1}AT$ is just the vector $\lambda_j E_j$, as required.   ∎

Example. Let

$$A = \begin{pmatrix} 1 & 2 & -1 \\ 0 & 3 & -2 \\ 0 & 2 & -2 \end{pmatrix}.$$

Expanding $\det(A - \lambda I)$ along the first column, we find that the characteristic equation of A is

$$\det(A - \lambda I) = (1 - \lambda) \det \begin{pmatrix} 3 - \lambda & -2 \\ 2 & -2 - \lambda \end{pmatrix}$$
$$= (1 - \lambda)((3 - \lambda)(-2 - \lambda) + 4)$$
$$= (1 - \lambda)(\lambda - 2)(\lambda + 1),$$

so the eigenvalues are 2, 1, and -1. The eigenvector corresponding to $\lambda = 2$ is given by solving the equations $(A - 2I)X = 0$, which yields

$$-x + 2y - z = 0$$
$$y - 2z = 0$$
$$2y - 4z = 0.$$

These equations reduce to

$$x - 3z = 0$$
$$y - 2z = 0.$$

Hence $V_1 = (3, 2, 1)$ is an eigenvector associated to $\lambda = 2$. In similar fashion we find that $(1, 0, 0)$ is an eigenvector associated to $\lambda = 1$, while $(0, 1, 2)$ is an eigenvector associated to $\lambda = -1$. Then we set

$$T = \begin{pmatrix} 3 & 1 & 0 \\ 2 & 0 & 1 \\ 1 & 0 & 2 \end{pmatrix}.$$

A simple calculation shows that

$$AT = T \begin{pmatrix} 2 & 0 & 0 \\ 0 & 1 & 0 \\ 0 & 0 & -1 \end{pmatrix}.$$

Since $\det T = -3$, T is invertible and we have

$$T^{-1}AT = \begin{pmatrix} 2 & 0 & 0 \\ 0 & 1 & 0 \\ 0 & 0 & -1 \end{pmatrix}. \qquad \blacksquare$$

5.3 Complex Eigenvalues

Now we treat the case where A has nonreal (complex) eigenvalues. Suppose $\alpha + i\beta$ is an eigenvalue of A with $\beta \neq 0$. Since the characteristic equation for A has real coefficients, it follows that if $\alpha + i\beta$ is an eigenvalue, then so is its complex conjugate $\overline{\alpha + i\beta} = \alpha - i\beta$.

Another way to see this is the following. Let V be an eigenvector associated to $\alpha + i\beta$. Then the equation

$$AV = (\alpha + i\beta)V$$

shows that V is a vector with complex entries. We write

$$V = \begin{pmatrix} x_1 + iy_1 \\ \vdots \\ x_n + iy_n \end{pmatrix}.$$

Let $\overline{V}$ denote the complex conjugate of V:

$$\overline{V} = \begin{pmatrix} x_1 - iy_1 \\ \vdots \\ x_n - iy_n \end{pmatrix}.$$

Then we have

$$A\overline{V} = \overline{AV} = \overline{(\alpha + i\beta)V} = (\alpha - i\beta)\overline{V},$$

which shows that $\overline{V}$ is an eigenvector associated to the eigenvalue $\alpha - i\beta$.

Notice that we have (temporarily) stepped out of the "real" world of $\mathbb{R}^n$ and into the world $\mathbb{C}^n$ of complex vectors. This is not really a problem, since all of the previous linear algebraic results hold equally well for complex vectors.

Now suppose that A is a $2n \times 2n$ matrix with distinct nonreal eigenvalues $\alpha_j \pm i\beta_j$ for $j = 1,\ldots,n$. Let V_j and $\overline{V}_j$ denote the associated eigenvectors. Then, just as in the previous proposition, this collection of eigenvectors is linearly independent. That is, if we have

$$\sum_{j=1}^{n}(c_j V_j + d_j \overline{V}_j) = 0$$

where the c_j and d_j are now complex numbers, then we must have $c_j = d_j = 0$ for each j.

Now we change coordinates to put A into canonical form. Let

$$W_{2j-1} = \frac{1}{2}(V_j + \overline{V}_j)$$

$$W_{2j} = \frac{-i}{2}(V_j - \overline{V}_j).$$

Note that W_{2j-1} and W_{2j} are both real vectors. Indeed, W_{2j-1} is just the real part of V_j while W_{2j} is its imaginary part. So working with the W_j brings us back home to $\mathbb{R}^n$.

Proposition. *The vectors $W_1,\ldots,W_{2n}$ are linearly independent.*

Proof: Suppose not. Then we can find real numbers c_j, d_j for $j = 1,\ldots,n$ such that

$$\sum_{j=1}^{n}\left(c_j W_{2j-1} + d_j W_{2j}\right) = 0$$

but not all of the c_j and d_j are zero. But then we have

$$\frac{1}{2} \sum_{j=1}^{n} \left(c_j(V_j + \overline{V_j}) - id_j(V_j - \overline{V_j}) \right) = 0$$

from which we find

$$\sum_{j=1}^{n} \left((c_j - id_j)V_j + (c_j + id_j)\overline{V_j} \right) = 0.$$

Since the V_j and $\overline{V_j}$'s are linearly independent, we must have $c_j \pm id_j = 0$, from which we conclude $c_j = d_j = 0$ for all j. This contradiction establishes the result.   ■

Note that we have

$$\begin{aligned}
AW_{2j-1} &= \frac{1}{2}(AV_j + A\overline{V_j}) \\
&= \frac{1}{2}\left((\alpha + i\beta)V_j + (\alpha - i\beta)\overline{V_j} \right) \\
&= \frac{\alpha}{2}(V_j + \overline{V_j}) + \frac{i\beta}{2}(V_j - \overline{V_j}) \\
&= \alpha W_{2j-1} - \beta W_{2j}.
\end{aligned}$$

Similarly, we compute

$$AW_{2j} = \beta W_{2j-1} + \alpha W_{2j}.$$

Now consider the linear map T for which $TE_j = W_j$ for $j = 1, \ldots, 2n$. That is, the matrix associated to T has columns $W_1, \ldots, W_{2n}$. Note that this matrix has real entries. Since the W_j are linearly independent, it follows from Section 5.1 that T is invertible. Now consider the matrix $T^{-1}AT$. We have

$$\begin{aligned}
(T^{-1}AT)E_{2j-1} &= T^{-1}AW_{2j-1} \\
&= T^{-1}(\alpha W_{2j-1} - \beta W_{2j}) \\
&= \alpha E_{2j-1} - \beta E_{2j}
\end{aligned}$$

and similarly

$$(T^{-1}AT)E_{2j} = \beta E_{2j-1} + \alpha E_{2j}.$$

Therefore the matrix associated to $T^{-1}AT$ is

$$T^{-1}AT = \begin{pmatrix} D_1 & & \\ & \ddots & \\ & & D_n \end{pmatrix}$$

where each D_j is a 2×2 matrix of the form

$$D_j = \begin{pmatrix} \alpha_j & \beta_j \\ -\beta_j & \alpha_j \end{pmatrix}.$$

This is our canonical form for matrices with distinct nonreal eigenvalues.

Combining the results of this and the previous section, we have:

Theorem. *Suppose that the $n \times n$ matrix A has distinct eigenvalues. Then there exists a linear map T so that*

$$T^{-1}AT = \begin{pmatrix} \lambda_1 & & & & & \\ & \ddots & & & & \\ & & \lambda_k & & & \\ & & & D_1 & & \\ & & & & \ddots & \\ & & & & & D_\ell \end{pmatrix}$$

where the D_j are 2×2 matrices in the form

$$D_j = \begin{pmatrix} \alpha_j & \beta_j \\ -\beta_j & \alpha_j \end{pmatrix}.$$
∎

5.4 Bases and Subspaces

To deal with the case of a matrix with repeated eigenvalues, we need some further algebraic concepts. Recall that the collection of all linear combinations of a given finite set of vectors is called a *subspace* of $\mathbb{R}^n$. More precisely, given $V_1, \ldots, V_k \in \mathbb{R}^n$, the set

$$\mathcal{S} = \{\alpha_1 V_1 + \cdots + \alpha_k V_k \mid \alpha_j \in \mathbb{R}\}$$

is a subspace of $\mathbb{R}^n$. In this case we say that $\mathcal{S}$ is *spanned* by $V_1, \ldots, V_k$.

Definition
Let S be a subspace of $\mathbb{R}^n$. A collection of vectors $V_1, \ldots, V_k$ is a *basis* of S if the V_j are linearly independent and span S.

Note that a subspace always has a basis, for if S is spanned by $V_1, \ldots, V_k$, we can always throw away certain of the V_j in order to reach a linearly independent subset of these vectors that spans S. More precisely, if the V_j are not linearly independent, then we may find one of these vectors, say, V_k, for which

$$V_k = \beta_1 V_1 + \cdots + \beta_{k-1} V_{k-1}.$$

Hence we can write any vector in S as a linear combination of the $V_1, \ldots, V_{k-1}$ alone; the vector V_k is extraneous. Continuing in this fashion, we eventually reach a linearly independent subset of the V_j that spans S.

More important for our purposes is:

Proposition. *Every basis of a subspace $S \subset \mathbb{R}^n$ has the same number of elements.*

Proof: We first observe that the system of k linear equations in $k+\ell$ unknowns given by

$$a_{11}x_1 + \cdots + a_{1\,k+\ell}x_{k+\ell} = 0$$

$$\vdots$$

$$a_{k1}x_1 + \cdots + a_{k\,k+\ell}x_{k+\ell} = 0$$

always has a nonzero solution if $\ell > 0$. Indeed, using row reduction, we may first solve for one unknown in terms of the others, and then we may eliminate this unknown to obtain a system of $k - 1$ equations in $k + \ell - 1$ unknowns. Thus we are finished by induction (the first case, $k = 1$, being obvious).

Now suppose that $V_1, \ldots, V_k$ is a basis for the subspace S. Suppose that $W_1, \ldots, W_{k+\ell}$ is also a basis of S, with $\ell > 0$. Then each W_j is a linear combination of the V_i, so we have constants a_{ij} such that

$$W_j = \sum_{i=1}^{k} a_{ij} V_i, \quad \text{for } j = 1, \ldots, k + \ell.$$

By the previous observation, the system of k equations

$$\sum_{j=1}^{k+\ell} a_{ij}x_j = 0, \quad \text{for } i = 1, \ldots, k$$

has a nonzero solution $(c_1, \ldots, c_{k+\ell})$. Then

$$\sum_{j=1}^{k+\ell} c_j W_j = \sum_{j=1}^{k+\ell} c_j \left(\sum_{i=1}^{k} a_{ij} V_i \right) = \sum_{i=1}^{k} \left(\sum_{j=1}^{k+\ell} a_{ij} c_j \right) V_i = 0$$

so that the W_j are linearly dependent. This contradiction completes the proof. ∎

As a consequence of this result, we may define the *dimension* of a subspace S as the number of vectors that form any basis for S. In particular, $\mathbb{R}^n$ is a subspace of itself, and its dimension is clearly n. Furthermore, any other subspace of $\mathbb{R}^n$ must have dimension less than n, for otherwise we would have a collection of more than n vectors in $\mathbb{R}^n$ that are linearly independent. This cannot happen by the previous proposition. The set consisting of only the 0 vector is also a subspace, and we define its dimension to be zero. We write $\dim S$ for the dimension of the subspace S.

Example. A straight line through the origin in $\mathbb{R}^n$ forms a one-dimensional subspace of $\mathbb{R}^n$, since any vector on this line may be written uniquely as tV where $V \in \mathbb{R}^n$ is a fixed nonzero vector lying on the line and $t \in \mathbb{R}$ is arbitrary. Clearly, the single vector V forms a basis for this subspace. ∎

Example. The plane P in $\mathbb{R}^3$ defined by

$$x + y + z = 0$$

is a two-dimensional subspace of $\mathbb{R}^3$. The vectors $(1, 0, -1)$ and $(0, 1, -1)$ both lie in P and are linearly independent. If $W \in P$, we may write

$$W = \begin{pmatrix} x \\ y \\ -y - x \end{pmatrix} = x \begin{pmatrix} 1 \\ 0 \\ -1 \end{pmatrix} + y \begin{pmatrix} 0 \\ 1 \\ -1 \end{pmatrix},$$

so these vectors also span P. ∎

As in the planar case, we say that a function $T: \mathbb{R}^n \to \mathbb{R}^n$ is linear if $T(X) = AX$ for some $n \times n$ matrix A. T is called a *linear map* or *linear transformation*. Using the properties of matrices discussed in Section 5.1, we have

$$T(\alpha X + \beta Y) = \alpha T(X) + \beta T(Y)$$

for any $\alpha, \beta \in \mathbb{R}$ and $X, Y \in \mathbb{R}^n$. We say that the linear map T is invertible if the matrix A associated to T has an inverse.

For the study of linear systems of differential equations, the most important types of subspaces are the kernels and ranges of linear maps. We define the *kernel* of T, denoted Ker T, to be the set of vectors mapped to 0 by T. The *range* of T consists of all vectors W for which there exists a vector V for which $TV = W$. This, of course, is a familiar concept from calculus. The difference here is that the range of T is always a subspace of $\mathbb{R}^n$.

Example. Consider the linear map

$$T(X) = \begin{pmatrix} 0 & 1 & 0 \\ 0 & 0 & 1 \\ 0 & 0 & 0 \end{pmatrix} X.$$

If $X = (x, y, z)$, then

$$T(X) = \begin{pmatrix} y \\ z \\ 0 \end{pmatrix}.$$

Hence Ker T consists of all vectors of the form $(\alpha, 0, 0)$ while Range T is the set of vectors of the form $(\beta, \gamma, 0)$, where $\alpha, \beta, \gamma \in \mathbb{R}$. Both sets are clearly subspaces. ■

Example. Let

$$T(X) = AX = \begin{pmatrix} 1 & 2 & 3 \\ 4 & 5 & 6 \\ 7 & 8 & 9 \end{pmatrix} X.$$

For Ker T, we seek vectors X that satisfy $AX = 0$. Using row reduction, we find that the reduced row echelon form of A is the matrix

$$\begin{pmatrix} 1 & 0 & -1 \\ 0 & 1 & 2 \\ 0 & 0 & 0 \end{pmatrix}.$$

Hence the solutions $X = (x, y, z)$ of $AX = 0$ satisfy $x = z$, $y = -2z$. Therefore any vector in Ker T is of the form $(z, -2z, z)$, so Ker T has dimension one. For Range T, note that the columns of A are vectors in Range T, since they are the images of $(1, 0, 0)$, $(0, 1, 0)$, and $(0, 0, 1)$, respectively. These vectors are not linearly independent since

$$-1 \begin{pmatrix} 1 \\ 4 \\ 7 \end{pmatrix} + 2 \begin{pmatrix} 2 \\ 5 \\ 8 \end{pmatrix} = \begin{pmatrix} 3 \\ 6 \\ 9 \end{pmatrix}.$$

However, $(1, 4, 7)$ and $(2, 5, 8)$ are linearly independent, so these two vectors give a basis of Range T.   ∎

Proposition.   *Let $T : \mathbb{R}^n \to \mathbb{R}^n$ be a linear map. Then* Ker T *and* Range T *are both subspaces of $\mathbb{R}^n$. Moreover,*

$$\dim \text{Ker } T + \dim \text{Range } T = n.$$

Proof: First suppose that Ker $T = \{0\}$. Let $E_1, \ldots, E_n$ be the standard basis of $\mathbb{R}^n$. Then we claim that $TE_1, \ldots, TE_n$ are linearly independent. If this is not the case, then we may find $\alpha_1, \ldots, \alpha_n$, not all 0, such that

$$\sum_{j=1}^{n} \alpha_j TE_j = 0.$$

But then we have

$$T \left(\sum_{j=1}^{n} \alpha_j E_j \right) = 0,$$

which implies that $\sum \alpha_j E_j \in$ Ker T, so that $\sum \alpha_j E_j = 0$. Hence each $\alpha_j = 0$, which is a contradiction. Thus the vectors TE_j are linearly independent. But then, given $V \in \mathbb{R}^n$, we may write

$$V = \sum_{j=1}^{n} \beta_j TE_j$$

for some $\beta_1, \ldots, \beta_n$. Hence

$$V = T\left(\sum_{j=1}^{n} \beta_j E_j\right)$$

which shows that Range $T = \mathbb{R}^n$. Hence both Ker T and Range T are subspaces of $\mathbb{R}^n$ and we have dim Ker $T = 0$ and dim Range $T = n$.

If Ker $T \neq \{0\}$, we may find a nonzero vector $V_1 \in$ Ker T. Clearly, $T(\alpha V_1) = 0$ for any $\alpha \in \mathbb{R}$, so all vectors of the form αV_1 lie in Ker T. If Ker T contains additional vectors, choose one and call it V_2. Then Ker T contains all linear combinations of V_1 and V_2, since

$$T(\alpha_1 V_1 + \alpha_2 V_2) = \alpha_1 TV_1 + \alpha_2 TV_2 = 0.$$

Continuing in this fashion we obtain a set of linearly independent vectors that span Ker T, thus showing that Ker T is a subspace. Note that this process must end, since every collection of more than n vectors in $\mathbb{R}^n$ is linearly dependent. A similar argument works to show that Range T is a subspace.

Now suppose that $V_1, \ldots, V_k$ form a basis of Ker T where $0 < k < n$ (the case where $k = n$ being obvious). Choose vectors $W_{k+1}, \ldots, W_n$ so that $V_1, \ldots, V_k, W_{k+1}, \ldots, W_n$ form a basis of $\mathbb{R}^n$. Let $Z_j = TW_j$ for each j. Then the vectors Z_j are linearly independent, for if we had

$$\alpha_{k+1} Z_{k+1} + \cdots + \alpha_n Z_n = 0,$$

then we would also have

$$T(\alpha_{k+1} W_{k+1} + \cdots + \alpha_n W_n) = 0.$$

This implies that

$$\alpha_{k+1} W_{k+1} + \cdots + \alpha_n W_n \in \text{Ker } T.$$

But this is impossible, since we cannot write any W_j (and hence any linear combination of the W_j) as a linear combination of the V_i. This proves that the sum of the dimensions of Ker T and Range T is n. ∎

We remark that it is easy to find a set of vectors that spans Range T; simply take the set of vectors that comprise the columns of the matrix associated to T. This works since the ith column vector of this matrix is the image of the standard basis vector E_i under T. In particular, if these column vectors are linearly independent, then Ker $T = \{0\}$ and there is a unique solution to the equation $T(X) = V$ for every $V \in \mathbb{R}^n$. Hence we have:

Corollary. *If $T: \mathbb{R}^n \to \mathbb{R}^n$ is a linear map with* dim Ker $T = 0$, *then T is invertible.* ∎

5.5 Repeated Eigenvalues

In this section we describe the canonical forms that arise when a matrix has repeated eigenvalues. Rather than spending an inordinate amount of time developing the general theory in this case, we will give the details only for 3×3 and 4×4 matrices with repeated eigenvalues. More general cases are relegated to the exercises. We justify this omission in the next section where we show that the "typical" matrix has distinct eigenvalues and hence can be handled as in the previous section. (If you happen to meet a random matrix while walking down the street, the chances are very good that this matrix will have distinct eigenvalues!) The most general result regarding matrices with repeated eigenvalues is given by:

Proposition. *Let A be an $n \times n$ matrix. Then there is a change of coordinates T for which*

$$
T^{-1}AT = \begin{pmatrix} B_1 & & \\ & \ddots & \\ & & B_k \end{pmatrix}
$$

where each of the B_j's is a square matrix (and all other entries are zero) of one of the following forms:

$$
\text{(i)} \quad \begin{pmatrix} \lambda & 1 & & & \\ & \lambda & 1 & & \\ & & \ddots & \ddots & \\ & & & \ddots & 1 \\ & & & & \lambda \end{pmatrix} \qquad \text{(ii)} \quad \begin{pmatrix} C_2 & I_2 & & & \\ & C_2 & I_2 & & \\ & & \ddots & \ddots & \\ & & & \ddots & I_2 \\ & & & & C_2 \end{pmatrix}
$$

where

$$
C_2 = \begin{pmatrix} \alpha & \beta \\ -\beta & \alpha \end{pmatrix}, \quad I_2 = \begin{pmatrix} 1 & 0 \\ 0 & 1 \end{pmatrix},
$$

and where $\alpha, \beta, \lambda \in \mathbb{R}$ with $\beta \neq 0$. The special cases where $B_j = (\lambda)$ or

$$B_j = \begin{pmatrix} \alpha & \beta \\ -\beta & \alpha \end{pmatrix}$$

are, of course, allowed. ∎

We first consider the case of $\mathbb{R}^3$. If A has repeated eigenvalues in $\mathbb{R}^3$, then all eigenvalues must be real. There are then two cases. Either there are two distinct eigenvalues, one of which is repeated, or else all eigenvalues are the same. The former case can be handled by a process similar to that described in Chapter 3, so we restrict our attention here to the case where A has a single eigenvalue λ of multiplicity 3.

Proposition. *Suppose A is a 3×3 matrix for which λ is the only eigenvalue. Then we may find a change of coordinates T such that $T^{-1}AT$ assumes one of the following three forms:*

$$
\text{(i)} \begin{pmatrix} \lambda & 0 & 0 \\ 0 & \lambda & 0 \\ 0 & 0 & \lambda \end{pmatrix}
\quad
\text{(ii)} \begin{pmatrix} \lambda & 1 & 0 \\ 0 & \lambda & 0 \\ 0 & 0 & \lambda \end{pmatrix}
\quad
\text{(iii)} \begin{pmatrix} \lambda & 1 & 0 \\ 0 & \lambda & 1 \\ 0 & 0 & \lambda \end{pmatrix}.
$$

Proof: Let K be the kernel of $A - \lambda I$. Any vector in K is an eigenvector of A. There are then three subcases depending on whether the dimension of K is 1, 2, or 3.

If the dimension of K is 3, then $(A - \lambda I)V = 0$ for any $V \in \mathbb{R}^3$. Hence $A = \lambda I$. This yields matrix (i).

Suppose the dimension of K is 2. Let R be the range of $A - \lambda I$. Then R has dimension 1 since $\dim K + \dim R = 3$, as we saw in the previous section. We claim that $R \subset K$. If this is not the case, let $V \in R$ be a nonzero vector. Since $(A - \lambda I)V \in R$ and R is one dimensional, we must have $(A - \lambda I)V = \mu V$ for some $\mu \neq 0$. But then $AV = (\lambda + \mu)V$, so we have found a new eigenvalue $\lambda + \mu$. This contradicts our assumption, so we must have $R \subset K$.

Now let $V_1 \in R$ be nonzero. Since $V_1 \in K$, V_1 is an eigenvector and so $(A - \lambda I)V_1 = 0$. Since V_1 also lies in R, we may find $V_2 \in \mathbb{R}^3 - K$ with $(A - \lambda I)V_2 = V_1$. Since K is two dimensional we may choose a second vector $V_3 \in K$ such that V_1 and V_3 are linearly independent. Note that V_3 is also an eigenvector. If we now choose the change of coordinates $TE_j = V_j$ for $j = 1, 2, 3$, then it follows easily that $T^{-1}AT$ assumes the form of case (ii).

Finally, suppose that K has dimension 1. Thus R has dimension 2. We claim that, in this case, $K \subset R$. If this is not the case, then $(A - \lambda I)R = R$ and so

$A - \lambda I$ is invertible on R. Thus, if $V \in R$, there is a unique $W \in R$ for which $(A - \lambda I)W = V$. In particular, we have

$$AV = A(A - \lambda I)W$$
$$= (A^2 - \lambda A)W$$
$$= (A - \lambda I)(AW).$$

This shows that, if $V \in R$, then so too is AV. Hence A also preserves the subspace R. It then follows immediately that A must have an eigenvector in R, but this then says that $K \subset R$ and we have a contradiction.

Next we claim that $(A - \lambda I)R = K$. To see this, note that $(A - \lambda I)R$ is one dimensional, since $K \subset R$. If $(A - \lambda I)R \neq K$, there is a nonzero vector $V \notin K$ for which $(A - \lambda I)R = \{tV\}$ where $t \in \mathbb{R}$. But then $(A - \lambda I)V = tV$ for some $t \in \mathbb{R}, t \neq 0$, and so $AV = (t + \lambda)V$ yields another new eigenvalue. Thus we must in fact have $(A - \lambda I)R = K$.

Now let $V_1 \in K$ be an eigenvector for A. As above there exists $V_2 \in R$ such that $(A - \lambda I)V_2 = V_1$. Since $V_2 \in R$ there exists V_3 such that $(A - \lambda I)V_3 = V_2$. Note that $(A - \lambda I)^2 V_3 = V_1$. The V_j are easily seen to be linearly independent. Moreover, the linear map defined by $TE_j = V_j$ finally puts A into canonical form (iii). This completes the proof. ■

Example. Suppose

$$A = \begin{pmatrix} 2 & 0 & -1 \\ 0 & 2 & 1 \\ -1 & -1 & 2 \end{pmatrix}.$$

Expanding along the first row, we find

$$\det(A - \lambda I) = (2 - \lambda)[(2 - \lambda)^2 + 1] - (2 - \lambda) = (2 - \lambda)^3$$

so the only eigenvalue is 2. Solving $(A - 2I)V = 0$ yields only one independent eigenvector $V_1 = (1, -1, 0)$, so we are in case (iii) of the proposition. We compute

$$(A - 2I)^2 = \begin{pmatrix} 1 & 1 & 0 \\ -1 & -1 & 0 \\ 0 & 0 & 0 \end{pmatrix}$$

so that the vector $V_3 = (1, 0, 0)$ solves $(A - 2I)^2 V_3 = V_1$. We also have

$$(A - 2I)V_3 = V_2 = (0, 0, -1).$$

As before, we let $TE_j = V_j$ for $j = 1, 2, 3$, so that

$$T = \begin{pmatrix} 1 & 0 & 1 \\ -1 & 0 & 0 \\ 0 & -1 & 0 \end{pmatrix}.$$

Then $T^{-1}AT$ assumes the canonical form

$$T^{-1}AT = \begin{pmatrix} 2 & 1 & 0 \\ 0 & 2 & 1 \\ 0 & 0 & 2 \end{pmatrix}.$$

∎

Example. Now suppose

$$A = \begin{pmatrix} 1 & 1 & 0 \\ -1 & 3 & 0 \\ -1 & 1 & 2 \end{pmatrix}.$$

Again expanding along the first row, we find

$$\det(A - \lambda I) = (1 - \lambda)[(3 - \lambda)(2 - \lambda)] + (2 - \lambda) = (2 - \lambda)^3$$

so again the only eigenvalue is 2. This time, however, we have

$$A - 2I = \begin{pmatrix} -1 & 1 & 0 \\ -1 & 1 & 0 \\ -1 & 1 & 0 \end{pmatrix}$$

so that we have two linearly independent eigenvectors (x, y, z) for which we must have $x = y$ while z is arbitrary. Note that $(A - 2I)^2$ is the zero matrix, so we may choose any vector that is not an eigenvector as V_2, say, $V_2 = (1, 0, 0)$. Then $(A - 2I)V_2 = V_1 = (-1, -1, -1)$ is an eigenvector. A second linearly independent eigenvector is then $V_3 = (0, 0, 1)$, for example. Defining $TE_j = V_j$ as usual then yields the canonical form

$$T^{-1}AT = \begin{pmatrix} 2 & 1 & 0 \\ 0 & 2 & 0 \\ 0 & 0 & 2 \end{pmatrix}.$$

∎

Now we turn to the 4×4 case. The case of all real eigenvalues is similar to the 3×3 case (though a little more complicated algebraically) and is left as an exercise. Thus we assume that A has repeated complex eigenvalues $\alpha \pm i\beta$ with $\beta \neq 0$.

There are just two cases; either we can find a pair of linearly independent eigenvectors corresponding to $\alpha + i\beta$, or we can find only one such eigenvector. In the former case, let V_1 and V_2 be the independent eigenvectors. The $\overline{V_1}$ and $\overline{V_2}$ are linearly independent eigenvectors for $\alpha - i\beta$. As before, choose the real vectors

$$W_1 = (V_1 + \overline{V_1})/2$$

$$W_2 = -i(V_1 - \overline{V_1})/2$$

$$W_3 = (V_2 + \overline{V_2})/2$$

$$W_4 = -i(V_2 - \overline{V_2})/2.$$

If we set $TE_j = W_j$ then changing coordinates via T puts A in canonical form

$$T^{-1}AT = \begin{pmatrix} \alpha & \beta & 0 & 0 \\ -\beta & \alpha & 0 & 0 \\ 0 & 0 & \alpha & \beta \\ 0 & 0 & -\beta & \alpha \end{pmatrix}.$$

If we find only one eigenvector V_1 for $\alpha + i\beta$, then we solve the system of equations $(A - (\alpha + i\beta)I)X = V_1$ as in the case of repeated real eigenvalues. The proof of the previous proposition shows that we can always find a nonzero solution V_2 of these equations. Then choose the W_j as above and set $TE_j = W_j$. Then T puts A into the canonical form

$$T^{-1}AT = \begin{pmatrix} \alpha & \beta & 1 & 0 \\ -\beta & \alpha & 0 & 1 \\ 0 & 0 & \alpha & \beta \\ 0 & 0 & -\beta & \alpha \end{pmatrix}.$$

For example, we compute

$$
\begin{aligned}
(T^{-1}AT)E_3 &= T^{-1}AW_3 \\
&= T^{-1}A(V_2 + \overline{V_2})/2 \\
&= T^{-1}\big((V_1 + (\alpha + i\beta)V_2)/2 + (\overline{V_1} + (\alpha - i\beta)\overline{V_2})/2\big) \\
&= T^{-1}\big((V_1 + \overline{V_1})/2 + \alpha(V_2 + \overline{V_2})/2 + i\beta(V_2 - \overline{V_2})/2\big) \\
&= E_1 + \alpha E_3 - \beta E_4.
\end{aligned}
$$

Example. Let

$$A = \begin{pmatrix} 1 & -1 & 0 & 1 \\ 2 & -1 & 1 & 0 \\ 0 & 0 & -1 & 2 \\ 0 & 0 & -1 & 1 \end{pmatrix}.$$

The characteristic equation, after a little computation, is

$$(\lambda^2 + 1)^2 = 0.$$

Hence A has eigenvalues $\pm i$, each repeated twice.

Solving the system $(A - iI)X = 0$ yields one linearly independent complex eigenvector $V_1 = (1, 1 - i, 0, 0)$ associated to i. Then $\overline{V_1}$ is an eigenvector associated to the eigenvalue $-i$.

Next we solve the system $(A - iI)X = V_1$ to find $V_2 = (0, 0, 1 - i, 1)$. Then $\overline{V_2}$ solves the system $(A - iI)X = \overline{V_1}$. Finally, choose

$$W_1 = \left(V_1 + \overline{V_1}\right)/2 = \text{Re } V_1$$

$$W_2 = -i\left(V_1 - \overline{V_1}\right)/2 = \text{Im } V_1$$

$$W_3 = \left(V_2 + \overline{V_2}\right)/2 = \text{Re } V_2$$

$$W_4 = -i\left(V_2 - \overline{V_2}\right)/2 = \text{Im } V_2$$

and let $TE_j = W_j$ for $j = 1, \ldots, 4$. We have

$$T = \begin{pmatrix} 1 & 0 & 0 & 0 \\ 1 & -1 & 0 & 0 \\ 0 & 0 & 1 & -1 \\ 0 & 0 & 1 & 0 \end{pmatrix}, \quad T^{-1} = \begin{pmatrix} 1 & 0 & 0 & 0 \\ 1 & -1 & 0 & 0 \\ 0 & 0 & 0 & 1 \\ 0 & 0 & -1 & 1 \end{pmatrix}$$

and we find the canonical form

$$T^{-1}AT = \begin{pmatrix} 0 & 1 & 1 & 0 \\ -1 & 0 & 0 & 1 \\ 0 & 0 & 0 & 1 \\ 0 & 0 & -1 & 0 \end{pmatrix}. \qquad \blacksquare$$

Example. Let

$$A = \begin{pmatrix} 2 & 0 & 1 & 0 \\ 0 & 2 & 0 & 1 \\ 0 & 0 & 2 & 0 \\ 0 & -1 & 0 & 2 \end{pmatrix}.$$

The characteristic equation for A is

$$(2 - \lambda)^2((2 - \lambda)^2 + 1) = 0$$

so the eigenvalues are $2 \pm i$ and 2 (with multiplicity 2).

Solving the equations $(A - (2 + i)I)X = 0$ yields an eigenvector $V = (0, -i, 0, 1)$ for $2 + i$. Let $W_1 = (0, 0, 0, 1)$ and $W_2 = (0, -1, 0, 0)$ be the real and imaginary parts of V.

Solving the equations $(A - 2I)X = 0$ yields only one eigenvector associated to 2, namely, $W_3 = (1, 0, 0, 0)$. Then we solve $(A - 2I)X = W_3$ to find $W_4 = (0, 0, 1, 0)$. Setting $TE_j = W_j$ as usual puts A into the canonical form

$$T^{-1}AT = \begin{pmatrix} 2 & 1 & 0 & 0 \\ -1 & 2 & 0 & 0 \\ 0 & 0 & 2 & 1 \\ 0 & 0 & 0 & 2 \end{pmatrix},$$

as is easily checked. ∎

5.6 Genericity

We have mentioned several times that "most" matrices have distinct eigenvalues. Our goal in this section is to make this precise.

Recall that a set $\mathcal{U} \subset \mathbb{R}^n$ is *open* if whenever $X \in \mathcal{U}$ there is an open ball about X contained in $\mathcal{U}$; that is, for some $a > 0$ (depending on X) the open ball about X of radius a,

$$\{Y \in \mathbb{R}^n \mid |Y - X| < a\},$$

is contained in $\mathcal{U}$. Using geometrical language we say that if X belongs to an open set $\mathcal{U}$, any point sufficiently near to X also belongs to $\mathcal{U}$.

Another kind of subset of $\mathbb{R}^n$ is a *dense* set: $\mathcal{U} \subset \mathbb{R}^n$ is dense if there are points in $\mathcal{U}$ arbitrarily close to each point in $\mathbb{R}^n$. More precisely, if $X \in \mathbb{R}^n$, then for every $\epsilon > 0$ there exists some $Y \in \mathcal{U}$ with $|X - Y| < \epsilon$. Equivalently, $\mathcal{U}$ is dense in $\mathbb{R}^n$ if $V \cap \mathcal{U}$ is nonempty for every nonempty open set $V \subset \mathbb{R}^n$. For example, the rational numbers form a dense subset of $\mathbb{R}$, as do the irrational numbers. Similarly

$$\{(x, y) \in \mathbb{R}^2 \mid \text{both } x \text{ and } y \text{ are rational}\}$$

is a dense subset of the plane.

An interesting kind of subset of $\mathbb{R}^n$ is a set that is both open and dense. Such a set $\mathcal{U}$ is characterized by the following properties: Every point in the complement of $\mathcal{U}$ can be approximated arbitrarily closely by points of $\mathcal{U}$ (since $\mathcal{U}$ is dense); but no point in $\mathcal{U}$ can be approximated arbitrarily closely by points in the complement (because $\mathcal{U}$ is open).

Here is a simple example of an open and dense subset of $\mathbb{R}^2$:

$$\mathcal{V} = \{(x, y) \in \mathbb{R}^2 \mid xy \neq 1\}.$$

This, of course, is the complement in $\mathbb{R}^2$ of the hyperbola defined by $xy = 1$. Suppose $(x_0, y_0) \in \mathcal{V}$. Then $x_0 y_0 \neq 1$ and if $|x - x_0|$, $|y - y_0|$ are small enough, then $xy \neq 1$; this proves that $\mathcal{V}$ is open. Given any $(x_0, y_0) \in \mathbb{R}^2$, we can find (x, y) as close as we like to (x_0, y_0) with $xy \neq 1$; this proves that $\mathcal{V}$ is dense.

An open and dense set is a very fat set, as the following proposition shows:

Proposition.   *Let $\mathcal{V}_1, \ldots, \mathcal{V}_m$ be open and dense subsets of $\mathbb{R}^n$. Then*

$$\mathcal{V} = \mathcal{V}_1 \cap \ldots \cap \mathcal{V}_m$$

is also open and dense.

Proof: It can be easily shown that the intersection of a finite number of open sets is open, so $\mathcal{V}$ is open. To prove that $\mathcal{V}$ is dense let $\mathcal{U} \subset \mathbb{R}^n$ be a nonempty open set. Then $\mathcal{U} \cap \mathcal{V}_1$ is nonempty since $\mathcal{V}_1$ is dense. Because $\mathcal{U}$ and $\mathcal{V}_1$ are open, $\mathcal{U} \cap \mathcal{V}_1$ is also open. Since $\mathcal{U} \cap \mathcal{V}_1$ is open and nonempty, $(\mathcal{U} \cap \mathcal{V}_1) \cap \mathcal{V}_2$ is nonempty because $\mathcal{V}_2$ is dense. Since $\mathcal{V}_1$ is open, $\mathcal{U} \cap \mathcal{V}_1 \cap \mathcal{V}_2$ is open. Thus $(\mathcal{U} \cap \mathcal{V}_1 \cap \mathcal{V}_2) \cap \mathcal{V}_3$ is nonempty, and so on. So $\mathcal{U} \cap \mathcal{V}$ is nonempty, which proves that $\mathcal{V}$ is dense in $\mathbb{R}^n$.   ∎

We therefore think of a subset of $\mathbb{R}^n$ as being large if this set contains an open and dense subset. To make precise what we mean by "most" matrices, we need to transfer the notion of an open and dense set to the set of all matrices.

Let $L(\mathbb{R}^n)$ denote the set of $n \times n$ matrices, or, equivalently, the set of linear maps of $\mathbb{R}^n$. To discuss open and dense sets in $L(\mathbb{R}^n)$, we need to have a notion of how far apart two given matrices in $L(\mathbb{R}^n)$ are. But we can do this by simply writing all of the entries of a matrix as one long vector (in a specified order) and thereby thinking of $L(\mathbb{R}^n)$ as $\mathbb{R}^{n^2}$.

Theorem.   *The set $\mathcal{M}$ of matrices in $L(\mathbb{R}^n)$ that have n distinct eigenvalues is open and dense in $L(\mathbb{R}^n)$.*

Proof: We first prove that $\mathcal{M}$ is dense. Let $A \in L(\mathbb{R}^n)$. Suppose that A has some repeated eigenvalues. The proposition from the previous section states that we can find a matrix T such that $T^{-1}AT$ assumes one of two forms. Either we have a canonical form with blocks along the diagonal of the form

(i)
$$\begin{pmatrix} \lambda & 1 & & & \\ & \lambda & 1 & & \\ & & \ddots & \ddots & \\ & & & \ddots & 1 \\ & & & & \lambda \end{pmatrix}$$
or (ii)
$$\begin{pmatrix} C_2 & I_2 & & & \\ & C_2 & I_2 & & \\ & & \ddots & \ddots & \\ & & & \ddots & I_2 \\ & & & & C_2 \end{pmatrix}$$

where $\alpha, \beta, \lambda \in \mathbb{R}$ with $\beta \neq 0$ and

$$C_2 = \begin{pmatrix} \alpha & \beta \\ -\beta & \alpha \end{pmatrix}, \quad I_2 = \begin{pmatrix} 1 & 0 \\ 0 & 1 \end{pmatrix},$$

or else we have a pair of separate diagonal blocks (λ) or C_2. Either case can be handled as follows.

Choose distinct values λ_j such that $|\lambda - \lambda_j|$ is as small as desired, and replace block (i) above by

$$\begin{pmatrix} \lambda_1 & 1 & & & \\ & \lambda_2 & 1 & & \\ & & \ddots & \ddots & \\ & & & \ddots & 1 \\ & & & & \lambda_j \end{pmatrix}.$$

This new block now has distinct eigenvalues. In block (ii) we may similarly replace each 2×2 block

$$\begin{pmatrix} \alpha & \beta \\ -\beta & \alpha \end{pmatrix}$$

with distinct α_i's. The new matrix thus has distinct eigenvalues $\alpha_i \pm \beta$. In this fashion, we find a new matrix B arbitrarily close to $T^{-1}AT$ with distinct eigenvalues. Then the matrix TBT^{-1} also has distinct eigenvalues and, moreover, this matrix is arbitrarily close to A. Indeed, the function $F: L(\mathbb{R}^n) \to L(\mathbb{R}^n)$ given by $F(M) = TMT^{-1}$ where T is a fixed invertible matrix is a continuous function on $L(\mathbb{R}^n)$ and hence takes matrices close to $T^{-1}AT$ to new matrices close to A. This shows that $\mathcal{M}$ is dense.

To prove that $\mathcal{M}$ is open, consider the characteristic polynomial of a matrix $A \in L(\mathbb{R}^n)$. If we vary the entries of A slightly, then the characteristic polynomial's coefficients vary only slightly. Hence the roots of this polynomial in $\mathbb{C}$

move only slightly as well. Thus, if we begin with a matrix whose eigenvalues are distinct, nearby matrices have this property as well. This proves that $\mathcal{M}$ is open.   ∎

A property $\mathcal{P}$ of matrices is a *generic property* if the set of matrices having property $\mathcal{P}$ contains an open and dense set in $L(\mathbb{R}^n)$. Thus a property is generic if it is shared by some open and dense set of matrices (and perhaps other matrices as well). Intuitively speaking, a generic property is one that "almost all" matrices have. Thus, having all distinct eigenvalues is a generic property of $n \times n$ matrices.

EXERCISES

1. Prove that the determinant of a 3×3 matrix can be computed by expanding along any row or column.
2. Find the eigenvalues and eigenvectors of the following matrices:

$$
\text{(a)} \begin{pmatrix} 0 & 0 & 1 \\ 0 & 1 & 0 \\ 1 & 0 & 0 \end{pmatrix} \quad
\text{(b)} \begin{pmatrix} 0 & 0 & 1 \\ 0 & 2 & 0 \\ 3 & 0 & 0 \end{pmatrix} \quad
\text{(c)} \begin{pmatrix} 1 & 1 & 1 \\ 1 & 1 & 1 \\ 1 & 1 & 1 \end{pmatrix}
$$

$$
\text{(d)} \begin{pmatrix} 0 & 0 & 2 \\ 0 & 2 & 0 \\ -2 & 0 & 0 \end{pmatrix} \quad
\text{(e)} \begin{pmatrix} 3 & 0 & 0 & 1 \\ 0 & 1 & 2 & 2 \\ 1 & -2 & -1 & -4 \\ -1 & 0 & 0 & 3 \end{pmatrix}
$$

3. Describe the regions in a, b, c–space where the matrix

$$
\begin{pmatrix} 0 & 0 & a \\ 0 & b & 0 \\ c & 0 & 0 \end{pmatrix}
$$

   has real, complex, and repeated eigenvalues.
4. Describe the regions in a, b, c–space where the matrix

$$
\begin{pmatrix} a & 0 & 0 & a \\ 0 & a & b & 0 \\ 0 & c & a & 0 \\ a & 0 & 0 & a \end{pmatrix}
$$

   has real, complex, and repeated eigenvalues.

5. Put the following matrices in canonical form

(a) $\begin{pmatrix} 0 & 0 & 1 \\ 0 & 1 & 0 \\ 1 & 0 & 0 \end{pmatrix}$
(b) $\begin{pmatrix} 1 & 0 & 1 \\ 0 & 1 & 0 \\ 0 & 0 & 1 \end{pmatrix}$
(c) $\begin{pmatrix} 0 & 1 & 0 \\ -1 & 0 & 0 \\ 1 & 1 & 1 \end{pmatrix}$

(d) $\begin{pmatrix} 0 & 1 & 0 \\ 1 & 0 & 0 \\ 1 & 1 & 1 \end{pmatrix}$
(e) $\begin{pmatrix} 1 & 0 & 1 \\ 0 & 1 & 0 \\ 1 & 0 & 1 \end{pmatrix}$
(f) $\begin{pmatrix} 1 & 1 & 0 \\ 1 & 1 & 1 \\ 0 & 1 & 1 \end{pmatrix}$

(g) $\begin{pmatrix} 1 & 0 & -1 \\ -1 & 1 & -1 \\ 0 & 0 & 1 \end{pmatrix}$
(h) $\begin{pmatrix} 1 & 0 & 0 & 1 \\ 0 & 1 & 1 & 0 \\ 0 & 0 & 1 & 0 \\ 1 & 0 & 0 & 0 \end{pmatrix}$

6. Suppose that a 5×5 matrix has eigenvalues 2 and $1 \pm i$. List all possible canonical forms for a matrix of this type.

7. Let L be the elementary matrix that interchanges the ith and jth rows of a given matrix. That is, L has 1's along the diagonal, with the exception that $\ell_{ii} = \ell_{jj} = 0$ but $\ell_{ij} = \ell_{ji} = 1$. Prove that $\det L = -1$.

8. Find a basis for both Ker T and Range T when T is the matrix

(a) $\begin{pmatrix} 1 & 2 \\ 2 & 4 \end{pmatrix}$
(b) $\begin{pmatrix} 1 & 1 & 1 \\ 1 & 1 & 1 \\ 1 & 1 & 1 \end{pmatrix}$
(c) $\begin{pmatrix} 1 & 9 & 6 \\ 1 & 4 & 1 \\ 2 & 7 & 1 \end{pmatrix}$

9. Suppose A is a 4×4 matrix that has a single real eigenvalue λ and only one independent eigenvector. Prove that A may be put in canonical form

$$\begin{pmatrix} \lambda & 1 & 0 & 0 \\ 0 & \lambda & 1 & 0 \\ 0 & 0 & \lambda & 1 \\ 0 & 0 & 0 & \lambda \end{pmatrix}.$$

10. Suppose A is a 4×4 matrix with a single real eigenvalue and two linearly independent eigenvectors. Describe the possible canonical forms for A and show that A may indeed be transformed into one of these canonical forms. Describe explicitly the conditions under which A is transformed into a particular form.

11. Show that if A and/or B are noninvertible matrices, then AB is also noninvertible.

12. Suppose that $\mathcal{S}$ is a subset of $\mathbb{R}^n$ having the following properties:

(a) If $X, Y \in \mathcal{S}$, then $X + Y \in \mathcal{S}$;

(b) If $X \in \mathcal{S}$ and $\alpha \in \mathbb{R}$, then $\alpha X \in \mathcal{S}$.

Prove that S may be written as the collection of all possible linear combinations of a finite set of vectors.

13. Which of the following subsets of $\mathbb{R}^n$ are open and/or dense? Give a brief reason in each case.

(a) $\mathcal{U}_1 = \{(x, y) \mid y > 0\}$;

(b) $\mathcal{U}_2 = \{(x, y) \mid x^2 + y^2 \neq 1\}$;

(c) $\mathcal{U}_3 = \{(x, y) \mid x \text{ is irrational}\}$;

(d) $\mathcal{U}_4 = \{(x, y) \mid x \text{ and } y \text{ are not integers}\}$;

(e) $\mathcal{U}_5$ is the complement of a set C_1 where C_1 is closed and not dense;

(f) $\mathcal{U}_6$ is the complement of a set C_2 which contains exactly 6 billion and two distinct points.

14. Each of the following properties defines a subset of real $n \times n$ matrices. Which of these sets are open and/or dense in the $L(\mathbb{R}^n)$? Give a brief reason in each case.

(a) $\det A \neq 0$.

(b) Trace A is rational.

(c) Entries of A are not integers.

(d) $3 \leq \det A < 4$.

(e) $-1 < |\lambda| < 1$ for every eigenvalue λ.

(f) A has no real eigenvalues.

(g) Each real eigenvalue of A has multiplicity one.

15. Which of the following properties of linear maps on $\mathbb{R}^n$ are generic?

(a) $|\lambda| \neq 1$ for every eigenvalue λ.

(b) $n = 2$; some eigenvalue is not real.

(c) $n = 3$; some eigenvalue is not real.

(d) No solution of $X' = AX$ is periodic (except the zero solution).

(e) There are n distinct eigenvalues, each with distinct imaginary parts.

(f) $AX \neq X$ and $AX \neq -X$ for all $X \neq 0$.

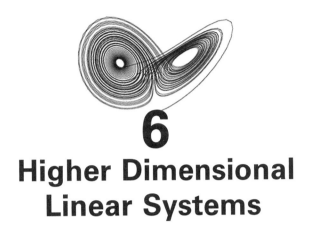

6

Higher Dimensional Linear Systems

After our little sojourn into the world of linear algebra, it's time to return to differential equations and, in particular, to the task of solving higher dimensional linear systems with constant coefficients. As in the linear algebra chapter, we have to deal with a number of different cases.

6.1 Distinct Eigenvalues

Consider first a linear system $X' = AX$ where the $n \times n$ matrix A has n distinct, real eigenvalues $\lambda_1, \ldots, \lambda_n$. By the results in Chapter 5, there is a change of coordinates T so that the new system $Y' = (T^{-1}AT)Y$ assumes the particularly simple form

$$y_1' = \lambda_1 y_1$$
$$\vdots$$
$$y_n' = \lambda_n y_n.$$

The linear map T is the map that takes the standard basis vector E_j to the eigenvector V_j associated to λ_j. Clearly, a function of the form

$$Y(t) = \begin{pmatrix} c_1 e^{\lambda_1 t} \\ \vdots \\ c_n e^{\lambda_n t} \end{pmatrix}$$

107

is a solution of $Y' = (T^{-1}AT)Y$ that satisfies the initial condition $Y(0) = (c_1, \ldots, c_n)$. As in Chapter 3, this is the only such solution, because if

$$W(t) = \begin{pmatrix} w_1(t) \\ \vdots \\ w_n(t) \end{pmatrix}$$

is another solution, then differentiating each expression $w_j(t) \exp(-\lambda_j t)$, we find

$$\frac{d}{dt} w_j(t) e^{-\lambda_j t} = (w_j' - \lambda_j w_j) e^{-\lambda_j t} = 0.$$

Hence $w_j(t) = c_j \exp(\lambda_j t)$ for each j. Therefore the collection of solutions $Y(t)$ yields the general solution of $Y' = (T^{-1}AT)Y$.

It then follows that $X(t) = TY(t)$ is the general solution of $X' = AX$, so this general solution may be written in the form

$$X(t) = \sum_{j=1}^{n} c_j e^{\lambda_j t} V_j.$$

Now suppose that the eigenvalues $\lambda_1, \ldots, \lambda_k$ of A are negative, while the eigenvalues $\lambda_{k+1}, \ldots, \lambda_n$ are positive. Since there are no zero eigenvalues, the system is hyperbolic. Then any solution that starts in the subspace spanned by the vectors $V_1, \ldots, V_k$ must first of all stay in that subspace for all time since $c_{k+1} = \cdots = c_n = 0$. Secondly, each such solution tends to the origin as $t \to \infty$. In analogy with the terminology introduced for planar systems, we call this subspace the *stable subspace*. Similarly, the subspace spanned by $V_{k+1}, \ldots, V_n$ contains solutions that move away from the origin. This subspace is the *unstable subspace*. All other solutions tend toward the stable subspace as time goes backward and toward the unstable subspace as time increases. Therefore this system is a higher dimensional analog of a *saddle*.

Example. Consider

$$X' = \begin{pmatrix} 1 & 2 & -1 \\ 0 & 3 & -2 \\ 0 & 2 & -2 \end{pmatrix} X.$$

In Section 5.2 in Chapter 5, we showed that this matrix has eigenvalues 2, 1, and -1 with associated eigenvectors $(3, 2, 1)$, $(1, 0, 0)$, and $(0, 1, 2)$, respectively. Therefore the matrix

$$T = \begin{pmatrix} 3 & 1 & 0 \\ 2 & 0 & 1 \\ 1 & 0 & 2 \end{pmatrix}$$

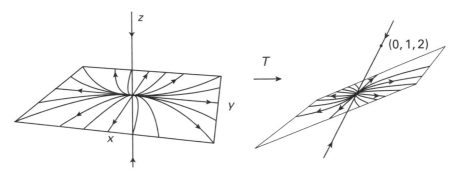

Figure 6.1 The stable and unstable subspaces of a saddle in dimension 3. On the left, the system is in canonical form.

converts $X' = AX$ to

$$Y' = (T^{-1}AT)Y = \begin{pmatrix} 2 & 0 & 0 \\ 0 & 1 & 0 \\ 0 & 0 & -1 \end{pmatrix} Y,$$

which we can solve immediately. Multiplying the solution by T then yields the general solution

$$X(t) = c_1 e^{2t} \begin{pmatrix} 3 \\ 2 \\ 1 \end{pmatrix} + c_2 e^t \begin{pmatrix} 1 \\ 0 \\ 0 \end{pmatrix} + c_3 e^{-t} \begin{pmatrix} 0 \\ 1 \\ 2 \end{pmatrix}$$

of $X' = AX$. The straight line through the origin and $(0, 1, 2)$ is the stable line, while the plane spanned by $(3, 2, 1)$ and $(1, 0, 0)$ is the unstable plane. A collection of solutions of this system as well as the system $Y' = (T^{-1}AT)Y$ is displayed in Figure 6.1. ■

Example. If the 3×3 matrix A has three real, distinct eigenvalues that are negative, then we may find a change of coordinates so that the system assumes the form

$$Y' = (T^{-1}AT)Y = \begin{pmatrix} \lambda_1 & 0 & 0 \\ 0 & \lambda_2 & 0 \\ 0 & 0 & \lambda_3 \end{pmatrix} Y$$

where $\lambda_3 < \lambda_2 < \lambda_1 < 0$. All solutions therefore tend to the origin and so we have a higher dimensional *sink*. See Figure 6.2. For an initial condition (x_0, y_0, z_0) with all three coordinates nonzero, the corresponding solution tends to the origin tangentially to the x-axis (see Exercise 2 at the end of the chapter). ■

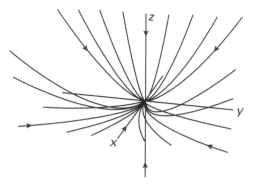

Figure 6.2 A sink in three dimensions.

Now suppose that the $n \times n$ matrix A has n distinct eigenvalues, of which k_1 are real and $2k_2$ are nonreal, so that $n = k_1 + 2k_2$. Then, as in Chapter 5, we may change coordinates so that the system assumes the form

$$x'_j = \lambda_j x_j$$
$$u'_\ell = \alpha_\ell u_\ell + \beta_\ell v_\ell$$
$$v'_\ell = -\beta_\ell u_\ell + \alpha_\ell v_\ell$$

for $j = 1, \ldots, k_1$ and $\ell = 1, \ldots, k_2$. As in Chapter 3, we therefore have solutions of the form

$$x_j(t) = c_j e^{\lambda_j t}$$
$$u_\ell(t) = p_\ell e^{\alpha_\ell t} \cos \beta_\ell t + q_\ell e^{\alpha_\ell t} \sin \beta_\ell t$$
$$v_\ell(t) = -p_\ell e^{\alpha_\ell t} \sin \beta_\ell t + q_\ell e^{\alpha_\ell t} \cos \beta_\ell t.$$

As before, it is straightforward to check that this is the general solution. We have therefore shown:

Theorem. *Consider the system $X' = AX$ where A has distinct eigenvalues $\lambda_1, \ldots, \lambda_{k_1} \in \mathbb{R}$ and $\alpha_1 + i\beta_1, \ldots, \alpha_{k_2} + i\beta_{k_2} \in \mathbb{C}$. Let T be the matrix that puts A in the canonical form*

$$T^{-1}AT = \begin{pmatrix} \lambda_1 & & & & & & \\ & \ddots & & & & & \\ & & \lambda_{k_1} & & & & \\ & & & B_1 & & & \\ & & & & \ddots & & \\ & & & & & B_{k_2} \end{pmatrix}$$

where

$$B_j = \begin{pmatrix} \alpha_j & \beta_j \\ -\beta_j & \alpha_j \end{pmatrix}.$$

Then the general solution of $X' = AX$ is $TY(t)$ where

$$Y(t) = \begin{pmatrix} c_1 e^{\lambda_1 t} \\ \vdots \\ c_{k_1} e^{\lambda_{k_1} t} \\ a_1 e^{\alpha_1 t} \cos \beta_1 t + b_1 e^{\alpha_1 t} \sin \beta_1 t \\ -a_1 e^{\alpha_1 t} \sin \beta_1 t + b_1 e^{\alpha_1 t} \cos \beta_1 t \\ \vdots \\ a_{k_2} e^{\alpha_{k_2} t} \cos \beta_{k_2} t + b_{k_2} e^{\alpha_{k_2} t} \sin \beta_{k_2} t \\ -a_{k_2} e^{\alpha_{k_2} t} \sin \beta_{k_2} t + b_{k_2} e^{\alpha_{k_2} t} \cos \beta_{k_2} t \end{pmatrix}$$ ■

As usual, the columns of the matrix T in this theorem are the eigenvectors (or the real and imaginary parts of the eigenvectors) corresponding to each eigenvalue. Also, as before, the subspace spanned by the eigenvectors corresponding to eigenvalues with negative (resp., positive) real parts is the stable (resp., unstable) subspace.

Example. Consider the system

$$X' = \begin{pmatrix} 0 & 1 & 0 \\ -1 & 0 & 0 \\ 0 & 0 & -1 \end{pmatrix} X$$

whose matrix is already in canonical form. The eigenvalues are $\pm i, -1$. The solution satisfying the initial condition (x_0, y_0, z_0) is given by

$$Y(t) = x_0 \begin{pmatrix} \cos t \\ -\sin t \\ 0 \end{pmatrix} + y_0 \begin{pmatrix} \sin t \\ \cos t \\ 0 \end{pmatrix} + z_0 e^{-t} \begin{pmatrix} 0 \\ 0 \\ 1 \end{pmatrix}$$

so this is the general solution. The phase portrait for this system is displayed in Figure 6.3. The stable line lies along the z-axis, whereas all solutions in the xy–plane travel around circles centered at the origin. In fact, each solution that does not lie on the stable line actually lies on a cylinder in $\mathbb{R}^3$ given by $x^2 + y^2 = $ constant. These solutions spiral toward the circular solution of radius $\sqrt{x_0^2 + y_0^2}$ in the xy–plane if $z_0 \neq 0$. ■

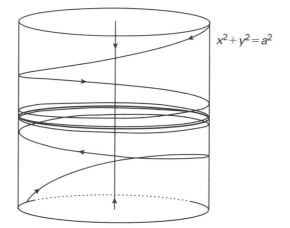

Figure 6.3   The phase portrait for a spiral center.

Example. Now consider $X' = AX$ where

$$A = \begin{pmatrix} -0.1 & 0 & 1 \\ -1 & 1 & -1.1 \\ -1 & 0 & -0.1 \end{pmatrix}.$$

The characteristic equation is

$$-\lambda^3 + 0.8\lambda^2 - 0.81\lambda + 1.01 = 0,$$

which we have kindly factored for you into

$$(1 - \lambda)(\lambda^2 + 0.2\lambda + 1.01) = 0.$$

Therefore the eigenvalues are the roots of this equation, which are 1 and $-0.1 \pm i$. Solving $(A - (-0.1 + i)I)X = 0$ yields the eigenvector $(-i, 1, 1)$ associated to $-0.1 + i$. Let $V_1 = \mathrm{Re}\,(-i, 1, 1) = (0, 1, 1)$ and $V_2 = \mathrm{Im}\,(-i, 1, 1) = (-1, 0, 0)$. Solving $(A - I)X = 0$ yields $V_3 = (0, 1, 0)$ as an eigenvector corresponding to $\lambda = 1$. Then the matrix whose columns are the V_i,

$$T = \begin{pmatrix} 0 & -1 & 0 \\ 1 & 0 & 1 \\ 1 & 0 & 0 \end{pmatrix},$$

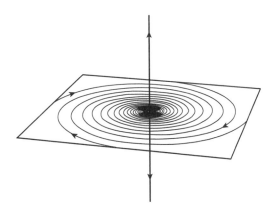

Figure 6.4 A spiral saddle in canonical
form.

converts $X' = AX$ into

$$Y' - \begin{pmatrix} -0.1 & 1 & 0 \\ -1 & -0.1 & 0 \\ 0 & 0 & 1 \end{pmatrix} Y.$$

This system has an unstable line along the z-axis, while the xy–plane is the stable plane. Note that solutions spiral into 0 in the stable plane. We call this system a *spiral saddle* (see Figure 6.4). Typical solutions of the stable plane spiral toward the z-axis while the z-coordinate meanwhile increases or decreases (see Figure 6.5). ∎

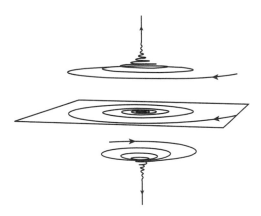

Figure 6.5 Typical solutions of the spiral
saddle tend to spiral toward the unstable
line.

6.2 Harmonic Oscillators

Consider a pair of undamped harmonic oscillators whose equations are

$$x_1'' = -\omega_1^2 x_1$$
$$x_2'' = -\omega_2^2 x_2.$$

We can almost solve these equations by inspection as visions of $\sin \omega t$ and $\cos \omega t$ pass through our minds. But let's push on a bit, first to illustrate the theorem in the previous section in the case of nonreal eigenvalues, but more importantly to introduce some interesting geometry.

We first introduce the new variables $y_j = x_j'$ for $j = 1, 2$ so that the equations can be written as a system

$$x_j' = y_j$$
$$y_j' = -\omega_j^2 x_j.$$

In matrix form, this system is $X' = AX$ where $X = (x_1, y_1, x_2, y_2)$ and

$$A = \begin{pmatrix} 0 & 1 & & \\ -\omega_1^2 & 0 & & \\ & & 0 & 1 \\ & & -\omega_2^2 & 0 \end{pmatrix}.$$

This system has eigenvalues $\pm i\omega_1$ and $\pm i\omega_2$. An eigenvector corresponding to $i\omega_1$ is $V_1 = (1, i\omega_1, 0, 0)$ while $V_2 = (0, 0, 1, i\omega_2)$ is associated to $i\omega_2$. Let W_1 and W_2 be the real and imaginary parts of V_1, and let W_3 and W_4 be the same for V_2. Then, as usual, we let $TE_j = W_j$ and the linear map T puts this system into canonical form with the matrix

$$T^{-1}AT = \begin{pmatrix} 0 & \omega_1 & & \\ -\omega_1 & 0 & & \\ & & 0 & \omega_2 \\ & & -\omega_2 & 0 \end{pmatrix}.$$

We then derive the general solution

$$Y(t) = \begin{pmatrix} x_1(t) \\ y_1(t) \\ x_2(t) \\ y_2(t) \end{pmatrix} = \begin{pmatrix} a_1 \cos \omega_1 t + b_1 \sin \omega_1 t \\ -a_1\omega_1 \sin \omega_1 t + b_1\omega_1 \cos \omega_1 t \\ a_2 \cos \omega_2 t + b_2 \sin \omega_2 t \\ -a_2\omega_2 \sin \omega_2 t + b_2\omega_2 \cos \omega_2 t \end{pmatrix}$$

just as we originally expected.

We could say that this is the end of the story and stop here since we have the formulas for the solution. However, let's push on a bit more.

Each pair of solutions $(x_j(t), y_j(t))$ for $j = 1, 2$ is clearly a periodic solution of the equation with period $2\pi/\omega_j$, but this does not mean that the full four-dimensional solution is a periodic function. Indeed, the full solution is a periodic function with period τ if and only if there exist integers m and n such that

$$\omega_1 \tau = m \cdot 2\pi \quad \text{and} \quad \omega_2 \tau = n \cdot 2\pi.$$

Thus, for periodicity, we must have

$$\tau = \frac{2\pi m}{\omega_1} = \frac{2\pi n}{\omega_2}$$

or, equivalently,

$$\frac{\omega_2}{\omega_1} = \frac{n}{m}.$$

That is, the ratio of the two frequencies of the oscillators must be a rational number. In Figure 6.6 we have plotted $(x_1(t), x_2(t))$ for particular solution of this system when the ratio of the frequencies is 5/2.

When the ratio of the frequencies is irrational, something very different happens. To understand this, we make another (and much more familiar) change of coordinates. In canonical form, our system currently is

$$x_j' = \omega_j y_j$$

$$y_j' = -\omega_j x_j.$$

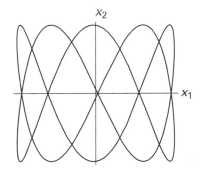

Figure 6.6   A solution with frequency ratio 5/2 projected into the $x_1 x_2$–plane. Note that $x_2(t)$ oscillates five times and $x_1(t)$ only twice before returning to the initial position.

Let's now introduce polar coordinates (r_j, θ_j) in place of the x_j and y_j variables. Differentiating

$$r_j^2 = x_j^2 + y_j^2,$$

we find

$$
\begin{aligned}
2r_j r_j' &= 2x_j x_j' + 2y_j y_j' \\
&= 2x_j y_j \omega_j - 2x_j y_j \omega_j \\
&= 0.
\end{aligned}
$$

Therefore $r_j' = 0$ for each j. Also, differentiating the equation

$$\tan \theta_j = \frac{y_j}{x_j}$$

yields

$$
\begin{aligned}
(\sec^2 \theta_j)\theta_j' &= \frac{y_j' x_j - y_j x_j'}{x_j^2} \\
&= \frac{-\omega_j r_j^2}{r_j^2 \cos^2 \theta_j}
\end{aligned}
$$

from which we find

$$\theta_j' = -\omega_j.$$

So, in polar coordinates, these equations really are quite simple:

$$
\begin{aligned}
r_j' &= 0 \\
\theta_j' &= -\omega_j.
\end{aligned}
$$

The first equation tells us that both r_1 and r_2 remain constant along any solution. Then, no matter what we pick for our initial r_1 and r_2 values, the θ_j equations remain the same. Hence we may as well restrict our attention to $r_1 = r_2 = 1$. The resulting set of points in $\mathbb{R}^4$ is a *torus*—the surface of a doughnut—although this is a little difficult to visualize in four-dimensional space. However, we know that we have two independent variables on this set, namely, θ_1 and θ_2, and both are periodic with period 2π. So this is akin to the two independent circular directions that parameterize the familiar torus in $\mathbb{R}^3$.

Restricted to this torus, the equations now read

$$\theta_1' = -\omega_1$$
$$\theta_2' = -\omega_2.$$

It is convenient to think of θ_1 and θ_2 as variables in a square of sidelength 2π where we glue together the opposite sides $\theta_j = 0$ and $\theta_j = 2\pi$ to make the torus. In this square our vector field now has constant slope

$$\frac{\theta_2'}{\theta_1'} = \frac{\omega_2}{\omega_1}.$$

Therefore solutions lie along straight lines with slope ω_2/ω_1 in this square. When a solution reaches the edge $\theta_1 = 2\pi$ (say, at $\theta_2 = c$), it instantly reappears on the edge $\theta_1 = 0$ with the θ_2 coordinate given by c, and then continues onward with slope ω_2/ω_1. A similar identification occurs when the solution meets $\theta_2 = 2\pi$.

So now we have a simplified geometric vision of what happens to these solutions on these tori. But what really happens? The answer depends on the ratio ω_2/ω_1. If this ratio is a rational number, say, n/m, then the solution starting at $(\theta_1(0), \theta_2(0))$ will pass horizontally through the torus exactly m times and vertically n times before returning to its starting point. This is the periodic solution we observed above. Incidentally, the picture of the straight line solutions in the $\theta_1\theta_2$–plane is not at all the same as our depiction of solutions in the x_1x_2–plane as shown in Figure 6.6.

In the irrational case, something quite different occurs. See Figure 6.7. To understand what is happening here, we return to the notion of a *Poincaré map* discussed in Chapter 1. Consider the circle $\theta_1 = 0$, the left-hand edge of our square representation of the torus. Given an initial point on this circle, say, $\theta_2 = x_0$, we follow the solution starting at this point until it next hits $\theta_1 = 2\pi$.

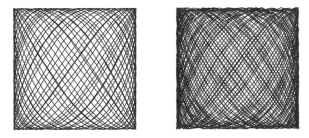

Figure 6.7   A solution with frequency ratio $\sqrt{2}$ projected into the $x_1 x_2$–plane, the left curve computed up to time 50π, the right to time 100π.

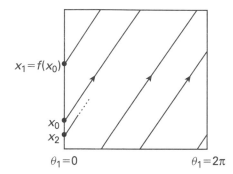

Figure 6.8   The Poincaré map on the
circle $\theta_1 = 0$ in the $\theta_1\theta_2$ torus.

By our identification, this solution has now returned to the circle $\theta_1 = 0$. The
solution may cross the boundary $\theta_2 = 2\pi$ several times in making this transit,
but it does eventually return to $\theta_1 = 0$. So we may define the Poincaré map
on $\theta_1 = 0$ by assigning to x_0 on this circle the corresponding coordinate of
the point of first return. Suppose that this first return occurs at the point $\theta_2(\tau)$
where τ is the time for which $\theta_1(\tau) = 2\pi$. Since $\theta_1(t) = \theta_1(0) - \omega_1 t$, we have
$\tau = 2\pi/\omega_1$. Hence $\theta_2(\tau) = x_0 - \omega_2(2\pi/\omega_1)$. Therefore the Poincaré map on
the circle may be written as

$$f(x_0) = x_0 + 2\pi(\omega_2/\omega_1) \bmod 2\pi$$

where $x_0 = \theta_2(0)$ is our initial θ_2 coordinate on the circle. See Figure 6.8. Thus
the Poincaré map on the circle is just the function that rotates points on the
circle by angle $2\pi(\omega_2/\omega_1)$. Since ω_2/ω_1 is irrational, this function is called an
irrational rotation of the circle.

Definition
The set of points $x_0, x_1 = f(x_0), x_2 = f(f(x_0)), \ldots, x_n = f(x_{n-1})$ is
called the *orbit* of x_0 under iteration of f.

The orbit of x_0 tracks how our solution successively crosses $\theta_1 = 2\pi$ as time
increases.

Proposition.   *Suppose ω_2/ω_1 is irrational. Then the orbit of any initial point
x_0 on the circle $\theta_1 = 0$ is dense in the circle.*

Proof: Recall from Section 5.6 that a subset of the circle is *dense* if there
are points in this subset that are arbitrarily close to any point whatsoever

in the circle. Therefore we must show that, given any point z on the circle and any $\epsilon > 0$, there is a point x_n on the orbit of x_0 such that $|z - x_n| < \epsilon$ where z and x_n are measured mod 2π. To see this, observe first that there must be n, m for which $m > n$ and $|x_n - x_m| < \epsilon$. Indeed, we know that the orbit of x_0 is not a finite set of points since ω_2/ω_1 is irrational. Hence there must be at least two of these points whose distance apart is less than ϵ since the circle has finite circumference. These are the points x_n and x_m (actually, there must be infinitely many such points). Now rotate these points in the reverse direction exactly n times. The points x_n and x_m are rotated to x_0 and x_{m-n}, respectively. We find, after this rotation, that $|x_0 - x_{m-n}| < \epsilon$. Now x_{m-n} is given by rotating the circle through angle $(m - n)2\pi(\omega_2/\omega_1)$, which, mod 2π, is therefore a rotation of angle less than ϵ. Hence, performing this rotation again, we find

$$|x_{2(m-n)} - x_{m-n}| < \epsilon$$

as well, and, inductively,

$$|x_{k(m-n)} - x_{(k-1)(m-n)}| < \epsilon$$

for each k. Thus we have found a sequence of points obtained by repeated rotation through angle $(m - n)2\pi(\omega_2/\omega_1)$, and each of these points is within ϵ of its predecessor. Hence there must be a point of this form within ϵ of z. ∎

Since the orbit of x_0 is dense in the circle $\theta_1 = 0$, it follows that the straight-line solutions connecting these points in the square are also dense, and so the original solutions are dense in the torus on which they reside. This accounts for the densely packed solution shown projected into the $x_1 x_2$–plane in Figure 6.7 when $\omega_2/\omega_1 = \sqrt{2}$.

Returning to the actual motion of the oscillators, we see that when ω_2/ω_1 is irrational, the masses do not move in periodic fashion. However, they do come back very close to their initial positions over and over again as time goes on due to the density of these solutions on the torus. These types of motions are called *quasi-periodic motions*. In Exercise 7 we investigate a related set of equations, namely, a pair of coupled oscillators.

6.3 Repeated Eigenvalues

As we saw in the previous chapter, the solution of systems with repeated real eigenvalues reduces to solving systems whose matrices contain blocks of the form

$$
\begin{pmatrix}
\lambda & 1 & & & \\
 & \lambda & 1 & & \\
 & & \ddots & \ddots & \\
 & & & \ddots & 1 \\
 & & & & \lambda
\end{pmatrix}.
$$

Example. Let

$$
X' = \begin{pmatrix} \lambda & 1 & 0 \\ 0 & \lambda & 1 \\ 0 & 0 & \lambda \end{pmatrix} X.
$$

The only eigenvalue for this system is λ, and its only eigenvector is $(1, 0, 0)$. We may solve this system as we did in Chapter 3, by first noting that $x_3' = \lambda x_3$, so we must have

$$
x_3(t) = c_3 e^{\lambda t}.
$$

Now we must have

$$
x_2' = \lambda x_2 + c_3 e^{\lambda t}.
$$

As in Chapter 3, we guess a solution of the form

$$
x_2(t) = c_2 e^{\lambda t} + \alpha t e^{\lambda t}.
$$

Substituting this guess into the differential equation for x_2', we determine that $\alpha = c_3$ and find

$$
x_2(t) = c_2 e^{\lambda t} + c_3 t e^{\lambda t}.
$$

Finally, the equation

$$
x_1' = \lambda x_1 + c_2 e^{\lambda t} + c_3 t e^{\lambda t}
$$

suggests the guess

$$
x_1(t) = c_1 e^{\lambda t} + \alpha t e^{\lambda t} + \beta t^2 e^{\lambda t}.
$$

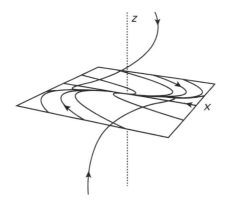

Figure 6.9 The phase portrait for repeated real eigenvalues.

Solving as above, we find

$$x_1(t) = c_1 e^{\lambda t} + c_2 t e^{\lambda t} + c_3 \frac{t^2}{2} e^{\lambda t}.$$

Altogether, we find

$$X(t) = c_1 e^{\lambda t} \begin{pmatrix} 1 \\ 0 \\ 0 \end{pmatrix} + c_2 e^{\lambda t} \begin{pmatrix} t \\ 1 \\ 0 \end{pmatrix} + c_3 e^{\lambda t} \begin{pmatrix} t^2/2 \\ t \\ 1 \end{pmatrix},$$

which is the general solution. Despite the presence of the polynomial terms in this solution, when $\lambda < 0$, the exponential term dominates and all solutions do tend to zero. Some representative solutions when $\lambda < 0$ are shown in Figure 6.9. Note that there is only one straight-line solution for this system; this solution lies on the x-axis. Also, the xy–plane is invariant and solutions there behave exactly as in the planar repeated eigenvalue case. ■

Example. Consider the following four-dimensional system:

$$x_1' = x_1 + x_2 - x_3$$
$$x_2' = x_2 + x_4$$
$$x_3' = x_3 + x_4$$
$$x_4' = x_4$$

We may write this system in matrix form as

$$X' = AX = \begin{pmatrix} 1 & 1 & -1 & 0 \\ 0 & 1 & 0 & 1 \\ 0 & 0 & 1 & 1 \\ 0 & 0 & 0 & 1 \end{pmatrix} X.$$

Because A is upper triangular, all of the eigenvalues are equal to 1. Solving $(A - I)X = 0$, we find two independent eigenvectors $V_1 = (1, 0, 0, 0)$ and $W_1 = (0, 1, 1, 0)$. This reduces the possible canonical forms for A to two possibilities. Solving $(A - I)X = V_1$ yields one solution $V_2 = (0, 1, 0, 0)$, and solving $(A - I)X = W_1$ yields another solution $W_2 = (0, 0, 0, 1)$. Thus we know that the system $X' = AX$ may be tranformed into

$$Y' = (T^{-1}AT)Y = \begin{pmatrix} 1 & 1 & 0 & 0 \\ 0 & 1 & 0 & 0 \\ 0 & 0 & 1 & 1 \\ 0 & 0 & 0 & 1 \end{pmatrix} Y$$

where the matrix T is given by

$$T = \begin{pmatrix} 1 & 0 & 0 & 0 \\ 0 & 1 & 1 & 0 \\ 0 & 0 & 1 & 0 \\ 0 & 0 & 0 & 1 \end{pmatrix}.$$

Solutions of $Y' = (T^{-1}AT)Y$ therefore are given by

$$y_1(t) = c_1 e^t + c_2 t e^t$$
$$y_2(t) = c_2 e^t$$
$$y_3(t) = c_3 e^t + c_4 t e^t$$
$$y_4(t) = c_4 e^t.$$

Applying the change of coordinates T, we find the general solution of the original system

$$x_1(t) = c_1 e^t + c_2 t e^t$$
$$x_2(t) = c_2 e^t + c_3 e^t + c_4 t e^t$$
$$x_3(t) = c_3 e^t + c_4 t e^t$$
$$x_4(t) = c_4 e^t.$$

■

6.4 The Exponential of a Matrix

We turn now to an alternative and elegant approach to solving linear systems using the exponential of a matrix. In a certain sense, this is the more natural way to attack these systems.

Recall from Section 1.1 in Chapter 1 how we solved the 1×1 "system" of linear equations $x' = ax$ where our matrix was now simply (a). We did not go through the process of finding eigenvalues and eigenvectors here. (Well, actually, we did, but the process was pretty simple.) Rather, we just exponentiated the matrix (a) to find the general solution $x(t) = c \exp(at)$. In fact, this process works in the general case where A is $n \times n$. All we need to know is how to exponentiate a matrix.

Here's how: Recall from calculus that the exponential function can be expressed as the infinite series

$$e^x = \sum_{k=0}^{\infty} \frac{x^k}{k!}.$$

We know that this series converges for every $x \in \mathbb{R}$. Now we can add matrices, we can raise them to the power k, and we can multiply each entry by $1/k!$. So this suggests that we can use this series to exponentiate them as well.

Definition
Let A be an $n \times n$ matrix. We define the *exponential* of A to be the matrix given by

$$\exp(A) = \sum_{k=0}^{\infty} \frac{A^k}{k!}.$$

Of course we have to worry about what it means for this sum of matrices to converge, but let's put that off and try to compute a few examples first.

Example. Let

$$A = \begin{pmatrix} \lambda & 0 \\ 0 & \mu \end{pmatrix}.$$

Then we have

$$A^k = \begin{pmatrix} \lambda^k & 0 \\ 0 & \mu^k \end{pmatrix}$$

so that

$$\exp(A) = \begin{pmatrix} \displaystyle\sum_{k=0}^{\infty} \lambda^k/k! & 0 \\ 0 & \displaystyle\sum_{k=0}^{\infty} \mu^k/k! \end{pmatrix} = \begin{pmatrix} e^\lambda & 0 \\ 0 & e^\mu \end{pmatrix}$$

as you may have guessed.   ∎

Example. For a slightly more complicated example, let

$$A = \begin{pmatrix} 0 & \beta \\ -\beta & 0 \end{pmatrix}.$$

We compute

$$A^0 = I, \quad A^2 = -\beta^2 I, \quad A^3 = -\beta^3 \begin{pmatrix} 0 & 1 \\ -1 & 0 \end{pmatrix},$$

$$A^4 = \beta^4 I, \quad A^5 = \beta^5 \begin{pmatrix} 0 & 1 \\ -1 & 0 \end{pmatrix}, \quad \cdots$$

so we find

$$\exp(A) = \begin{pmatrix} \displaystyle\sum_{k=0}^{\infty}(-1)^k \frac{\beta^{2k}}{(2k)!} & \displaystyle\sum_{k=0}^{\infty}(-1)^k \frac{\beta^{2k+1}}{(2k+1)!} \\ -\displaystyle\sum_{k=0}^{\infty}(-1)^k \frac{\beta^{2k+1}}{(2k+1)!} & \displaystyle\sum_{k=0}^{\infty}(-1)^k \frac{\beta^{2k}}{(2k)!} \end{pmatrix}$$

$$= \begin{pmatrix} \cos\beta & \sin\beta \\ -\sin\beta & \cos\beta \end{pmatrix}.$$ ∎

Example. Now let

$$A = \begin{pmatrix} \lambda & 1 \\ 0 & \lambda \end{pmatrix}$$

with $\lambda \neq 0$. With an eye toward what comes later, we compute, not exp A, but rather $\exp(tA)$. We have

$$(tA)^k = \begin{pmatrix} (t\lambda)^k & kt^k\lambda^{k-1} \\ 0 & (t\lambda)^k \end{pmatrix}.$$

Hence we find

$$\exp(tA) = \begin{pmatrix} \displaystyle\sum_{k=0}^{\infty} \frac{(t\lambda)^k}{k!} & \displaystyle t\sum_{k=0}^{\infty} \frac{(t\lambda)^k}{k!} \\ 0 & \displaystyle\sum_{k=0}^{\infty} \frac{(t\lambda)^k}{k!} \end{pmatrix} = \begin{pmatrix} e^{t\lambda} & te^{t\lambda} \\ 0 & e^{t\lambda} \end{pmatrix}. \qquad \blacksquare$$

Note that, in each of these three examples, the matrix $\exp(A)$ is a matrix whose entries are infinite series. We therefore say that the infinite series of matrices $\exp(A)$ converges absolutely if each of its individual terms does so. In each of the previous cases, this convergence was clear. Unfortunately, in the case of a general matrix A, this is not so clear. To prove convergence here, we need to work a little harder.

Let $a_{ij}(k)$ denote the ij entry of A^k. Let $a = \max |a_{ij}|$. We have

$$|a_{ij}(2)| = \left| \sum_{k=1}^{n} a_{ik}a_{kj} \right| < na^2$$

$$|a_{ij}(3)| = \left| \sum_{k,\ell=1}^{n} a_{ik}a_{k\ell}a_{\ell j} \right| \leq n^2 a^3$$

$$\vdots$$

$$|a_{ij}(k)| \leq n^{k-1}a^k.$$

Hence we have a bound for the ij entry of the $n \times n$ matrix $\exp(A)$:

$$\left| \sum_{k=0}^{\infty} \frac{a_{ij}(k)}{k!} \right| \leq \sum_{k=0}^{\infty} \frac{|a_{ij}(k)|}{k!} \leq \sum_{k=0}^{\infty} \frac{n^{k-1}a^k}{k!} \leq \sum_{k=0}^{\infty} \frac{(na)^k}{k!} \leq \exp na$$

so that this series converges absolutely by the comparison test. Therefore the matrix $\exp A$ makes sense for any $A \in L(\mathbb{R}^n)$.

The following result shows that matrix exponentiation shares many of the familiar properties of the usual exponential function.

Proposition. *Let A, B, and T be n × n matrices. Then:*

1. *If* $B = T^{-1}AT$, *then* $\exp(B) = T^{-1}\exp(A)T$.
2. *If* $AB = BA$, *then* $\exp(A + B) = \exp(A)\exp(B)$.
3. $\exp(-A) = (\exp(A))^{-1}$.

Proof: The proof of (1) follows from the identities $T^{-1}(A+B)T = T^{-1}AT + T^{-1}BT$ and $(T^{-1}AT)^k = T^{-1}A^kT$. Therefore

$$T^{-1}\left(\sum_{k=0}^{n}\frac{A^k}{k!}\right)T = \sum_{k=0}^{n}\frac{\left(T^{-1}AT\right)^k}{k!}$$

and (1) follows by taking limits.

To prove (2), observe that because $AB = BA$ we have, by the binomial theorem,

$$(A + B)^n = n!\sum_{j+k=n}\frac{A^j\,B^k}{j!\;k!}.$$

Therefore we must show that

$$\sum_{n=0}^{\infty}\left(\sum_{j+k=n}\frac{A^j\,B^k}{j!\;k!}\right) = \left(\sum_{j=0}^{\infty}\frac{A^j}{j!}\right)\left(\sum_{k=0}^{\infty}\frac{B^k}{k!}\right).$$

This is not as obvious as it may seem, since we are dealing here with series of matrices, not series of real numbers. So we will prove this in the following lemma, which then proves (2). Putting $B = -A$ in (2) gives (3). ∎

Lemma. *For any n × n matrices A and B, we have:*

$$\sum_{n=0}^{\infty}\left(\sum_{j+k=n}\frac{A^j\,B^k}{j!\;k!}\right) = \left(\sum_{j=0}^{\infty}\frac{A^j}{j!}\right)\left(\sum_{k=0}^{\infty}\frac{B^k}{k!}\right).$$

Proof: We know that each of these infinite series of matrices converges. We just have to check that they converge to each other. To do this, consider the partial sums

$$\gamma_{2m} = \sum_{n=0}^{2m}\left(\sum_{j+k=n}\frac{A^j\,B^k}{j!\;k!}\right)$$

and

$$\alpha_m = \left(\sum_{j=0}^{m} \frac{A^j}{j!}\right) \text{ and } \beta_m = \left(\sum_{k=0}^{m} \frac{B^k}{k!}\right).$$

We need to show that the matrices $\gamma_{2m} - \alpha_m\beta_m$ tend to the zero matrix as $m \to \infty$. Toward that end, for a matrix $M = [m_{ij}]$, we let $\|M\| = \max |m_{ij}|$. We will show that $\|\gamma_{2m} - \alpha_m\beta_m\| \to 0$ as $m \to \infty$.

A computation shows that

$$\gamma_{2m} - \alpha_m\beta_m = {\sum}' \frac{A^j}{j!} \frac{B^k}{k!} + {\sum}'' \frac{A^j}{j!} \frac{B^k}{k!}$$

where $\sum'$ denotes the sum over terms with indices satisfying

$$j + k \le 2m, \quad 0 \le j \le m, \quad m + 1 \le k \le 2m$$

while $\sum''$ is the sum corresponding to

$$j + k \le 2m, \quad m + 1 \le j \le 2m, \quad 0 \le k \le m.$$

Therefore

$$\|\gamma_{2m} - \alpha_m\beta_m\| \le {\sum}' \|\frac{A^j}{j!}\| \cdot \|\frac{B^k}{k!}\| + {\sum}'' \|\frac{A^j}{j!}\| \cdot \|\frac{B^k}{k!}\|.$$

Now

$$\sum{}' \|\frac{A^j}{j!}\| \cdot \|\frac{B^k}{k!}\| \le \left(\sum_{j=0}^{m} \|\frac{A^j}{j!}\|\right)\left(\sum_{k=m+1}^{2m} \|\frac{B^k}{k!}\|\right).$$

This tends to 0 as $m \to \infty$ since, as we saw above,

$$\sum_{j=0}^{\infty} \|\frac{A^j}{j!}\| \le \exp(n\|A\|) < \infty.$$

Similarly,

$$\sum{}'' \|\frac{A^j}{j!}\| \cdot \|\frac{B^k}{k!}\| \to 0$$

as $m \to \infty$. Therefore $\lim_{m\to\infty}(\gamma_{2m} - \alpha_m\beta_m) = 0$, proving the lemma. ∎

Observe that statement (3) of the proposition implies that $\exp(A)$ is invertible for every matrix A. This is analogous to the fact that $e^a \neq 0$ for every real number a.

There is a very simple relationship between the eigenvectors of A and those of $\exp(A)$.

Proposition. *If $V \in \mathbb{R}^n$ is an eigenvector of A associated to the eigenvalue λ, then V is also an eigenvector of $\exp(A)$ associated to e^λ.*

Proof: From $AV = \lambda V$, we obtain

$$\exp(A)V = \lim_{n\to\infty} \left(\sum_{k=0}^{n} \frac{A^k V}{k!} \right)$$

$$= \lim_{n\to\infty} \left(\sum_{k=0}^{n} \frac{\lambda^k}{k!} V \right)$$

$$= \left(\sum_{k=0}^{\infty} \frac{\lambda^k}{k!} \right) V$$

$$= e^\lambda V. \qquad\blacksquare$$

Now let's return to the setting of systems of differential equations. Let A be an $n \times n$ matrix and consider the system $X' = AX$. Recall that $L(\mathbb{R}^n)$ denotes the set of all $n \times n$ matrices. We have a function $\mathbb{R} \to L(\mathbb{R}^n)$, which assigns the matrix $\exp(tA)$ to $t \in \mathbb{R}$. Since $L(\mathbb{R}^n)$ is identified with $\mathbb{R}^{n^2}$, it makes sense to speak of the derivative of this function.

Proposition.

$$\frac{d}{dt} \exp(tA) = A \exp(tA) = \exp(tA)A.$$

In other words, the derivative of the matrix-valued function $t \to \exp(tA)$ is another matrix-valued function $A \exp(tA)$.

Proof: We have

$$\frac{d}{dt} \exp(tA) = \lim_{h\to 0} \frac{\exp((t+h)A) - \exp(tA)}{h}$$

$$= \lim_{h\to 0} \frac{\exp(tA)\exp(hA) - \exp(tA)}{h}$$

$$= \exp(tA) \lim_{h \to 0} \left(\frac{\exp(hA) - I}{h} \right)$$

$$= \exp(tA)A;$$

that the last limit equals A follows from the series definition of $\exp(hA)$. Note that A commutes with each term of the series for $\exp(tA)$, hence with $\exp(tA)$. This proves the proposition.   ∎

Now we return to solving systems of differential equations. The following may be considered as the fundamental theorem of linear differential equations with constant coefficients.

Theorem. *Let A be an $n \times n$ matrix. Then the solution of the initial value problem $X' = AX$ with $X(0) = X_0$ is $X(t) = \exp(tA)X_0$. Moreover, this is the only such solution.*

Proof: The preceding proposition shows that

$$\frac{d}{dt}(\exp(tA)X_0) = \left(\frac{d}{dt} \exp(tA) \right) X_0 = A \exp(tA)X_0.$$

Moreover, since $\exp(0A)X_0 = X_0$, it follows that this is a solution of the initial value problem. To see that there are no other solutions, let $Y(t)$ be another solution satisfying $Y(0) = X_0$ and set

$$Z(t) = \exp(-tA)Y(t).$$

Then

$$\begin{aligned}
Z'(t) &= \left(\frac{d}{dt} \exp(-tA) \right) Y(t) + \exp(-tA)Y'(t) \\
&= -A \exp(-tA)Y(t) + \exp(-tA)AY(t) \\
&= \exp(-tA)(-A + A)Y(t) \\
&\equiv 0.
\end{aligned}$$

Therefore $Z(t)$ is a constant. Setting $t = 0$ shows $Z(t) = X_0$, so that $Y(t) = \exp(tA)X_0$. This completes the proof of the theorem.   ∎

Note that this proof is identical to that given way back in Section 1.1. Only the meaning of the letter A has changed.

Example. Consider the system

$$X' = \begin{pmatrix} \lambda & 1 \\ 0 & \lambda \end{pmatrix} X.$$

By the theorem, the general solution is

$$X(t) = \exp(tA)X_0 = \exp\begin{pmatrix} t\lambda & t \\ 0 & t\lambda \end{pmatrix} X_0.$$

But this is precisely the matrix whose exponential we computed earlier. We find

$$X(t) = \begin{pmatrix} e^{t\lambda} & te^{t\lambda} \\ 0 & e^{t\lambda} \end{pmatrix} X_0.$$

Note that this agrees with our computations in Chapter 3. ■

6.5 Nonautonomous Linear Systems

Up to this point, virtually all of the linear systems of differential equations that we have encountered have been autonomous. There are, however, certain types of nonautonomous systems that often arise in applications. One such system is of the form

$$X' = A(t)X$$

where $A(t) = [a_{ij}(t)]$ is an $n \times n$ matrix that depends continuously on time. We will investigate these types of systems further when we encounter the variational equation in subsequent chapters.

Here we restrict our attention to a different type of nonautonomous linear system given by

$$X' = AX + G(t)$$

where A is a constant $n \times n$ matrix and $G : \mathbb{R} \to \mathbb{R}^n$ is a *forcing term* that depends explicitly on t. This is an example of a first-order, linear, nonhomogeneous system of equations.

Example. (Forced Harmonic Oscillator) If we apply an external force to the harmonic oscillator system, the differential equation governing the motion becomes

$$x'' + bx' + kx = f(t)$$

where $f(t)$ measures the external force. An important special case occurs when this force is a periodic function of time, which corresponds, for example, to moving the table on which the mass-spring apparatus resides back and forth periodically. As a system, the forced harmonic oscillator equation becomes

$$X' = \begin{pmatrix} 0 & 1 \\ -k & -b \end{pmatrix} X + G(t) \quad \text{where} \quad G(t) = \begin{pmatrix} 0 \\ f(t) \end{pmatrix}. \qquad \blacksquare$$

For a nonhomogeneous system, the equation that results from dropping the time-dependent term, namely, $X' = AX$, is called the *homogeneous equation*. We know how to find the general solution of this system. Borrowing the notation from the previous section, the solution satisfying the initial condition $X(0) = X_0$ is

$$X(t) = \exp(tA)X_0,$$

so this is the general solution of the homogeneous equation.

To find the general solution of the nonhomogeneous equation, suppose that we have one particular solution $Z(t)$ of this equation. So $Z'(t) = AZ(t) + G(t)$. If $X(t)$ is any solution of the homogeneous equation, then the function $Y(t) = X(t) + Z(t)$ is another solution of the nonhomogeneous equation. This follows since we have

$$\begin{aligned} Y' &= X' + Z' = AX + AZ + G(t) \\ &= A(X + Z) + G(t) \\ &= AY + G(t). \end{aligned}$$

Therefore, since we know all solutions of the homogeneous equation, we can now find the general solution to the nonhomogeneous equation, provided that we can find just one particular solution of this equation. Often one gets such a solution by simply guessing that solution (in calculus, this method is usually called the *method of undetermined coefficients*). Unfortunately, guessing a solution does not always work. The following method, called *variation of parameters*, does work in all cases. However, there is no guarantee that we can actually evaluate the required integrals.

Theorem. (Variation of Parameters) *Consider the nonhomogeneous equation*

$$X' = AX + G(t)$$

where A is an n × n matrix and G(t) is a continuous function of t. Then

$$X(t) = \exp(tA) \left(X_0 + \int_0^t \exp(-sA)\, G(s)\, ds \right)$$

is a solution of this equation satisfying $X(0) = X_0$.

Proof: Differentiating $X(t)$, we obtain

$$X'(t) = A \exp(tA) \left(X_0 + \int_0^t \exp(-sA)\, G(s)\, ds \right)$$

$$+ \exp(tA)\, \frac{d}{dt} \int_0^t \exp(-sA)\, G(s)\, ds$$

$$= A \exp(tA) \left(X_0 + \int_0^t \exp(-sA)\, G(s)\, ds \right) + G(t)$$

$$= AX(t) + G(t). \qquad \blacksquare$$

We now give several applications of this result in the case of the periodically forced harmonic oscillator. Assume first that we have a damped oscillator that is forced by $\cos t$, so the period of the forcing term is 2π. The system is

$$X' = AX + G(t)$$

where $G(t) = (0, \cos t)$ and A is the matrix

$$A = \begin{pmatrix} 0 & 1 \\ -k & -b \end{pmatrix}$$

with $b, k > 0$. We claim that there is a unique periodic solution of this system that has period 2π. To prove this, we must first find a solution $X(t)$ satisfying $X(0) = X_0 = X(2\pi)$. By variation of parameters, we need to find X_0 such that

$$X_0 = \exp(2\pi A)X_0 + \exp(2\pi A) \int_0^{2\pi} \exp(-sA)\, G(s)\, ds.$$

Now the term

$$\exp(2\pi A) \int_0^{2\pi} \exp(-sA)\, G(s)\, ds$$

is a constant vector, which we denote by W. Therefore we must solve the equation

$$\left(\exp(2\pi A) - I\right) X_0 = -W.$$

There is a unique solution to this equation, since the matrix $\exp(2\pi A) - I$ is invertible. For if this matrix were not invertible, we would have a nonzero vector V with

$$\left(\exp(2\pi A) - I\right) V = 0,$$

or, in other words, the matrix $\exp(2\pi A)$ would have an eigenvalue of 1. But, from the previous section, the eigenvalues of $\exp(2\pi A)$ are given by $\exp(2\pi \lambda_j)$ where the λ_j are the eigenvalues of A. But each λ_j has real part less than 0, so the magnitude of $\exp(2\pi \lambda_j)$ is smaller than 1. Thus the matrix $\exp(2\pi A) - I$ is indeed invertible, and the unique initial value leading to a 2π-periodic solution is

$$X_0 = \left(\exp(2\pi A) - I\right)^{-1} (-W).$$

So let $X(t)$ be this periodic solution with $X(0) = X_0$. This solution is called the *steady-state* solution. If Y_0 is any other initial condition, then we may write $Y_0 = (Y_0 - X_0) + X_0$, so the solution through Y_0 is given by

$$Y(t) = \exp(tA)(Y_0 - X_0) + \exp(tA)X_0 + \exp(tA) \int_0^t \exp(-sA)\, G(s)\, ds$$

$$= \exp(tA)(Y_0 - X_0) + X(t).$$

The first term in this expression tends to 0 as $t \to \infty$, since it is a solution of the homogeneous equation. Hence every solution of this system tends to the steady-state solution as $t \to \infty$. Physically, this is clear: The motion of the damped (and unforced) oscillator tends to equilibrium, leaving only the motion due to the periodic forcing. We have proved:

Theorem. *Consider the forced, damped harmonic oscillator equation*

$$x'' + bx' + kx = \cos t$$

with $k, b > 0$. Then all solutions of this equation tend to the steady-state solution, which is periodic with period 2π. ■

Now consider a particular example of a forced, undamped harmonic oscillator

$$X' = \begin{pmatrix} 0 & 1 \\ -1 & 0 \end{pmatrix} X + \begin{pmatrix} 0 \\ \cos \omega t \end{pmatrix}$$

where the period of the forcing is now $2\pi/\omega$ with $\omega \neq \pm 1$. Let

$$A = \begin{pmatrix} 0 & 1 \\ -1 & 0 \end{pmatrix}.$$

The solution of the homogeneous equation is

$$X(t) = \exp(tA)X_0 = \begin{pmatrix} \cos t & \sin t \\ -\sin t & \cos t \end{pmatrix} X_0.$$

Variation of parameters provides a solution of the nonhomogeneous equation starting at the origin:

$$Y(t) = \exp(tA) \int_0^t \exp(-sA) \begin{pmatrix} 0 \\ \cos \omega s \end{pmatrix} ds$$

$$= \exp(tA) \int_0^t \begin{pmatrix} \cos s & -\sin s \\ \sin s & \cos s \end{pmatrix} \begin{pmatrix} 0 \\ \cos \omega s \end{pmatrix} ds$$

$$= \exp(tA) \int_0^t \begin{pmatrix} -\sin s \cos \omega s \\ \cos s \cos \omega s \end{pmatrix} ds$$

$$= \frac{1}{2} \exp(tA) \int_0^t \begin{pmatrix} \sin(\omega - 1)s - \sin(\omega + 1)s \\ \cos(\omega - 1)s + \cos(\omega + 1)s \end{pmatrix} ds.$$

Recalling that

$$\exp(tA) = \begin{pmatrix} \cos t & \sin t \\ -\sin t & \cos t \end{pmatrix}$$

and using the fact that $\omega \neq \pm 1$, evaluation of this integral plus a long computation yields

$$Y(t) = \frac{1}{2} \exp(tA) \begin{pmatrix} \dfrac{-\cos(\omega - 1)t}{\omega - 1} + \dfrac{\cos(\omega + 1)t}{\omega + 1} \\[2mm] \dfrac{\sin(\omega - 1)t}{\omega - 1} + \dfrac{\sin(\omega + 1)t}{\omega + 1} \end{pmatrix}$$

$$+ \exp(tA) \begin{pmatrix} (\omega^2 - 1)^{-1} \\ 0 \end{pmatrix}$$

$$= \frac{1}{\omega^2 - 1} \begin{pmatrix} -\cos \omega t \\ \omega \sin \omega t \end{pmatrix} + \exp(tA) \begin{pmatrix} (\omega^2 - 1)^{-1} \\ 0 \end{pmatrix}.$$

Thus the general solution of this equation is

$$Y(t) = \exp(tA) \left(X_0 + \begin{pmatrix} (\omega^2 - 1)^{-1} \\ 0 \end{pmatrix} \right) + \frac{1}{\omega^2 - 1} \begin{pmatrix} -\cos \omega t \\ \omega \sin \omega t \end{pmatrix}.$$

The first term in this expression is periodic with period 2π while the second has period $2\pi/\omega$. Unlike the damped case, this solution does not necessarily yield a periodic motion. Indeed, this solution is periodic if and only if ω is a rational number. If ω is irrational, the motion is quasiperiodic, just as we saw in Section 6.2.

EXERCISES

1. Find the general solution for $X' = AX$ where A is given by:

(a) $\begin{pmatrix} 0 & 0 & 1 \\ 0 & 1 & 0 \\ 1 & 0 & 0 \end{pmatrix}$ (b) $\begin{pmatrix} 1 & 0 & 1 \\ 0 & 1 & 0 \\ 1 & 0 & 1 \end{pmatrix}$ (c) $\begin{pmatrix} 0 & 1 & 0 \\ -1 & 0 & 0 \\ 1 & 1 & 1 \end{pmatrix}$

(d) $\begin{pmatrix} 0 & 1 & 0 \\ 1 & 0 & 0 \\ 1 & 1 & 1 \end{pmatrix}$ (e) $\begin{pmatrix} 1 & 0 & 1 \\ 0 & 1 & 0 \\ 0 & 0 & 1 \end{pmatrix}$ (f) $\begin{pmatrix} 1 & 1 & 0 \\ 1 & 1 & 1 \\ 0 & 1 & 1 \end{pmatrix}$

(g) $\begin{pmatrix} 1 & 0 & -1 \\ -1 & 1 & -1 \\ 0 & 0 & 1 \end{pmatrix}$ (h) $\begin{pmatrix} 1 & 0 & 0 & 1 \\ 0 & 1 & 1 & 0 \\ 0 & 0 & 1 & 0 \\ 1 & 0 & 0 & 0 \end{pmatrix}$

2. Consider the linear system

$$X' = \begin{pmatrix} \lambda_1 & 0 & 0 \\ 0 & \lambda_2 & 0 \\ 0 & 0 & \lambda_3 \end{pmatrix} X$$

where $\lambda_3 < \lambda_2 < \lambda_1 < 0$. Describe how the solution through an arbitrary initial value tends to the origin.

3. Give an example of a 3×3 matrix A for which all nonequilibrium solutions of $X' = AX$ are periodic with period 2π. Sketch the phase portrait.

4. Find the general solution of

$$X' = \begin{pmatrix} 0 & 1 & 1 & 0 \\ -1 & 0 & 0 & 1 \\ 0 & 0 & 0 & 1 \\ 0 & 0 & -1 & 0 \end{pmatrix} X.$$

5. Consider the system

$$X' = \begin{pmatrix} 0 & 0 & a \\ 0 & b & 0 \\ a & 0 & 0 \end{pmatrix} X$$

depending on the two parameters a and b.

(a) Find the general solution of this system.

(b) Sketch the region in the ab–plane where this system has different types of phase portraits.

6. Consider the system

$$X' = \begin{pmatrix} a & 0 & b \\ 0 & b & 0 \\ -b & 0 & a \end{pmatrix} X$$

depending on the two parameters a and b.

(a) Find the general solution of this system.

(b) Sketch the region in the ab–plane where this system has different types of phase portraits.

7. Coupled Harmonic Oscillators. In this series of exercises you are asked to generalize the material on harmonic oscillators in Section 6.2 to the case where the oscillators are *coupled*. Suppose there are two masses m_1

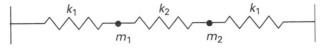

Figure 6.10   A coupled oscillator.

and m_2 attached to springs and walls as shown in Figure 6.10. The springs connecting m_j to the walls both have spring constants k_1, while the spring connecting m_1 and m_2 has spring constant k_2. This coupling means that the motion of either mass affects the behavior of the other.

Let x_j denote the displacement of each mass from its rest position, and assume that both masses are equal to 1. The differential equations for these coupled oscillators are then given by

$$x_1'' = -(k_1 + k_2)x_1 + k_2 x_2$$
$$x_2'' = k_2 x_1 - (k_1 + k_2)x_2.$$

These equations are derived as follows. If m_1 is moved to the right ($x_1 > 0$), the left spring is stretched and exerts a restorative force on m_1 given by $-k_1 x_1$. Meanwhile, the central spring is compressed, so it exerts a restorative force on m_1 given by $-k_2 x_1$. If the right spring is stretched, then the central spring is compressed and exerts a restorative force on m_1 given by $k_2 x_2$ (since $x_2 < 0$). The forces on m_2 are similar.

(a) Write these equations as a first-order linear system.
(b) Determine the eigenvalues and eigenvectors of the corresponding matrix.
(c) Find the general solution.
(d) Let $\omega_1 = \sqrt{k_1}$ and $\omega_2 = \sqrt{k_1 + 2k_2}$. What can be said about the periodicity of solutions relative to the ω_j? Prove this.

8. Suppose $X' = AX$ where A is a 4×4 matrix whose eigenvalues are $\pm i\sqrt{2}$ and $\pm i\sqrt{3}$. Describe this flow.

9. Suppose $X' = AX$ where A is a 4×4 matrix whose eigenvalues are $\pm i$ and $-1 \pm i$. Describe this flow.

10. Suppose $X' = AX$ where A is a 4×4 matrix whose eigenvalues are $\pm i$ and ± 1. Describe this flow.

11. Consider the system $X' = AX$ where $X = (x_1, \ldots, x_6)$ and

$$A = \begin{pmatrix} 0 & \omega_1 & & & & \\ -\omega_1 & 0 & & & & \\ & & 0 & \omega_2 & & \\ & & -\omega_2 & 0 & & \\ & & & & -1 & \\ & & & & & 1 \end{pmatrix}$$

and ω_1/ω_2 is irrational. Describe qualitatively how a solution behaves when, at time 0, each x_j is nonzero with the exception that

(a) $x_6 = 0$;

(b) $x_5 = 0$;

(c) $x_3 = x_4 = x_5 = 0$;

(d) $x_3 = x_4 = x_5 = x_6 = 0$.

12. Compute the exponentials of the following matrices:

(a) $\begin{pmatrix} 5 & -6 \\ 3 & -4 \end{pmatrix}$ (b) $\begin{pmatrix} 2 & -1 \\ 1 & 2 \end{pmatrix}$ (c) $\begin{pmatrix} 2 & -1 \\ 0 & 2 \end{pmatrix}$ (d) $\begin{pmatrix} 0 & 1 \\ 1 & 0 \end{pmatrix}$

(e) $\begin{pmatrix} 0 & 1 & 2 \\ 0 & 0 & 3 \\ 0 & 0 & 0 \end{pmatrix}$ (f) $\begin{pmatrix} 2 & 0 & 0 \\ 0 & 3 & 0 \\ 0 & 1 & 3 \end{pmatrix}$ (g) $\begin{pmatrix} \lambda & 0 & 0 \\ 1 & \lambda & 0 \\ 0 & 1 & \lambda \end{pmatrix}$

(h) $\begin{pmatrix} i & 0 \\ 0 & -i \end{pmatrix}$ (i) $\begin{pmatrix} 1+i & 0 \\ 2 & 1+i \end{pmatrix}$ (j) $\begin{pmatrix} 1 & 0 & 0 & 0 \\ 1 & 0 & 0 & 0 \\ 1 & 0 & 0 & 0 \\ 1 & 0 & 0 & 0 \end{pmatrix}$

13. Find an example of two matrices A, B such that

$$\exp(A + B) \neq \exp(A)\exp(B).$$

14. Show that if $AB = BA$, then

(a) $\exp(A)\exp(B) = \exp(B)\exp(A)$

(b) $\exp(A)B = B\exp(A)$.

15. Consider the triplet of harmonic oscillators

$$x_1'' = -x_1$$
$$x_2'' = -2x_2$$
$$x_3'' = -\omega^2 x_3$$

where ω is irrational. What can you say about the qualitative behavior of solutions of this six-dimensional system?

7
Nonlinear Systems

In this chapter we begin the study of nonlinear differential equations. Unlike linear (constant coefficient) systems, where we can always find the explicit solution of any initial value problem, this is rarely the case for nonlinear systems. In fact, basic properties such as the existence and uniqueness of solutions, which was so obvious in the linear case, no longer hold for nonlinear systems. As we shall see, some nonlinear systems have no solutions whatsoever to a given initial value problem. On the other hand, some systems have infinitely many different such solutions. Even if we do find a solution of such a system, this solution need not be defined for all time; for example, the solution may tend to ∞ in finite time. And other questions arise: For example, what happens if we vary the initial condition of a system ever so slightly? Does the corresponding solution vary continuously? All of this is clear for linear systems, but not at all clear in the nonlinear case. This means that the underlying theory behind nonlinear systems of differential equations is quite a bit more complicated than that for linear systems.

In practice, most nonlinear systems that arise are "nice" in the sense that we do have existence and uniqueness of solutions as well as continuity of solutions when initial conditions are varied, and other "natural" properties. Thus we have a choice: Given a nonlinear system, we could simply plunge ahead and either hope that or, if possible, verify that, in each specific case, the system's solutions behave nicely. Alternatively, we could take a long pause at this stage to develop the necessary hypotheses that guarantee that solutions of a given nonlinear system behave nicely.

In this book we pursue a compromise route. In this chapter, we will spell out in precise detail many of the theoretical results that govern the behavior of solutions of differential equations. We will present examples of how and when these results fail, but we will not prove these theorems here. Rather, we will postpone all of the technicalities until Chapter 17, primarily because understanding this material demands a firm and extensive background in the principles of real analysis. In subsequent chapters, we will make use of the results stated here, but readers who are primarily interested in applications of differential equations or in understanding how specific nonlinear systems may be analyzed need not get bogged down in these details here. Readers who want the technical details may take a detour to Chapter 17 now.

7.1 Dynamical Systems

As mentioned previously, most nonlinear systems of differential equations are impossible to solve analytically. One reason for this is the unfortunate fact that we simply do not have enough functions with specific names that we can use to write down explicit solutions of these systems. Equally problematic is the fact that, as we shall see, higher dimensional systems may exhibit chaotic behavior, a property that makes knowing a particular explicit solution essentially worthless in the larger scheme of understanding the behavior of the system. Hence we are forced to resort to different means in order to understand these systems. These are the techniques that arise in the field of dynamical systems. We will use a combination of analytic, geometric, and topological techniques to derive rigorous results about the behavior of solutions of these equations.

We begin by collecting some of the terminology regarding dynamical systems that we have introduced at various points in the preceding chapters. A *dynamical system* is a way of describing the passage in time of all points of a given space S. The space S could be thought of, for example, as the space of states of some physical system. Mathematically, S might be a Euclidean space or an open subset of Euclidean space or some other space such as a surface in $\mathbb{R}^3$. When we consider dynamical systems that arise in mechanics, the space S will be the set of possible positions and velocities of the system. For the sake of simplicity, we will assume throughout that the space S is Euclidean space $\mathbb{R}^n$, although in certain cases the important dynamical behavior will be confined to a particular subset of $\mathbb{R}^n$.

Given an initial position $X \in \mathbb{R}^n$, a dynamical system on $\mathbb{R}^n$ tells us where X is located 1 unit of time later, 2 units of time later, and so on. We denote these new positions of X by X_1, X_2, and so forth. At time zero, X is located at position X_0. One unit before time zero, X was at X_{-1}. In general, the "trajectory" of X is given by X_t. If we measure the positions X_t using only integer time values, we have an example of a *discrete* dynamical system, which we shall study in Chapter 15. If time is measured continuously with $t \in \mathbb{R}$, we have a *continuous* dynamical system. If the system depends on time in a continuously differentiable manner, we have a *smooth* dynamical system. These are the three principal types of dynamical systems that arise in the study of systems of differential equations, and they will form the backbone of Chapters 8 through 14.

The function that takes t to X_t yields either a sequence of points or a curve in $\mathbb{R}^n$ that represents the life history of X as time runs from $-\infty$ to ∞. Different branches of dynamical systems make different assumptions about how the function X_t depends on t. For example, ergodic theory deals with such functions under the assumption that they preserve a measure on $\mathbb{R}^n$. Topological dynamics deals with such functions under the assumption that X_t varies only continuously. In the case of differential equations, we will usually assume that the function X_t is continuously differentiable. The map $\phi_t \colon \mathbb{R}^n \to \mathbb{R}^n$ that takes X into X_t is defined for each t and, from our interpretation of X_t as a state moving in time, it is reasonable to expect ϕ_t to have ϕ_{-t} as its inverse. Also, ϕ_0 should be the identity function $\phi_0(X) = X$ and $\phi_t(\phi_s(X)) = \phi_{t+s}(X)$ is also a natural condition. We formalize all of this in the following definition:

Definition
A *smooth dynamical system* on $\mathbb{R}^n$ is a continuously differentiable function $\phi \colon \mathbb{R} \times \mathbb{R}^n \to \mathbb{R}^n$ where $\phi(t, X) = \phi_t(X)$ satisfies

1. $\phi_0 \colon \mathbb{R}^n \to \mathbb{R}^n$ is the identity function: $\phi_0(X_0) = X_0$;
2. The composition $\phi_t \circ \phi_s = \phi_{t+s}$ for each $t, s \in \mathbb{R}$.

Recall that a function is continuously differentiable if all of its partial derivatives exist and are continuous throughout its domain. It is traditional to call a continuously differentiable function a C^1 function. If the function is k times continuously differentiable, it is called a C^k function. Note that the definition above implies that the map $\phi_t \colon \mathbb{R}^n \to \mathbb{R}^n$ is C^1 for each t and has a C^1 inverse ϕ_{-t} (take $s = -t$ in part 2).

Example. For the first-order differential equation $x' = ax$, the function $\phi_t(x_0) = x_0 \exp(at)$ gives the solutions of this equation and also defines a smooth dynamical system on $\mathbb{R}$. ∎

Example. Let A be an $n \times n$ matrix. Then the function $\phi_t(X_0) = \exp(tA)X_0$ defines a smooth dynamical system on $\mathbb{R}^n$. Clearly, $\phi_0 = \exp(0) = I$ and, as we saw in the previous chapter, we have

$$\phi_{t+s} = \exp((t+s)A) = (\exp(tA))(\exp(sA)) = \phi_t \circ \phi_s. \qquad \blacksquare$$

Note that these examples are intimately related to the system of differential equations $X' = AX$. In general, a smooth dynamical system always yields a vector field on $\mathbb{R}^n$ via the following rule: Given ϕ_t, let

$$F(X) = \left.\frac{d}{dt}\right|_{t=0} \phi_t(X).$$

Then ϕ_t is just the time t map associated to the flow of $X' = F(X)$.

Conversely, the differential equation $X' = F(X)$ generates a smooth dynamical system provided the time t map of the flow is well defined and continuously differentiable for all time. Unfortunately, this is not always the case.

7.2 The Existence and Uniqueness Theorem

We turn now to the fundamental theorem of differential equations, the existence and uniqueness theorem. Consider the system of differential equations

$$X' = F(X)$$

where $F: \mathbb{R}^n \to \mathbb{R}^n$. Recall that a solution of this system is a function $X: J \to \mathbb{R}^n$ defined on some interval $J \subset \mathbb{R}$ such that, for all $t \in J$,

$$X'(t) = F(X(t)).$$

Geometrically, $X(t)$ is a curve in $\mathbb{R}^n$ whose tangent vector $X'(t)$ exists for all $t \in J$ and equals $F(X(t))$. As in previous chapters, we think of this vector as being based at $X(t)$, so that the map $F: \mathbb{R}^n \to \mathbb{R}^n$ defines a vector field on $\mathbb{R}^n$. An *initial condition* or *initial value* for a solution $X: J \to \mathbb{R}^n$ is a specification of the form $X(t_0) = X_0$ where $t_0 \in J$ and $X_0 \in \mathbb{R}^n$. For simplicity, we usually take $t_0 = 0$. The main problem in differential equations is to find the solution of any *initial value problem*; that is, to determine the solution of the system that satisfies the initial condition $X(0) = X_0$ for each $X_0 \in \mathbb{R}^n$.

Unfortunately, nonlinear differential equations may have no solutions satisfying certain initial conditions.

Example. Consider the simple first-order differential equation

$$x' = \begin{cases} 1 & \text{if } x < 0 \\ -1 & \text{if } x \geq 0. \end{cases}$$

This vector field on $\mathbb{R}$ points to the left when $x \geq 0$ and to the right if $x < 0$. Consequently, there is no solution that satisfies the initial condition $x(0) = 0$. Indeed, such a solution must initially decrease since $x'(0) = -1$, but for all negative values of x, solutions must increase. This cannot happen. Note further that solutions are never defined for all time. For example, if $x_0 > 0$, then the solution through x_0 is given by $x(t) = x_0 - t$, but this solution is only valid for $-\infty < t < x_0$ for the same reason as above.

The problem in this example is that the vector field is not continuous at 0; whenever a vector field is discontinuous we face the possibility that nearby vectors may point in "opposing" directions, thereby causing solutions to halt at these bad points.   ■

Beyond the problem of existence of solutions of nonlinear differential equations, we also must confront the fact that certain equations may have many different solutions to the same initial value problem.

Example. Consider the differential equation

$$x' = 3x^{2/3}.$$

The identically zero function $u: \mathbb{R} \to \mathbb{R}$ given by $u(t) \equiv 0$ is clearly a solution with initial condition $u(0) = 0$. But $u_0(t) = t^3$ is also a solution satisfying this initial condition. Moreover, for any $\tau > 0$, the function given by

$$u_\tau(t) = \begin{cases} 0 & \text{if } t \leq \tau \\ (t - \tau)^3 & \text{if } t > \tau \end{cases}$$

is also a solution satisfying the initial condition $u_\tau(0) = 0$. While the differential equation in this example is continuous at $x_0 = 0$, the problems arise because $3x^{2/3}$ is not differentiable at this point.   ■

From these two examples it is clear that, to ensure existence and uniqueness of solutions, certain conditions must be imposed on the function F. In the first example, F was not continuous at the problematic point 0, while in the second example, F failed to be differentiable at 0. It turns out that the assumption that F is continuously differentiable is sufficient to guarantee both existence and uniqueness of solutions, as we shall see. Fortunately, differential equations that are not continuously differentiable rarely arise in applications, so the phenomenon of nonexistence or nonuniqueness of solutions with given initial conditions is quite exceptional.

The following is the fundamental local theorem of ordinary differential equations.

The Existence and Uniqueness Theorem.   *Consider the initial value problem*

$$X' = F(X),\ X(t_0) = X_0$$

where $X_0 \in \mathbb{R}^n$. Suppose that $F\colon \mathbb{R}^n \to \mathbb{R}^n$ is C^1. Then, first of all, there exists a solution of this initial value problem and, secondly, this is the only such solution. More precisely, there exists an $a > 0$ and a unique solution

$$X\colon (t_0 - a, t_0 + a) \to \mathbb{R}^n$$

of this differential equation satisfying the initial condition $X(t_0) = X_0$.   ■

The important proof of this theorem is contained in Chapter 17.

Without dwelling on the details here, the proof of this theorem depends on an important technique known as *Picard iteration*. Before moving on, we illustrate how the Picard iteration scheme used in the proof of the theorem works in several special examples. The basic idea behind this iterative process is to construct a sequence of functions that converges to the solution of the differential equation. The sequence of functions $u_k(t)$ is defined inductively by $u_0(t) = x_0$ where x_0 is the given initial condition, and then

$$u_{k+1}(t) = x_0 + \int_0^t (u_k(s))\, ds.$$

Example.   Consider the simple differential equation $x' = x$. We will produce the solution of this equation satisfying $x(0) = x_0$. We know, of course, that this solution is given by $x(t) = x_0 e^t$. We will construct a sequence of functions $u_k(t)$, one for each k, which converges to the actual solution $x(t)$ as $k \to \infty$.

We start with

$$u_0(t) = x_0,$$

the given initial value. Then we set

$$u_1(t) = x_0 + \int_0^t u_0(s)\, ds = x_0 + \int_0^t x_0\, ds,$$

so that $u_1(t) = x_0 + tx_0$. Given u_1 we define

$$u_2(t) = x_0 + \int_0^t u_1(s)\,ds = x_0 + \int_0^t (x_0 + sx_0)\,ds,$$

so that $u_2(t) = x_0 + tx_0 + \frac{t^2}{2}x_0$. You can probably see where this is heading. Inductively, we set

$$u_{k+1}(t) = x_0 + \int_0^t u_k(s)\,ds,$$

and so

$$u_{k+1}(t) = x_0 \sum_{i=0}^{k+1} \frac{t^i}{i!}.$$

As $k \to \infty$, $u_k(t)$ converges to

$$x_0 \sum_{i=0}^{\infty} \frac{t^i}{i!} = x_0 e^t = x(t),$$

which is the solution of our original equation.   ∎

Example. For an example of Picard iteration applied to a system of differential equations, consider the linear system

$$X' = F(X) = \begin{pmatrix} 0 & 1 \\ -1 & 0 \end{pmatrix} X$$

with initial condition $X(0) = (1, 0)$. As we have seen, the solution of this initial value problem is

$$X(t) = \begin{pmatrix} \cos t \\ -\sin t \end{pmatrix}.$$

Using Picard iteration, we have

$$U_0(t) = \begin{pmatrix} 1 \\ 0 \end{pmatrix}$$

$$U_1(t) = \begin{pmatrix} 1 \\ 0 \end{pmatrix} + \int_0^t F\begin{pmatrix} 1 \\ 0 \end{pmatrix} ds = \begin{pmatrix} 1 \\ 0 \end{pmatrix} + \int_0^t \begin{pmatrix} 0 \\ -1 \end{pmatrix} ds = \begin{pmatrix} 1 \\ -t \end{pmatrix}$$

$$U_2(t) = \begin{pmatrix} 1 \\ 0 \end{pmatrix} + \int_0^t \begin{pmatrix} -s \\ -1 \end{pmatrix} ds = \begin{pmatrix} 1 - t^2/2 \\ -t \end{pmatrix}$$

$$U_3(t) = \begin{pmatrix} 1 \\ 0 \end{pmatrix} + \int_0^t \begin{pmatrix} -s \\ -1 + s^2/2 \end{pmatrix} ds = \begin{pmatrix} 1 - t^2/2 \\ -t + t^3/3! \end{pmatrix}$$

$$U_4(t) = \begin{pmatrix} 1 - t^2/2 + t^4/4! \\ -t + t^3/3! \end{pmatrix}$$

and we see the infinite series for the cosine and sine functions emerging from this iteration. ∎

Now suppose that we have two solutions $Y(t)$ and $Z(t)$ of the differential equation $X' = F(X)$, and that $Y(t)$ and $Z(t)$ satisfy $Y(t_0) = Z(t_0)$. Suppose that both solutions are defined on an interval J. The existence and uniqueness theorem guarantees that $Y(t) = Z(t)$ for all t in an interval about t_0, which may *a priori* be smaller than J. However, this is not the case. To see this, suppose that J^* is the largest interval on which $Y(t) = Z(t)$. Let t_1 be an endpoint of J^*. By continuity, we have $Y(t_1) = Z(t_1)$. The theorem then guarantees that, in fact, $Y(t)$ and $Z(t)$ agree on an open interval containing t_1. This contradicts the assertion that J^* is the largest interval on which the two solutions agree.

Thus we can always assume that we have a unique solution defined on a maximal time domain. There is, however, no guarantee that a solution $X(t)$ can be defined for all time, no matter how "nice" $F(X)$ is.

Example. Consider the differential equation in $\mathbb{R}$ given by

$$x' = 1 + x^2.$$

This equation has as solutions the functions $x(t) = \tan(t + c)$ where c is a constant. Such a function cannot be extended over an interval larger than

$$-c - \frac{\pi}{2} < t < -c + \frac{\pi}{2}$$

since $x(t) \to \pm\infty$ as $t \to -c \pm \pi/2$. ∎

This example is typical, for we have

Theorem. *Let $U \subset \mathbb{R}^n$ be an open set, and let $F : U \to \mathbb{R}^n$ be C^1. Let $X(t)$ be a solution of $X' = F(X)$ defined on a maximal open interval $J = (\alpha, \beta) \subset \mathbb{R}$ with $\beta < \infty$. Then given any closed and bounded set $K \subset U$, there is some $t \in (\alpha, \beta)$ with $X(t) \notin K$.* ∎

This theorem says that if a solution $X(t)$ cannot be extended to a larger time interval, then this solution leaves any closed and bounded set in U. This implies that $X(t)$ must come arbitrarily close to the boundary of U as $t \to \beta$. Similar results hold as $t \to \alpha$.

7.3 Continuous Dependence of Solutions

For the existence and uniqueness theorem to be at all interesting in any physical (or even mathematical) sense, this result needs to be complemented by the property that the solution $X(t)$ depends continuously on the initial condition $X(0)$. The next theorem gives a precise statement of this property.

Theorem. *Consider the differential equation $X' = F(X)$ where $F: \mathbb{R}^n \to \mathbb{R}^n$ is C^1. Suppose that $X(t)$ is a solution of this equation which is defined on the closed interval $[t_0, t_1]$ with $X(t_0) = X_0$. Then there is a neighborhood $U \subset \mathbb{R}^n$ of X_0 and a constant K such that if $Y_0 \in U$, then there is a unique solution $Y(t)$ also defined on $[t_0, t_1]$ with $Y(t_0) = Y_0$. Moreover $Y(t)$ satisfies*

$$|Y(t) - X(t)| \le K|Y_0 - X_0| \exp(K(t - t_0))$$

for all $t \in [t_0, t_1]$. ■

This result says that, if the solutions $X(t)$ and $Y(t)$ start out close together, then they remain close together for t close to t_0. While these solutions may separate from each other, they do so no faster than exponentially. In particular, since the right-hand side of this inequality depends on $|Y_0 - X_0|$, which we may assume is small, we have:

Corollary. (Continuous Dependence on Initial Conditions) *Let $\phi(t, X)$ be the flow of the system $X' = F(X)$ where F is C^1. Then ϕ is a continuous function of X.* ■

Example. Let $k > 0$. For the system

$$X' = \begin{pmatrix} -1 & 0 \\ 0 & k \end{pmatrix} X,$$

we know that the solution $X(t)$ satisfying $X(0) = (-1, 0)$ is given by

$$X(t) = (-e^{-t}, 0).$$

For any $\eta \neq 0$, let $Y_\eta(t)$ be the solution satisfying $Y_\eta(0) = (-1, \eta)$. Then

$$Y_\eta(t) = (-e^{-t}, \eta e^{kt}).$$

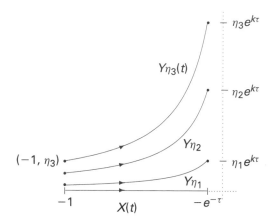

Figure 7.1   The solutions $Y_\eta(t)$ separate exponentially from $X(t)$ but nonetheless are continuous in their initial conditions.

As in the theorem, we have

$$|Y_\eta(t) - X(t)| = |\eta e^{kt} - 0| = |\eta - 0|e^{kt} = |Y_\eta(0) - X(0)|e^{kt}.$$

The solutions Y_η do indeed separate from $X(t)$, as we see in Figure 7.1, but they do so at most exponentially. Moreover, for any fixed time t, we have $Y_\eta(t) \to X(t)$ as $\eta \to 0$.   ∎

Differential equations often depend on parameters. For example, the harmonic oscillator equations depend on the parameters b (the damping constant) and k (the spring constant). Then the natural question is how do solutions of these equations depend on these parameters? As in the previous case, solutions depend continuously on these parameters provided that the system depends on the parameters in a continuously differentiable fashion. We can see this easily by using a special little trick. Suppose the system

$$X' = F_a(X)$$

depends on the parameter a in a C^1 fashion. Let's consider an "artificially" augmented system of differential equations given by

$$x_1' = f_1(x_1, \ldots, x_n, a)$$
$$\vdots$$
$$x_n' = f_n(x_1, \ldots, x_n, a)$$
$$a' = 0.$$

This is now an autonomous system of $n + 1$ differential equations. While this expansion of the system may seem trivial, we may now invoke the previous result about continuous dependence of solutions on initial conditions to verify that solutions of the original system depend continuously on a as well.

Theorem. (Continuous Dependence on Parameters) *Let $X' = F_a(X)$ be a system of differential equations for which F_a is continuously differentiable in both X and a. Then the flow of this system depends continuously on a as well.* ■

7.4 The Variational Equation

Consider an autonomous system $X' = F(X)$ where, as usual, F is assumed to be C^1. The flow $\phi(t, X)$ of this system is a function of both t and X. From the results of the previous section, we know that ϕ is continuous in the variable X. We also know that ϕ is differentiable in the variable t, since $t \to \phi(t, X)$ is just the solution curve through X. In fact, ϕ is also differentiable in the variable X, for we shall prove in Chapter 17:

Theorem. (Smoothness of Flows). *Consider the system $X' = F(X)$ where F is C^1. Then the flow $\phi(t, X)$ of this system is a C^1 function; that is, $\partial \phi / \partial t$ and $\partial \phi / \partial X$ exist and are continuous in t and X.* ■

Note that we can compute $\partial \phi / \partial t$ for any value of t as long as we know the solution passing through X_0, for we have

$$\frac{\partial \phi}{\partial t}(t, X_0) = F(\phi(t, X_0)).$$

We also have

$$\frac{\partial \phi}{\partial X}(t, X_0) = D\phi_t(X_0)$$

where $D\phi_t$ is the Jacobian of the function $X \to \phi_t(X)$. To compute $\partial \phi / \partial X$, however, it appears that we need to know the solution through X_0 as well as the solutions through all nearby initial positions, since we need to compute the partial derivatives of the various components of ϕ_t. However, we can get around this difficulty by introducing the variational equation along the solution through X_0.

To accomplish this, we need to take another brief detour into the world of nonautonomous differential equations. Let $A(t)$ be a family of $n \times n$ matrices that depends continuously on t. The system

$$X' = A(t)X$$

is a linear, nonautonomous system. We have an existence and uniqueness theorem for these types of equations:

Theorem. *Let $A(t)$ be a continuous family of $n \times n$ matrices defined for $t \in [\alpha, \beta]$. Then the initial value problem*

$$X' = A(t)X, \ X(t_0) = X_0$$

has a unique solution that is defined on the entire interval $[\alpha, \beta]$. ∎

Note that there is some additional content to this theorem: We do not assume that the right-hand side is a C^1 function in t. Continuity of $A(t)$ suffices to guarantee existence and uniqueness of solutions.

Example. Consider the first-order, linear, nonautonomous differential equation

$$x' = a(t)x.$$

The unique solution of this equation satisfying $x(0) = x_0$ is given by

$$x(t) = x_0 \exp\left(\int_0^t a(s)\, ds\right),$$

as is easily checked using the methods of Chapter 1. All we need is that $a(t)$ is continuous in order that $x'(t) = a(t)x(t)$; we do not need differentiability of $a(t)$ for this to be true. ∎

Note that solutions of linear, nonautonomous equations satisfy the linearity principle. That is, if $Y(t)$ and $Z(t)$ are two solutions of such a system, then so too is $\alpha Y(t) + \beta Z(t)$ for any constants α and β.

Now we return to the autonomous, nonlinear system $X' = F(X)$. Let $X(t)$ be a particular solution of this system defined for t in some interval $J = [\alpha, \beta]$. Fix $t_0 \in J$ and set $X(t_0) = X_0$. For each $t \in J$ let

$$A(t) = DF_{X(t)}$$

where $DF_{X(t)}$ denotes the Jacobian matrix of F at the point $X(t) \in \mathbb{R}^n$. Since F is C^1, $A(t) = DF_{X(t)}$ is a continuous family of $n \times n$ matrices. Consider the nonautonomous linear equation

$$U' = A(t)U.$$

This equation is known as the *variational equation* along the solution $X(t)$. By the previous theorem, we know that this variational equation has a solution defined on all of J for every initial condition $U(t_0) = U_0$.

The significance of this equation is that, if $U(t)$ is the solution of the variational equation that satisfies $U(t_0) = U_0$, then the function

$$t \to X(t) + U(t)$$

is a good approximation to the solution $Y(t)$ of the autonomous equation with initial value $Y(t_0) = X_0 + U_0$, provided U_0 is sufficiently small. This is the content of the following result.

Proposition.   *Consider the system $X' = F(X)$ where F is C^1. Suppose*

1. *$X(t)$ is a solution of $X' = F(X)$, which is defined for all $t \in [\alpha, \beta]$ and satisfies $X(t_0) = X_0$;*
2. *$U(t)$ is the solution to the variational equation along $X(t)$ that satisfies $U(t_0) = U_0$;*
3. *$Y(t)$ is the solution of $X' = F(X)$ that satisfies $Y(t_0) = X_0 + U_0$.*

Then

$$\lim_{U_0 \to 0} \frac{|Y(t) - (X(t) + U(t))|}{|U_0|}$$

converges to 0 uniformly in $t \in [\alpha, \beta]$.   ∎

Technically, this means that for every $\epsilon > 0$ there exists $\delta > 0$ such that if $|U_0| \le \delta$, then

$$|Y(t) - (X(t) + U(t))| \le \epsilon |U_0|$$

for all $t \in [\alpha, \beta]$. Thus, as $U_0 \to 0$, the curve $t \to X(t) + U(t)$ is a better and better approximation to $Y(t)$. In many applications, the solution of the variational equation $X(t) + U(t)$ is used in place of $Y(t)$; this is convenient because $U(t)$ depends linearly on U_0 by the linearity principle.

Example. Consider the nonlinear system of equations

$$x' = x + y^2$$
$$y' = -y.$$

We will discuss this system in more detail in the next chapter. For now, note that we know one solution of this system explicitly, namely, the equilibrium solution at the origin $X(t) \equiv (0,0)$. The variational equation along this solution is given by

$$U' = DF_0(U) = \begin{pmatrix} 1 & 0 \\ 0 & -1 \end{pmatrix} U,$$

which is an autonomous linear system. We obtain the solutions of this equation immediately: They are given by

$$U(t) = \begin{pmatrix} x_0 e^t \\ y_0 e^{-t} \end{pmatrix}.$$

The result above then guarantees that the solution of the nonlinear equation through (x_0, y_0) and defined on the interval $[-\tau, \tau]$ is as close as we wish to $U(t)$, provided (x_0, y_0) is sufficiently close to the origin. ∎

Note that the arguments in this example are perfectly general. Given any nonlinear system of differential equations $X' = F(X)$ with an equilibrium point at X_0, we may consider the variational equation along this solution. But DF_{X_0} is a constant matrix A. The variational equation is then $U' = AU$, which is an autonomous linear system. This system is called the *linearized system* at X_0. We know that flow of the linearized system is $\exp(tA)U_0$, so the result above says that near an equilibrium point of a nonlinear system, the phase portrait resembles that of the corresponding linearized system. We will make the term *resembles* more precise in the next chapter.

Using the previous proposition, we may now compute $\partial \phi / \partial X$, assuming we know the solution $X(t)$:

Theorem. *Let $X' = F(X)$ be a system of differential equations where F is C^1. Let $X(t)$ be a solution of this system satisfying the initial condition $X(0) = X_0$ and defined for $t \in [\alpha, \beta]$, and let $U(t, U_0)$ be the solution to the variational equation along $X(t)$ that satisfies $U(0, U_0) = U_0$. Then*

$$D\phi_t(X_0) \, U_0 = U(t, U_0).$$

That is, $\partial \phi / \partial X$ applied to U_0 is given by solving the corresponding variational equation starting at U_0.

Proof: Using the proposition, we have for all $t \in [\alpha, \beta]$:

$$D\phi_t(X_0)U_0 = \lim_{h \to 0} \frac{\phi_t(X_0 + hU_0) - \phi_t(X_0)}{h} = \lim_{h \to 0} \frac{U(t, hU_0)}{h} = U(t, U_0).$$

∎

Example. As an illustration of these ideas, consider the differential equation $x' = x^2$. An easy integration shows that the solution $x(t)$ satisfying the initial condition $x(0) = x_0$ is

$$x(t) = \frac{-x_0}{x_0 t - 1}.$$

Hence we have

$$\frac{\partial \phi}{\partial x}(t, x_0) = \frac{1}{(x_0 t - 1)^2}.$$

On the other hand, the variational equation for $x(t)$ is

$$u' = 2x(t)\, u = \frac{-2x_0}{x_0 t - 1}\, u.$$

The solution of this equation satisfying the initial condition $u(0) = u_0$ is given by

$$u(t) = u_0 \left(\frac{1}{x_0 t - 1} \right)^2$$

as required. ■

7.5 Exploration: Numerical Methods

In this exploration, we first describe three different methods for approximating the solutions of first-order differential equations. Your task will be to evaluate the effectiveness of each method.

Each of these methods involves an iterative process whereby we find a sequence of points (t_k, x_k) that approximates selected points $(t_k, x(t_k))$ along the graph of a solution of the first-order differential equation $x' = f(t, x)$. In each case we will begin with an initial value $x(0) = x_0$. So $t_0 = 0$ and x_0 is our given initial value. We need to produce t_k and x_k.

In each of the three methods we will generate the t_k recursively by choosing a step size Δt and simply incrementing t_k at each stage by Δt. Hence

$$t_{k+1} = t_k + \Delta t$$

in each case. Choosing Δt small will (hopefully) improve the accuracy of the method.

Therefore, to describe each method, we only need to determine the values of x_k. In each case, x_{k+1} will be the x-coordinate of the point that sits directly over t_{k+1} on a certain straight line through (t_k, x_k) in the tx–plane. Thus all we need to do is to provide you with the slope of this straight line and then x_{k+1} is determined. Each of the three methods involves a different straight line.

1. *Euler's method.* Here x_{k+1} is generated by moving Δt time units along the straight line generated by the slope field at the point (t_k, x_k). Since the slope at this point is $f(t_k, x_k)$, taking this short step puts us at

$$x_{k+1} = x_k + f(t_k, x_k)\,\Delta t.$$

2. *Improved Euler's method.* In this method, we use the average of two slopes to move from (t_k, x_k) to (t_{k+1}, x_{k+1}). The first slope is just that of the slope field at (t_k, x_k), namely,

$$m_k = f(t_k, x_k).$$

The second is the slope of the slope field at the point (t_{k+1}, y_k), where y_k is the terminal point determined by Euler's method applied at (t_k, x_k). That is,

$$n_k = f(t_{k+1}, y_k) \quad \text{where } y_k = x_k + f(t_k, x_k)\,\Delta t.$$

Then we have

$$x_{k+1} = x_k + \left(\frac{m_k + n_k}{2}\right)\Delta t.$$

3. *(Fourth-order) Runge-Kutta method.* This method is the one most often used to solve differential equations. There are more sophisticated numerical methods that are specifically designed for special situations, but this method has served as a general-purpose solver for decades. In this method, we will determine four slopes, m_k, n_k, p_k, and q_k. The step from (t_k, x_k) to (t_{k+1}, x_{k+1}) will be given by moving along a straight line whose slope is a weighted average of these four values:

$$x_{k+1} = x_k + \left(\frac{m_k + 2n_k + 2p_k + q_k}{6}\right)\Delta t.$$

These slopes are determined as follows:

(a) m_k is given as in Euler's method:

$$m_k = f(t_k, x_k).$$

(b) n_k is the slope at the point obtained by moving halfway along the slope field line at (t_k, x_k) to the intermediate point $(t_k + (\Delta t)/2, y_k)$, so that

$$n_k = f(t_k + \frac{\Delta t}{2}, y_k) \quad \text{where } y_k = x_k + m_k \frac{\Delta t}{2}.$$

(c) p_k is the slope at the point obtained by moving halfway along a different straight line at (t_k, x_k), where the slope is now n_k rather than m_k as before. Hence

$$p_k = f(t_k + \frac{\Delta t}{2}, z_k) \quad \text{where } z_k = x_k + n_k \frac{\Delta t}{2}.$$

(d) Finally, q_k is the slope at the point (t_{k+1}, w_k) where we use a line with slope p_k at (t_k, x_k) to determine this point. Hence

$$q_k = f(t_{k+1}, w_k) \quad \text{where } w_k = x_k + p_k \Delta t.$$

Your goal in this exploration is to compare the effectiveness of these three methods by evaluating the errors made in carrying out this procedure in several examples. We suggest that you use a spreadsheet to make these lengthy calculations.

1. First, just to make sure you comprehend the rather terse descriptions of the three methods above, draw a picture in the tx–plane that illustrates the process of moving from (t_k, x_k) to (t_{k+1}, x_{k+1}) in each of the three cases.

2. Now let's investigate how the various methods work when applied to an especially simple differential equation, $x' = x$.

   (a) First find the explicit solution $x(t)$ of this equation satisfying the initial condition $x(0) = 1$ (now there's a free gift from the math department...).

   (b) Now use Euler's method to approximate the value of $x(1) = e$ using the step size $\Delta t = 0.1$. That is, recursively determine t_k and x_k for $k = 1, \ldots, 10$ using $\Delta t = 0.1$ and starting with $t_0 = 0$ and $x_0 = 1$.

   (c) Repeat the previous step with Δt half the size, namely, 0.05.

   (d) Again use Euler's method, this time reducing the step size by a factor of 5, so that $\Delta t = 0.01$, to approximate $x(1)$.

   (e) Now repeat the previous three steps using the improved Euler's method with the same step sizes.

   (f) And now repeat using Runge-Kutta.

   (g) You now have nine different approximations for the value of $x(1) = e$, three for each method. Now calculate the error in each case. For the record, use the value $e = 2.71828182845235360287\ldots$ in calculating the error.

(h) Finally calculate how the error changes as you change the step size from 0. 1 to 0. 05 and then from 0. 05 to 0. 01. That is, if ρ_Δ denotes the error made using step size Δ, compute both $\rho_{0.1}/\rho_{0.05}$ and $\rho_{0.05}/\rho_{0.01}$.

3. Now repeat the previous exploration, this time for the nonautonomous equation $x' = 2t(1 + x^2)$. Use the value $\tan 1 = 1.557407724654\ldots$.

4. Discuss how the errors change as you shorten the step size by a factor of 2 or a factor of 5. Why, in particular, is the Runge-Kutta method called a "fourth-order" method?

EXERCISES

1. Write out the first few terms of the Picard iteration scheme for each of the following initial value problems. Where possible, find explicit solutions and describe the domain of this solution.

   (a) $x' = x + 2; x(0) = 2$
   (b) $x' = x^{4/3}; x(0) = 0$
   (c) $x' = x^{4/3}; x(0) = 1$
   (d) $x' = \cos x; x(0) = 0$
   (e) $x' = 1/(2x); x(1) = 1$

2. Let A be an $n \times n$ matrix. Show that the Picard method for solving $X' = AX, X(0) = X_0$ gives the solution $\exp(tA)X_0$.

3. Derive the Taylor series for $\sin 2t$ by applying the Picard method to the first-order system corresponding to the second-order initial value problem

$$x'' = -4x; \quad x(0) = 0, \quad x'(0) = 2.$$

4. Verify the linearity principle for linear, nonautonomous systems of differential equations.

5. Consider the first-order equation $x' = x/t$. What can you say about "solutions" that satisfy $x(0) = 0$? $x(0) = a \neq 0$?

6. Discuss the existence and uniqueness of solutions of the equation $x' = x^a$ where $a > 0$ and $x(0) = 0$.

7. Let $A(t)$ be a continuous family of $n \times n$ matrices and let $P(t)$ be the matrix solution to the initial value problem $P' = A(t)P$, $P(0) = P_0$. Show that

$$\det P(t) = (\det P_0) \exp \left(\int_0^t \operatorname{Tr} A(s) \, ds \right).$$

8. Construct an example of a first-order differential equation on $\mathbb{R}$ for which there are no solutions to any initial value problem.

9. Construct an example of a differential equation depending on a parameter a for which some solutions do not depend continuously on a.

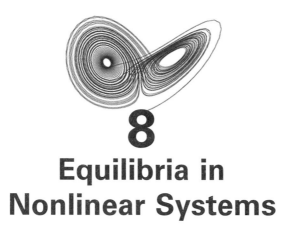

8

Equilibria in Nonlinear Systems

To avoid some of the technicalities that we encountered in the previous chapter, we henceforth assume that our differential equations are C^∞, except when specifically noted. This means that the right-hand side of the differential equation is k times continuously differentiable for all k. This will at the very least allow us to keep the number of hypotheses in our theorems to a minimum.

As we have seen, it is often impossible to write down explicit solutions of nonlinear systems of differential equations. The one exception to this occurs when we have equilibrium solutions. Provided we can solve the algebraic equations, we can write down the equilibria explicitly. Often, these are the most important solutions of a particular nonlinear system. More importantly, given our extended work on linear systems, we can usually use the technique of *linearization* to determine the behavior of solutions near equilibrium points. We describe this process in detail in this chapter.

8.1 Some Illustrative Examples

In this section we consider several planar nonlinear systems of differential equations. Each will have an equilibrium point at the origin. Our goal is to see that the solutions of the nonlinear system near the origin resemble those of the *linearized* system, at least in certain cases.

As a first example, consider the system:

$$x' = x + y^2$$
$$y' = -y.$$

There is a single equilibrium point at the origin. To picture nearby solutions, we note that, when y is small, y^2 is much smaller. Hence, near the origin at least, the differential equation $x' = x + y^2$ is very close to $x' = x$. In Section 7.4 in Chapter 7 we showed that the flow of this system near the origin is also "close" to that of the linearized system $X' = DF_0 X$. This suggests that we consider instead the linearized equation

$$x' = x$$
$$y' = -y,$$

derived by simply dropping the higher order term. We can, of course, solve this system immediately. We have a saddle at the origin with a stable line along the y-axis and an unstable line along the x-axis.

Now let's go back to the original nonlinear system. Luckily, we can also solve this system explicitly. For the second equation $y' = -y$ yields $y(t) = y_0 e^{-t}$. Inserting this into the first equation, we must solve

$$x' = x + y_0^2 e^{-2t}.$$

This is a first-order, nonautonomous equation whose solutions may be determined as in calculus by "guessing" a particular solution of the form ce^{-2t}. Inserting this guess into the equation yields a particular solution:

$$x(t) = -\frac{1}{3} y_0^2 e^{-2t}.$$

Hence any function of the form

$$x(t) = ce^t - \frac{1}{3} y_0^2 e^{-2t}$$

is a solution of this equation, as is easily checked. The general solution is then

$$x(t) = \left(x_0 + \frac{1}{3} y_0^2 \right) e^t - \frac{1}{3} y_0^2 e^{-2t}$$
$$y(t) = y_0 e^{-t}.$$

If $y_0 = 0$, we find a straight-line solution $x(t) = x_0 e^t$, $y(t) = 0$, just as in the linear case. However, unlike the linear case, the y-axis is no longer home

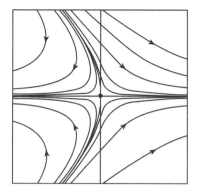

Figure 8.1 The phase plane
for $x' = x + y^2$, $y' = -y$.
Note the stable curve tangent
to the y-axis.

to a solution that tends to the origin. Indeed, the vector field along the y-axis is given by $(y^2, -y)$, which is not tangent to the axis; rather, all nonzero vectors point to the right along this axis.

On the other hand, there is a curve through the origin on which solutions tend to $(0, 0)$. Consider the curve $x + \frac{1}{3}y^2 = 0$ in $\mathbb{R}^2$. Suppose (x_0, y_0) lies on this curve, and let $(x(t), y(t))$ be the solution satisfying this initial condition. Since $x_0 + \frac{1}{3}y_0^2 = 0$ this solution becomes

$$x(t) = -\frac{1}{3}y_0^2 e^{-2t}$$

$$y(t) = y_0 e^{-t}.$$

Note that we have $x(t) + \frac{1}{3}(y(t))^2 = 0$ for all t, so this solution remains for all time on this curve. Moreover, as $t \to \infty$, this solution tends to the equilibrium point. That is, we have found a *stable curve* through the origin on which all solutions tend to $(0, 0)$. Note that this curve is tangent to the y-axis at the origin. See Figure 8.1.

Can we just "drop" the nonlinear terms in a system? The answer is, as we shall see below, it depends! In this case, however, doing so is perfectly legal, for we can find a change of variables that actually converts the original system to the linear system.

To see this, we introduce new variables u and v via

$$u = x + \frac{1}{3}y^2$$

$$v = y.$$

Then, in these new coordinates, the system becomes

$$u' = x' + \frac{2}{3}yy' = x + \frac{1}{3}y^2 = u$$
$$v' = y' = -y = -v.$$

That is to say, the nonlinear change of variables $F(x, y) = (x + \frac{1}{3}y^2, y)$ converts the original nonlinear system to a linear one, in fact, to the linearized system above.

Example. In general, it is impossible to convert a nonlinear system to a linear one as in the previous example, since the nonlinear terms almost always make huge changes in the system far from the equilibrium point at the origin. For example, consider the nonlinear system

$$x' = \frac{1}{2}x - y - \frac{1}{2}\left(x^3 + y^2x\right)$$
$$y' = x + \frac{1}{2}y - \frac{1}{2}\left(y^3 + x^2y\right).$$

Again we have an equilibrium point at the origin. The linearized system is now

$$X' = \begin{pmatrix} \frac{1}{2} & -1 \\ 1 & \frac{1}{2} \end{pmatrix} X,$$

which has eigenvalues $\frac{1}{2} \pm i$. All solutions of this system spiral away from the origin and toward ∞ in the counterclockwise direction, as is easily checked.

Solving the nonlinear system looks formidable. However, if we change to polar coordinates, the equations become much simpler. We compute

$$r' \cos\theta - r(\sin\theta)\theta' = x' = \frac{1}{2}\left(r - r^3\right)\cos\theta - r\sin\theta$$
$$r' \sin\theta + r(\cos\theta)\theta' = y' = \frac{1}{2}\left(r - r^3\right)\sin\theta + r\cos\theta$$

from which we conclude, after equating the coefficients of $\cos\theta$ and $\sin\theta$,

$$r' = r(1 - r^2)/2$$
$$\theta' = 1.$$

We can now solve this system explicitly, since the equations are decoupled. Rather than do this, we will proceed in a more geometric fashion. From the equation $\theta' = 1$, we conclude that all nonzero solutions spiral around the

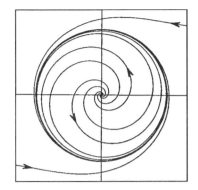

Figure 8.2   The phase plane
for $r' = \frac{1}{2}(r - r^3), \theta' = 1$.

origin in the counterclockwise direction. From the first equation, we see that solutions do not spiral toward ∞. Indeed, we have $r' = 0$ when $r = 1$, so all solutions that start on the unit circle stay there forever and move periodically around the circle. Since $r' > 0$ when $0 < r < 1$, we conclude that nonzero solutions inside the circle spiral away from the origin and toward the unit circle. Since $r' < 0$ when $r > 1$, solutions outside the circle spiral toward it. See Figure 8.2.   ■

In this example, there is no way to find a global change of coordinates that puts the system into a linear form, since no linear system has this type of spiraling toward a circle. However, near the origin this is still possible.

To see this, first note that if r_0 satisfies $0 < r_0 < 1$, then the nonlinear vector field points outside the circle of radius r_0. This follows from the fact that, on any such circle, $r' = r_0(1 - r_0^2)/2 > 0$. Consequently, in backward time, all solutions of the nonlinear system tend to the origin, and in fact spiral as they do so.

We can use this fact to define a conjugacy between the linear and nonlinear system in the disk $r \leq r_0$, much as we did in Chapter 4. Let ϕ_t denote the flow of the nonlinear system. In polar coordinates, the linearized system above becomes

$$r' = r/2$$
$$\theta' = 1.$$

Let ψ_t denote the flow of this system. Again, all solutions of the linear system tend toward the origin in backward time. We will now define a conjugacy between these two flows in the disk D given by $r < 1$. Fix r_0 with $0 < r_0 < 1$. For any point (r, θ) in D with $r > 0$, there is a unique

$t = t(r, \theta)$ for which $\phi_t(r, \theta)$ belongs to the circle $r = r_0$. We then define the function

$$h(r, \theta) = \psi_{-t}\phi_t(r, \theta)$$

where $t = t(r, \theta)$. We stipulate also that h takes the origin to the origin. Then it is straightforward to check that

$$h \circ \phi_s(r, \theta) = \psi_s \circ h(r, \theta)$$

for any point (r, θ) in D, so that h gives a conjugacy between the nonlinear and linear systems. It is also easy to check that h takes D onto all of $\mathbb{R}^2$.

Thus we see that, while it may not always be possible to linearize a system *globally*, we may sometimes accomplish this *locally*. Unfortunately, such a local linearization may not provide any useful information about the behavior of solutions.

Example.   Now consider the system

$$x' = -y + \epsilon x(x^2 + y^2)$$
$$y' = x + \epsilon y(x^2 + y^2).$$

Here ϵ is a parameter that we may take to be either positive or negative.

The linearized system is

$$x' = -y$$
$$y' = x,$$

so we see that the origin is a center and all solutions travel in the counterclockwise direction around circles centered at the origin with unit angular speed.

This is hardly the case for the nonlinear system. In polar coordinates, this system reduces to

$$r' = \epsilon r^3$$
$$\theta' = 1.$$

Thus when $\epsilon > 0$, all solutions spiral away from the origin, whereas when $\epsilon < 0$, all solutions spiral toward the origin. The addition of the nonlinear terms, no matter how small near the origin, changes the linearized phase portrait dramatically; we cannot use linearization to determine the behavior of this system near the equilibrium point.   ∎

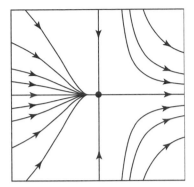

Figure 8.3 The phase plane
for $x' = x^2$, $y' = -y$.

Example. Now consider one final example:

$$x' = x^2$$
$$y' = -y.$$

The only equilibrium solution for this system is the origin. All other solutions (except those on the y-axis) move to the right and toward the x-axis. On the y-axis, solutions tend along this straight line to the origin. Hence the phase portrait is as shown in Figure 8.3. ∎

Note that the picture in Figure 8.3 is quite different from the corresponding picture for the linearized system

$$x' = 0$$
$$y' = -y,$$

for which all points on the x-axis are equilibrium points, and all other solutions lie on vertical lines $x =$ constant.

The problem here, as in the previous example, is that the equilibrium point for the linearized system at the origin is not hyperbolic. When a linear planar system has a zero eigenvalue or a center, the addition of nonlinear terms often completely changes the phase portrait.

8.2 Nonlinear Sinks and Sources

As we saw in the examples of the previous section, solutions of planar nonlinear systems near equilibrium points resemble those of their linear parts only in the case where the linearized system is hyperbolic; that is, when neither of the eigenvalues of the system has zero real part. In this section we begin to describe the situation in the general case of a hyperbolic equilibrium point in a nonlinear system by considering the special case of a sink. For simplicity, we will prove the results below in the planar case, although all of the results hold in $\mathbb{R}^n$.

Let $X' = F(X)$ and suppose that $F(X_0) = 0$. Let DF_{X_0} denote the Jacobian matrix of F evaluated at X_0. Then, as in Chapter 7, the linear system of differential equations

$$Y' = DF_{X_0} Y$$

is called the *linearized system near* X_0. Note that, if $X_0 = 0$, the linearized system is obtained by simply dropping all of the nonlinear terms in F, just as we did in the previous section.

In analogy with our work with linear systems, we say that an equilibrium point X_0 of a nonlinear system is *hyperbolic* if all of the eigenvalues of DF_{X_0} have nonzero real parts.

We now specialize the discussion to the case of an equilibrium of a planar system for which the linearized system has a sink at 0. Suppose our system is

$$x' = f(x, y)$$
$$y' = g(x, y)$$

with $f(x_0, y_0) = 0 = g(x_0, y_0)$. If we make the change of coordinates $u = x - x_0, v = y - y_0$ then the new system has an equilibrium point at $(0, 0)$. Hence we may as well assume that $x_0 = y_0 = 0$ at the outset. We then make a further linear change of coordinates that puts the linearized system in canonical form. For simplicity, let us assume at first that the linearized system has distinct eigenvalues $-\lambda < -\mu < 0$. Thus after these changes of coordinates, our system become

$$x' = -\lambda x + h_1(x, y)$$
$$y' = -\mu y + h_2(x, y)$$

where $h_j = h_j(x, y)$ contains all of the "higher order terms." That is, in terms of its Taylor expansion, each h_j contains terms that are quadratic or higher order in x and/or y. Equivalently, we have

$$\lim_{(x,y) \to (0,0)} \frac{h_j(x, y)}{r} = 0$$

where $r^2 = x^2 + y^2$.

The linearized system is now given by

$$x' = -\lambda x$$
$$y' = -\mu y.$$

For this linearized system, recall that the vector field always points inside the circle of radius r centered at the origin. Indeed, if we take the dot product of the linear vector field with the outward normal vector (x, y), we find

$$(-\lambda x, -\mu y) \cdot (x, y) = -\lambda x^2 - \mu y^2 < 0$$

for any nonzero vector (x, y). As we saw in Chapter 4, this forces all solutions to tend to the origin with strictly decreasing radial components.

The same thing happens for the nonlinear system, at least close to $(0, 0)$. Let $q(x, y)$ denote the dot product of the vector field with (x, y). We have

$$
\begin{aligned}
q(x, y) &= (-\lambda x + h_1(x, y), -\mu y + h_2(x, y)) \cdot (x, y) \\
&= -\lambda x^2 + x h_1(x, y) - \mu y^2 + y h_2(x, y) \\
&= -\mu(x^2 + y^2) + (\mu - \lambda)x^2 + x h_1(x, y) + y h_2(x, y) \\
&\leq -\mu r^2 + x h_1(x, y) + y h_2(x, y)
\end{aligned}
$$

since $(\mu - \lambda)x^2 \leq 0$. Therefore we have

$$\frac{q(x, y)}{r^2} \leq -\mu + \frac{x h_1(x, y) + y h_2(x, y)}{r^2}.$$

As $r \to 0$, the right-hand side tends to $-\mu$. Thus it follows that $q(x, y)$ is negative, at least close to the origin. As a consequence, the nonlinear vector field points into the interior of circles of small radius about 0, and so all solutions whose initial conditions lie inside these circles must tend to the origin. Thus we are justified in calling this type of equilibrium point a *sink*, just as in the linear case.

It is straightforward to check that the same result holds if the linearized system has eigenvalues $\alpha + i\beta$ with $\alpha < 0$, $\beta \neq 0$. In the case of repeated negative eigenvalues, we first need to change coordinates so that the linearized system is

$$x' = -\lambda x + \epsilon y$$
$$y' = -\lambda y$$

where ϵ is sufficiently small. We showed how to do this in Chapter 4. Then again the vector field points inside circles of sufficiently small radius.

We can now conjugate the flow of a nonlinear system near a hyperbolic equilibrium point that is a sink to the flow of its linearized system. Indeed, the argument used in the second example of the previous section goes over essentially unchanged. In similar fashion, nonlinear systems near a hyperbolic source are also conjugate to the corresponding linearized system.

This result is a special case of the following more general theorem.

The Linearization Theorem. *Suppose the n-dimensional system* $X' = F(X)$ *has an equilibrium point at* X_0 *that is hyperbolic. Then the nonlinear flow is conjugate to the flow of the linearized system in a neighborhood of* X_0. ∎

We will not prove this theorem here, since the proof requires analytical techniques beyond the scope of this book when there are eigenvalues present with both positive and negative real parts.

8.3 Saddles

We turn now to the case of an equilibrium for which the linearized system has a saddle at the origin in $\mathbb{R}^2$. As in the previous section, we may assume that this system is in the form

$$x' = \lambda x + f_1(x, y)$$
$$y' = -\mu y + f_2(x, y)$$

where $-\mu < 0 < \lambda$ and $f_j(x, y)/r$ tends to 0 as $r \to 0$. As in the case of a linear system, we call this type of equilibrium point a *saddle*.

For the linearized system, the y-axis serves as the stable line, with all solutions on this line tending to 0 as $t \to \infty$. Similarly, the x-axis is the unstable line. As we saw in Section 8.1, we cannot expect these stable and unstable straight lines to persist in the nonlinear case. However, there do exist a pair of curves through the origin that have similar properties.

Let $W^s(0)$ denote the set of initial conditions whose solutions tend to the origin as $t \to \infty$. Let $W^u(0)$ denote the set of points whose solutions tend to the origin as $t \to -\infty$. $W^s(0)$ and $W^u(0)$ are called the *stable curve* and *unstable curve*, respectively.

The following theorem shows that solutions near nonlinear saddles behave much the same as in the linear case.

The Stable Curve Theorem. *Suppose the system*

$$x' = \lambda x + f_1(x, y)$$
$$y' = -\mu y + f_2(x, y)$$

satisfies $-\mu < 0 < \lambda$ and $f_j(x, y)/r \to 0$ as $r \to 0$. Then there is an $\epsilon > 0$ and a curve $x = h^s(y)$ that is defined for $|y| < \epsilon$ and satisfies $h^s(0) = 0$. Furthermore:

1. All solutions whose initial conditions lie on this curve remain on this curve for all $t \geq 0$ and tend to the origin as $t \to \infty$;
2. The curve $x = h^s(y)$ passes through the origin tangent to the y-axis;
3. All other solutions whose initial conditions lie in the disk of radius ϵ centered at the origin leave this disk as time increases. ∎

Some remarks are in order. The curve $x = h^s(y)$ is called the *local stable curve* at 0. We can find the complete stable curve $W^s(0)$ by following solutions that lie on the local stable curve backward in time. The function $h^s(y)$ is actually C^∞ at all points, though we will not prove this result here.

A similar unstable curve theorem provides us with a *local unstable curve* of the form $y = h^u(x)$. This curve is tangent to the x-axis at the origin. All solutions on this curve tend to the origin as $t \to -\infty$.

We begin with a brief sketch of the proof of the stable curve theorem. Consider the square bounded by the lines $|x| = \epsilon$ and $|y| = \epsilon$ for $\epsilon > 0$ sufficiently small. The nonlinear vector field points into the square along the interior of the top and the bottom boundaries $y = \pm\epsilon$ since the system is close to the linear system $x' = \lambda x$, $y' = -\mu y$, which clearly has this property. Similarly, the vector field points outside the square along the left and right boundaries $x = \pm\epsilon$.

Now consider the initial conditions that lie along the top boundary $y = \epsilon$. Some of these solutions will exit the square to the left, while others will exit to the right. Solutions cannot do both, so these sets are disjoint. Moreover, these sets are open. So there must be some initial conditions whose solutions do not exit at all. We will show first of all that each of these nonexiting solutions tends to the origin. Secondly, we will show that there is only one initial condition on the top and bottom boundary whose solution behaves in this way. Finally we will show that this solution lies along some graph of the form $x = h^s(y)$, which has the required properties.

Now we fill in the details of the proof. Let B_ϵ denote the square bounded by $x = \pm\epsilon$ and $y = \pm\epsilon$. Let $S_\epsilon^\pm$ denote the top and bottom boundaries of B_ϵ. Let C_M denote the conical region given by $|y| \geq M|x|$ inside B_ϵ. Here we think of the slopes $\pm M$ of the boundary of C_M as being large in absolute value. See Figure 8.4.

Lemma. *Given $M > 0$, there exists $\epsilon > 0$ such that the vector field points outside C_M for points on the boundary of $C_M \cap B_\epsilon$ (except, of course, at the origin).*

Proof: Given M, choose $\epsilon > 0$ so that

$$|f_1(x, y)| \leq \frac{\lambda}{2\sqrt{M^2 + 1}} r$$

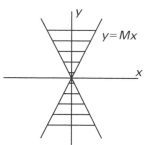

Figure 8.4   The cone C_M.

for all $(x, y) \in B_\epsilon$. Now suppose $x > 0$. Then along the right boundary of C_M we have

$$
\begin{aligned}
x' &= \lambda x + f_1(x, Mx) \\
&\geq \lambda x - |f_1(x, Mx)| \\
&\geq \lambda x - \frac{\lambda}{2\sqrt{M^2 + 1}} r \\
&= \lambda x - \frac{\lambda}{2\sqrt{M^2 + 1}} (x\sqrt{M^2 + 1}) \\
&= \frac{\lambda}{2} x > 0.
\end{aligned}
$$

Hence $x' > 0$ on this side of the boundary of the cone.

Similarly, if $y > 0$, we may choose $\epsilon > 0$ smaller if necessary so that we have $y' < 0$ on the edges of C_M where $y > 0$. Indeed, choosing ϵ so that

$$
|f_2(x, y)| \leq \frac{\mu}{2\sqrt{M^2 + 1}} r
$$

guarantees this exactly as above. Hence on the edge of C_M that lies in the first quadrant, we have shown that the vector field points down and to the right and hence out of C_M. Similar calculations show that the vector field points outside C_M on all other edges of C_M. This proves the lemma. ∎

It follows from this lemma that there is a set of initial conditions in $S_\epsilon^\pm \cap C_M$ whose solutions eventually exit from C_M to the right, and another set in $S_\epsilon^\pm \cap C_M$ whose solutions exit left. These sets are open because of continuity of solutions with respect to initial conditions (see Section 7.3). We next show that each of these sets is actually a single open interval.

Let C_M^+ denote the portion of C_M lying above the x-axis, and let C_M^- denote the portion lying below this axis.

Lemma. *Suppose $M > 1$. Then there is an $\epsilon > 0$ such that $y' < 0$ in C_M^+ and $y' > 0$ in C_M^-.*

Proof: In C_M^+ we have $|Mx| \leq y$ so that

$$r^2 \leq \frac{y^2}{M^2} + y^2$$

or

$$r \leq \frac{y}{M}\sqrt{1 + M^2}.$$

As in the previous lemma, we choose ϵ so that

$$|f_2(x, y)| \leq \frac{\mu}{2\sqrt{M^2 + 1}}\, r$$

for all $(x, y) \in B_\epsilon$. We then have in C_M^+

$$
\begin{aligned}
y' &\leq -\mu y + |f_2(x, y)| \\
&\leq -\mu y + \frac{\mu}{2\sqrt{M^2 + 1}}\, r \\
&\leq -\mu y + \frac{\mu}{2M}\, y \\
&\leq -\frac{\mu}{2} y
\end{aligned}
$$

since $M > 1$. This proves the result for C_M^+; the proof for C_M^- is similar. ∎

From this result we see that solutions that begin on $S_\epsilon^+ \cap C_M$ decrease in the y-direction while they remain in C_M^+. In particular, no solution can remain in C_M^+ for all time unless that solution tends to the origin. By the existence and uniqueness theorem, the set of points in S_ϵ^+ that exit to the right (or left) must then be a single open interval. The complement of these two intervals in S_ϵ^+ is therefore a nonempty closed interval on which solutions do not leave C_M and therefore tend to 0 as $t \to \infty$. We have similar behavior in C_M^-.

We now claim that the interval of initial conditions in $S_\epsilon^\pm$ whose solutions tend to 0 is actually a single point. To see this, recall the variational equation for this system discussed in Section 7.4. This is the nonautonomous linear system

$$U' = DF_{X(t)}U$$

where $X(t)$ is a given solution of the nonlinear system and

$$F\begin{pmatrix} x \\ y \end{pmatrix} = \begin{pmatrix} \lambda x + f_1(x, y) \\ -\mu y + f_2(x, y) \end{pmatrix}.$$

We showed in Chapter 7 that, given $\tau > 0$, if $|U_0|$ is sufficiently small, then the solution of the variational equation $U(t)$ that satisfies $U(0) = U_0$ and the solution $Y(t)$ of the nonlinear system satisfying $Y(0) = X_0 + U_0$ remain close to each other for $0 \leq t \leq \tau$.

Now let $X(t) = (x(t), y(t))$ be one of the solutions of the system that never leaves C_M. Suppose $x(0) = x_0$ and $y(0) = \epsilon$.

Lemma. *Let $U(t) = (u_1(t), u_2(t))$ be any solution of the variational equation*

$$U' = DF_{X(t)}U$$

that satisfies $|u_2(0)/u_1(0)| \leq 1$. Then we may choose $\epsilon > 0$ sufficiently small so that $|u_2(t)| \leq |u_1(t)|$ for all $t \geq 0$.

Proof: The variational equation may be written

$$u_1' = \lambda u_1 + f_{11}u_1 + f_{12}u_2$$
$$u_2' = -\mu u_2 + f_{21}u_1 + f_{22}u_2$$

where the $|f_{ij}(t)|$ may be made as small as we like by choosing ϵ smaller. Given any solution $(u_1(t), u_2(t))$ of this equation, let $\eta = \eta(t)$ be the slope $u_2(t)/u_1(t)$ of this solution. Using the variational equation and the quotient rule, we find that η satisfies

$$\eta' = \alpha(t) - (\mu + \lambda + \beta(t))\eta + \gamma(t)\eta^2$$

where $\alpha(t) = f_{21}(t)$, $\beta(t) = f_{11}(t) - f_{22}(t)$, and $\gamma(t) = -f_{12}(t)$. In particular, α, β, and γ are all small. Recall that $\mu + \lambda > 0$, so that

$$-(\mu + \lambda + \beta(t)) < 0.$$

It follows that, if $\eta(t) > 0$, there are constants a_1 and a_2 with $0 < a_1 < a_2$ such that

$$-a_2\eta + \alpha(t) \leq \eta' \leq -a_1\eta + \alpha(t).$$

Now, if we choose ϵ small enough and set $u_1(0) = u_2(0)$ so that $\eta(0) = 1$, then we have $\eta'(0) < 0$ so that $\eta(t)$ initially decreases. Arguing similarly, if $\eta(0) = -1$, then $\eta'(0) > 0$ so $\eta(t)$ increases. In particular, any solution that begins with $|\eta(0)| < 1$ can never satisfy $\eta(t) = \pm 1$. Since our solution satisfies $|\eta(0)| \leq 1$, it follows that this solution is trapped in the region where $|\eta(t)| < 1$, and so we have $|u_2(t)| \leq |u_1(t)|$ for all $t > 0$. This proves the lemma. ∎

To complete the proof of the theorem, suppose that we have a second solution $Y(t)$ that lies in C_M for all time. So $Y(t) \to 0$ as $t \to \infty$. Suppose that $Y(0) = (x_0', \epsilon)$ where $x_0' \neq x_0$. Note that any solution with initial value (x, ϵ) with $x_0 < x < x_0'$ must also tend to the origin, so we may assume that x_0' is as close as we like to x_0. We will approximate $Y(t) - X(t)$ by a solution of the variational equation. Let $U(t)$ denote the solution of the variational equation

$$U' = DF_{X(t)} U$$

satisfying $u_1(0) = x_0' - x_0, u_2(0) = 0$. The above lemma applies to $U(t)$ so that $|u_2(t)| \leq |u_1(t)|$ for all t.

We have

$$|u_1'| = |\lambda u_1 + f_{11}(t)u_1 + f_{12}(t)u_2|$$
$$\geq \lambda |u_1| - (|f_{11}(t)| + |f_{12}(t)|)|u_1|.$$

But $|f_{11} + f_{12}|$ tends to 0 as $\epsilon \to 0$, so we have

$$|u_1'| \geq K|u_1|$$

for some $K > 0$. Hence $|u_1(t)| \geq Ce^{Kt}$ for $t \geq 0$. Now the solution $Y(t)$ tends to 0 by assumption. But then the distance between $Y(t)$ and $U(t)$ must become large if $u_1(0) \neq 0$. This yields a contradiction and shows that $Y(t) = X(t)$.

We finally claim that $X(t)$ tends to the origin tangentially to the y-axis. This follows immediately from the fact that we may choose ϵ small enough so that $X(t)$ lies in C_M no matter how large M is. Hence the slope of $X(t)$ tends to $\pm\infty$ as $X(t)$ approaches $(0,0)$. ∎

We conclude this section with a brief discussion of higher dimensional saddles. Suppose $X' = F(X)$ where $X \in \mathbb{R}^n$. Suppose that X_0 is an equilibrium solution for which the linearized system has k eigenvalues with negative real parts and $n - k$ eigenvalues with positive real parts. Then the local stable and unstable sets are not generally curves. Rather, they are "submanifolds" of dimension k and $n - k$, respectively. Without entering the realm of manifold theory, we simply note that this means there is a linear change of coordinates in which the local stable set is given near the origin by the graph of a C^∞ function $g : B_r \to \mathbb{R}^{n-k}$ that satisfies $g(0) = 0$, and all partial derivatives of g vanish at the origin. Here B_r is the disk of radius r centered at the origin in $\mathbb{R}^k$. The local unstable set is a similar graph over an $n - k$-dimensional disk. Each of these graphs is tangent at the equilibrium point to the stable and unstable subspaces at X_0. Hence they meet only at X_0.

Example. Consider the system

$$x' = -x$$
$$y' = -y$$
$$z' = z + x^2 + y^2.$$

The linearized system at the origin has eigenvalues 1 and -1 (repeated). The change of coordinates

$$u = x$$
$$v = y$$
$$w = z + \frac{1}{3}(x^2 + y^2).$$

converts the nonlinear system to the linear system

$$u' = -u$$
$$v' = -v$$
$$w' = w.$$

The plane $w = 0$ for the linear system is the stable plane. Under the change of coordinates this plane is transformed to the surface

$$z = -\frac{1}{3}(x^2 + y^2),$$

which is a paraboloid passing through the origin in $\mathbb{R}^3$ and opening downward. All solutions tend to the origin on this surface; we call this the *stable surface* for the nonlinear system. See Figure 8.5.   ∎

8.4 Stability

The study of equilibria plays a central role in ordinary differential equations and their applications. An equilibrium point, however, must satisfy a certain

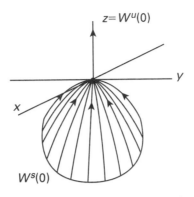

Figure 8.5   The phase portrait
for $x' = -x$, $y' = -y$, $z' =$
$z + x^2 + y^2$.

stability criterion in order to be significant physically. (Here, as in several other
places in this book, we use the word *physical* in a broad sense; in some contexts,
physical could be replaced by *biological, chemical,* or even *economic.*)

An equilibrium is said to be *stable* if nearby solutions stay nearby for all
future time. In applications of dynamical systems one cannot usually pinpoint
positions exactly, but only approximately, so an equilibrium must be stable to
be physically meaningful.

More precisely, suppose $X^* \in \mathbb{R}^n$ is an equilibrium point for the differential
equation

$$X' = F(X).$$

Then X^* is a *stable* equilibrium if for every neighborhood $\mathcal{O}$ of X^* in $\mathbb{R}^n$ there
is a neighborhood $\mathcal{O}_1$ of X^* in $\mathcal{O}$ such that every solution $X(t)$ with $X(0) = X_0$
in $\mathcal{O}_1$ is defined and remains in $\mathcal{O}$ for all $t > 0$.

A different form of stability is *asymptotic stability.* If $\mathcal{O}_1$ can be chosen above
so that, in addition to the properties for stability, we have $\lim_{t \to \infty} X(t) = X^*$,
then we say that X^* is *asymptotically stable.*

An equilibrium X^* that is not stable is called *unstable.* This means there is a
neighborhood $\mathcal{O}$ of X^* such that for *every* neighborhood $\mathcal{O}_1$ of X^* in $\mathcal{O}$, there
is at least one solution $X(t)$ starting at $X(0) \in \mathcal{O}_1$ that does not lie entirely in
$\mathcal{O}$ for all $t > 0$.

A sink is asymptotically stable and therefore stable. Sources and saddles
are examples of unstable equilibria. An example of an equilibrium that is
stable but not asymptotically stable is the origin in $\mathbb{R}^2$ for a linear equation
$X' = AX$ where A has pure imaginary eigenvalues. The importance of this
example in applications is limited (despite the famed harmonic oscillator)
because the slightest nonlinear perturbation will destroy its character, as we

saw in Section 8.1. Even a small linear perturbation can make a center into a sink or a source.

Thus when the linearization of the system at an equilibrium point is hyperbolic, we can immediately determine the stability of that point. Unfortunately, many important equilibrium points that arise in applications are nonhyperbolic. It would be wonderful to have a technique that determined the stability of an equilibrium point that works in all cases. Unfortunately, as yet, we have no universal way of determining stability except by actually finding all solutions of the system, which is usually difficult if not impossible. We will present some techniques that allow us to determine stability in certain special cases in the next chapter.

8.5 Bifurcations

In this section we will describe some simple examples of bifurcations that occur for nonlinear systems. We consider a family of systems

$$X' = F_a(X)$$

where a is a real parameter. We assume that F_a depends on a in a C^∞ fashion. A bifurcation occurs when there is a "significant" change in the structure of the solutions of the system as a varies. The simplest types of bifurcations occur when the number of equilibrium solutions changes as a varies.

Recall the elementary bifurcations we encountered in Chapter 1 for first-order equations $x' = f_a(x)$. If x_0 is an equilibrium point, then we have $f_a(x_0) = 0$. If $f_a'(x_0) \neq 0$, then small changes in a do not change the local structure near x_0: that is, the differential equation

$$x' = f_{a+\epsilon}(x)$$

has an equilibrium point $x_0(\epsilon)$ that varies continuously with ϵ for ϵ small. A glance at the (increasing or decreasing) graphs of $f_{a+\epsilon}(x)$ near x_0 shows why this is true. More rigorously, this is an immediate consequence of the implicit function theorem (see Exercise 3 at the end of this chapter). Thus bifurcations for first-order equations only occur in the nonhyperbolic case where $f_a'(x_0) = 0$.

Example. The first-order equation

$$x' = f_a(x) = x^2 + a$$

has a single equilibrium point at $x = 0$ when $a = 0$. Note $f_0'(0) = 0$, but $f_0''(0) \neq 0$. For $a > 0$ this equation has no equilibrium points since $f_a(x) > 0$ for all x, but for $a < 0$ this equation has a pair of equilibria. Thus a bifurcation occurs as the parameter passes through $a = 0$. ■

This kind of bifurcation is called a *saddle-node bifurcation* (we will see the "saddle" in this bifurcation a little later). In a saddle-node bifurcation, there is an interval about the bifurcation value a_0 and another interval I on the x-axis in which the differential equation has

1. Two equilibrium points in I if $a < a_0$;
2. One equilibrium point in I if $a = a_0$;
3. No equilibrium points in I if $a > a_0$.

Of course, the bifurcation could take place "the other way," with no equilibria when $a < a_0$. The example above is actually the typical type of bifurcation for first-order equations.

Theorem. (Saddle-Node Bifurcation) *Suppose $x' = f_a(x)$ is a first-order differential equation for which*

1. $f_{a_0}(x_0) = 0$;
2. $f_{a_0}'(x_0) = 0$;
3. $f_{a_0}''(x_0) \neq 0$;
4. $\dfrac{\partial f_{a_0}}{\partial a}(x_0) \neq 0$.

Then this differential equation undergoes a saddle-node bifurcation at $a = a_0$.

Proof: Let $G(x, a) = f_a(x)$. We have $G(x_0, a_0) = 0$. Also,

$$\frac{\partial G}{\partial a}(x_0, a_0) = \frac{\partial f_{a_0}}{\partial a}(x_0) \neq 0,$$

so we may apply the implicit function theorem to conclude that there is a smooth function $a = a(x)$ such that $G(x, a(x)) = 0$. In particular, if x^* belongs to the domain of $a(x)$, then x^* is an equilibrium point for the equation $x' = f_{a(x^*)}(x)$, since $f_{a(x^*)}(x^*) = 0$. Differentiating $G(x, a(x)) = 0$ with respect to x, we find

$$a'(x) = \frac{-\partial G/\partial x}{\partial G/\partial a}.$$

Now $(\partial G/\partial x)(x_0, a_0) = f'_{a_0}(x_0) = 0$, while $(\partial G/\partial a)(x_0, a_0) \neq 0$ by assumption. Hence $a'(x_0) = 0$. Differentiating once more, we find

$$a''(x) = \dfrac{-\dfrac{\partial^2 G}{\partial x^2}\dfrac{\partial G}{\partial a} + \dfrac{\partial G}{\partial x}\dfrac{\partial^2 G}{\partial x \partial a}}{\left(\dfrac{\partial G}{\partial a}\right)^2}.$$

Since $(\partial G/\partial x)(x_0, a_0) = 0$, we have

$$a''(x_0) = \dfrac{-\dfrac{\partial^2 G}{\partial x^2}(x_0, a_0)}{\dfrac{\partial G}{\partial a}(x_0, a_0)} \neq 0$$

since $(\partial^2 G/\partial x^2)(x_0, a_0) = f''_{a_0}(x_0) \neq 0$. This implies that the graph of $a = a(x)$ is either concave up or concave down, so we have two equilibria near x_0 for a-values on one side of a_0 and no equilibria for a-values on the other side. ∎

We said earlier that such saddle-node bifurcations were the "typical" bifurcations involving equilibrium points for first-order equations. The reason for this is that we must have both

1. $f_{a_0}(x_0) = 0$
2. $f'_{a_0}(x_0) = 0$

if $x' = f_a(x)$ is to undergo a bifurcation when $a = a_0$. Generically (in the sense of Section 5.6), the next higher order derivatives at (x_0, a_0) will be nonzero. That is, we typically have

3. $f''_{a_0}(x_0) \neq 0$
4. $\dfrac{\partial f_a}{\partial a}(x_0, a_0) \neq 0$

at such a bifurcation point. But these are precisely the conditions that guarantee a saddle-node bifurcation.

Recall that the bifurcation diagram for $x' = f_a(x)$ is a plot of the various phase lines of the equation versus the parameter a. The bifurcation diagram for a typical saddle-node bifurcation is displayed in Figure 8.6. (The directions of the arrows and the curve of equilibria may change.)

Example. (Pitchfork Bifurcation) Consider

$$x' = x^3 - ax.$$

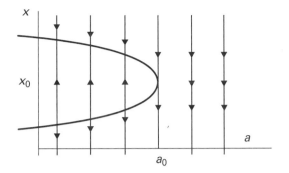

Figure 8.6   The bifurcation diagram for a saddle-node bifurcation.

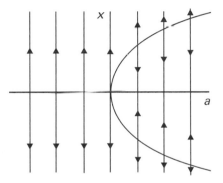

Figure 8.7   The bifurcation diagram for a pitchfork bifurcation.

There are three equilibria for this equation, at $x = 0$ and $x = \pm\sqrt{a}$ when $a > 0$. When $a \leq 0$, $x = 0$ is the only equilibrium point. The bifurcation diagram shown in Figure 8.7 explains why this bifurcation is so named. ▪

Now we turn to some bifurcations in higher dimensions. The saddle-node bifurcation in the plane is similar to its one-dimensional cousin, only now we see where the "saddle" comes from.

Example. Consider the system

$$x' = x^2 + a$$
$$y' = -y.$$

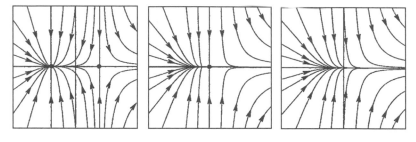

Figure 8.8   The saddle-node bifurcation when, from left to right,
$a < 0$, $a = 0$, and $a > 0$.

When $a = 0$, this is one of the systems considered in Section 8.1. There is a unique equilibrium point at the origin, and the linearized system has a zero eigenvalue.

When a passes through $a = 0$, a *saddle-node* bifurcation occurs. When $a > 0$, we have $x' > 0$ so all solutions move to the right; the equilibrium point disappears. When $a < 0$ we have a pair of equilibria, at the points $(\pm\sqrt{-a}, 0)$. The linearized equation is

$$X' = \begin{pmatrix} 2x & 0 \\ 0 & -1 \end{pmatrix} X.$$

So we have a sink at $(-\sqrt{-a}, 0)$ and a saddle at $(\sqrt{-a}, 0)$. Note that solutions on the lines $x = \pm\sqrt{-a}$ remain for all time on these lines since $x' = 0$ on these lines. Solutions tend directly to the equilibria on these lines since $y' = -y$. This bifurcation is sketched in Figure 8.8.   ■

A saddle-node bifurcation may have serious global implications for the behavior of solutions, as the following example shows.

Example.   Consider the system given in polar coordinates by

$$r' = r - r^3$$
$$\theta' = \sin^2(\theta) + a$$

where a is again a parameter. The origin is always an equilibrium point since $r' = 0$ when $r = 0$. There are no other equilibria when $a > 0$ since, in that case, $\theta' > 0$. When $a = 0$ two additional equilibria appear at $(r, \theta) = (1, 0)$ and $(r, \theta) = (1, \pi)$. When $-1 < a < 0$, there are four equilibria on the circle $r = 1$. These occur at the roots of the equation

$$\sin^2(\theta) = -a.$$

We denote these roots by $\theta_\pm$ and $\theta_\pm + \pi$, where we assume that $0 < \theta_+ < \pi/2$ and $-\pi/2 < \theta_- < 0$.

Note that the flow of this system takes the straight rays through the origin $\theta = $ constant to other straight rays. This occurs since θ' depends only on θ, not on r. Also, the unit circle is *invariant* in the sense that any solution that starts on the circle remains there for all time. This follows since $r' = 0$ on this circle. All other nonzero solutions tend to this circle, since $r' > 0$ if $0 < r < 1$ whereas $r' < 0$ if $r > 1$.

Now consider the case $a = 0$. In this case the x-axis is invariant and all nonzero solutions on this line tend to the equilibrium points at $x = \pm 1$. In the upper half-plane we have $\theta' > 0$, so all other solutions in this region wind counterclockwise about 0 and tend to $x = -1$; the θ-coordinate increases to $\theta = \pi$ while r tends monotonically to 1. No solution winds more than angle π about the origin, since the x-axis acts as a barrier. The system behaves symmetrically in the lower half-plane.

When $a > 0$ two things happen. First of all, the equilibrium points at $x = \pm 1$ disappear and now $\theta' > 0$ everywhere. Thus the barrier on the x-axis has been removed and all solutions suddenly are free to wind forever about the origin. Secondly, we now have a periodic solution on the circle $r = 1$, and all nonzero solutions are attracted to it.

This dramatic change is caused by a pair of saddle-node bifurcations at $a = 0$. Indeed, when $-1 < a < 0$ we have two pair of equilibria on the unit circle. The rays $\theta = \theta_\pm$ and $\theta = \theta_\pm + \pi$ are invariant, and all solutions on these rays tend to the equilibria on the circle. Consider the half-plane $\theta_- < \theta < \theta_- + \pi$. For θ-values in the interval $\theta_- < \theta < \theta_+$, we have $\theta' < 0$, while $\theta' > 0$ in the interval $\theta_+ < \theta < \theta_- + \pi$. Solutions behave symmetrically in the complementary half-plane. Therefore all solutions that do not lie on the rays $\theta = \theta_+$ or $\theta = \theta_+ + \pi$ tend to the equilibrium points at $r = 1$, $\theta = \theta_-$ or at $r = 1$, $\theta = \theta_- + \pi$. These equilibria are therefore sinks. At the other equilibria we have saddles. The stable curves of these saddles lie on the rays $\theta = \theta_+$ and $\theta = \pi + \theta_+$ and the unstable curves of the saddles are given by the unit circle minus the sinks. See Figure 8.9. ∎

The previous examples all featured bifurcations that occur when the linearized system has a zero eigenvalue. Another case where the linearized system fails to be hyperbolic occurs when the system has pure imaginary eigenvalues.

Example. (Hopf Bifurcation) Consider the system

$$x' = ax - y - x(x^2 + y^2)$$
$$y' = x + ay - y(x^2 + y^2).$$

Figure 8.9   Global effects of saddle-node bifurcations when, from left to right, $a < 0$, $a = 0$, and $a > 0$.

There is an equilibrium point at the origin and the linearized system is

$$X' = \begin{pmatrix} a & -1 \\ 1 & a \end{pmatrix} X.$$

The eigenvalues are $a \pm i$, so we expect a bifurcation when $a = 0$.

To see what happens as a passes through 0, we change to polar coordinates. The system becomes

$$r' = ar - r^3$$
$$\theta' = 1.$$

Note that the origin is the only equilibrium point for this system, since $\theta' \neq 0$. For $a < 0$ the origin is a sink since $ar - r^3 < 0$ for all $r > 0$. Thus all solutions tend to the origin in this case. When $a > 0$, the equilibrium becomes a source. So what else happens? When $a > 0$ we have $r' = 0$ if $r = \sqrt{a}$. So the circle of radius $\sqrt{a}$ is a periodic solution with period 2π. We also have $r' > 0$ if $0 < r < \sqrt{a}$, while $r' < 0$ if $r > \sqrt{a}$. Thus, all nonzero solutions spiral toward this circular solution as $t \to \infty$.

This type of bifurcation is called a *Hopf bifurcation*. Thus at a Hopf bifurcation, no new equilibria arise. Instead, a periodic solution is born at the equilibrium point as a passes through the bifurcation value. See Figure 8.10.   ■

8.6 Exploration: Complex Vector Fields

In this exploration, you will investigate the behavior of systems of differential equations in the complex plane of the form $z' = F(z)$. Throughout this section z will denote the complex number $z = x + iy$ and $F(z)$ will be a

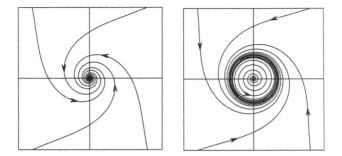

Figure 8.10   The Hopf bifurcation for $a < 0$ and $a > 0$.

polynomial with complex coefficients. Solutions of the differential equation will be expressed as curves $z(t) = x(t) + iy(t)$ in the complex plane.

You should be familiar with complex functions such as the exponential, sine, and cosine as well as with the process of taking complex square roots in order to comprehend fully what you see below. Theoretically, you should also have a grasp of complex analysis as well. However, all of the routine tricks from integration of functions of real variables work just as well when integrating with respect to z. You need not prove this, because you can always check the validity of your solutions when you have completed the integrals.

1. Solve the equation $z' = az$ where a is a complex number. What kind of equilibrium points do you find at the origin for these differential equations?
2. Solve each of the following complex differential equations and sketch the phase portrait.

   (a) $z' = z^2$

   (b) $z' = z^2 - 1$

   (c) $z' = z^2 + 1$

3. For a complex polynomial $F(z)$, the complex derivative is defined just as the real derivative

$$F'(z_0) = \lim_{z \to z_0} \frac{F(z) - F(z_0)}{z - z_0},$$

only this limit is evaluated along any smooth curve in $\mathbb{C}$ that passes through z_0. This limit must exist and yield the same (complex) number for each such curve. For polynomials, this is easily checked. Now write

$$F(x + iy) = u(x, y) + iv(x, y).$$

Evaluate $F'(z_0)$ in terms of the derivatives of u and v by first taking the limit along the horizontal line $z_0 + t$, and secondly along the vertical line $z_0 + it$. Use this to conclude that, if the derivative exists, then we must have

$$\frac{\partial u}{\partial x} = \frac{\partial v}{\partial y} \quad \text{and} \quad \frac{\partial u}{\partial y} = -\frac{\partial v}{\partial x}$$

at every point in the plane. The equations are called the *Cauchy-Riemann equations*.

4. Use the above observation to determine all possible types of equilibrium points for complex vector fields.

5. Solve the equation

$$z' = (z - z_0)(z - z_1)$$

where $z_0, z_1 \in \mathbb{C}$, and $z_0 \neq z_1$. What types of equilibrium points occur for different values of z_0 and z_1?

6. Find a nonlinear change of variables that converts the previous system to $w' = \alpha w$ with $\alpha \in \mathbb{C}$. *Hint:* Since the original system has two equilibrium points and the linear system only one, the change of variables must send one of the equilibrium points to ∞.

7. Classify all complex quadratic systems of the form

$$z' = z^2 + az + b$$

where $a, b \in \mathbb{C}$.

8. Consider the equation

$$z' = z^3 + az$$

with $a \in \mathbb{C}$. First use a computer to describe the phase portraits for these systems. Then prove as much as you can about these systems and classify them with respect to a.

9. Choose your own (nontrivial) family of complex functions depending on a parameter $a \in \mathbb{C}$ and provide a complete analysis of the phase portraits for each a. Some neat families to consider include $a \exp z$, $a \sin z$, or $(z^2 + a)(z^2 - a)$.

EXERCISES

1. For each of the following nonlinear systems,

    (a) Find all of the equilibrium points and describe the behavior of the associated linearized system.

(b) Describe the phase portrait for the nonlinear system.

(c) Does the linearized system accurately describe the local behavior near the equilibrium points?

(i) $x' = \sin x$, $y' = \cos y$
(ii) $x' = x(x^2 + y^2)$, $y' = y(x^2 + y^2)$
(iii) $x' = x + y^2$, $y' = 2y$
(iv) $x' = y^2$, $y' = y$
(v) $x' = x^2$, $y' = y^2$

2. Find a global change of coordinates that linearizes the system

$$x' = x + y^2$$
$$y' = -y$$
$$z' = -z + y^2.$$

3. Consider a first-order differential equation

$$x' = f_a(x)$$

for which $f_a(x_0) = 0$ and $f_a'(x_0) \neq 0$. Prove that the differential equation

$$x' = f_{a+\epsilon}(x)$$

has an equilibrium point $x_0(\epsilon)$ where $\epsilon \rightarrow x_0(\epsilon)$ is a smooth function satisfying $x_0(0) = x_0$ for ϵ sufficiently small.

4. Find general conditions on the derivatives of $f_a(x)$ so that the equation

$$x' = f_a(x)$$

undergoes a pitchfork bifurcation at $a = a_0$. Prove that your conditions lead to such a bifurcation.

5. Consider the system

$$x' = x^2 + y$$
$$y' = x - y + a$$

where a is a parameter.

(a) Find all equilibrium points and compute the linearized equation at each.

(b) Describe the behavior of the linearized system at each equilibrium point.

(c) Describe any bifurcations that occur.

6. Given an example of a family of differential equations $x' = f_a(x)$ for which there are no equilibrium points if $a < 0$; a single equilibrium if $a = 0$; and four equilibrium points if $a > 0$. Sketch the bifurcation diagram for this family.

7. Discuss the local and global behavior of solutions of

$$r' = r - r^3$$
$$\theta' = \sin^2(\theta) + a$$

at the bifurcation value $a = -1$.

8. Discuss the local and global behavior of solutions of

$$r' = r - r^2$$
$$\theta' = \sin^2(\theta/2) + a$$

at all of the bifurcation values.

9. Consider the system

$$r' = r - r^2$$
$$\theta' = \sin\theta + a$$

(a) For which values of a does this system undergo a bifurcation?

(b) Describe the local behavior of solutions near the bifurcation values (at, before, and after the bifurcation).

(c) Sketch the phase portrait of the system for all possible different cases.

(d) Discuss any global changes that occur at the bifurcations.

10. Let $X' = F(X)$ be a nonlinear system in $\mathbb{R}^n$. Suppose that $F(0) = 0$ and that DF_0 has n distinct eigenvalues with negative real parts. Describe the construction of a conjugacy between this system and its linearization.

11. Consider the system $X' = F(X)$ where $X \in \mathbb{R}^n$. Suppose that F has an equilibrium point at X_0. Show that there is a change of coordinates that moves X_0 to the origin and converts the system to

$$X' = AX + G(X)$$

where A is an $n \times n$ matrix, which is the canonical form of DF_{X_0} and where $G(X)$ satisfies

$$\lim_{|X| \to 0} \frac{|G(X)|}{|X|} = 0.$$

12. In the definition of an asymptotically stable equilibrium point, we required that the equilibrium point also be stable. This requirement is not vacuous. Give an example of a phase portrait (a sketch is sufficient) that has an equilibrium point toward which all nearby solution curves (eventually) tend, but that is not stable.

9

Global Nonlinear Techniques

In this chapter we present a variety of qualitative techniques for analyzing the behavior of nonlinear systems of differential equations. The reader should be forewarned that none of these techniques works for all nonlinear systems; most work only in specialized situations, which, as we shall see in the ensuing chapters, nonetheless occur in many important applications of differential equations.

9.1 Nullclines

One of the most useful tools for analyzing nonlinear systems of differential equations (especially planar systems) are the nullclines. For a system in the form

$$x_1' = f_1(x_1, \ldots, x_n)$$
$$\vdots$$
$$x_n' = f_n(x_1, \ldots, x_n),$$

the x_j-nullcline is the set of points where x_j' vanishes, so the x_j-nullcline is the set of points determined by setting $f_j(x_1, \ldots, x_n) = 0$.

The x_j-nullclines usually separate $\mathbb{R}^n$ into a collection of regions in which the x_j-components of the vector field point in either the positive or negative direction. If we determine all of the nullclines, then this allows us to decompose

$\mathbb{R}^n$ into a collection of open sets, in each of which the vector field points in a "certain direction."

This is easiest to understand in the case of a planar system

$$x' = f(x, y)$$
$$y' = g(x, y).$$

On the x-nullclines, we have $x' = 0$, so the vector field points straight up or down, and these are the only points at which this happens. Therefore the x-nullclines divide $\mathbb{R}^2$ into regions where the vector field points either to the left or to the right. Similarly, on the y-nullclines, the vector field is horizontal, so the y-nullclines separate $\mathbb{R}^2$ into regions where the vector field points either upward or downward. The intersections of the x- and y-nullclines yield the equilibrium points. In any of the regions between the nullclines, the vector field is neither vertical nor horizontal, so it must point in one of four directions: northeast, northwest, southeast, or southwest. We call such regions *basic regions*. Often, a simple sketch of the basic regions allows us to understand the phase portrait completely, at least from a qualitative point of view.

Example. For the system

$$x' = y - x^2$$
$$y' = x - 2,$$

the x-nullcline is the parabola $y = x^2$ and the y-nullcline is the vertical line $x = 2$. These nullclines meet at $(2, 4)$ so this is the only equilibrium point. The nullclines divide $\mathbb{R}^2$ into four basic regions labeled A through D in Figure 9.1(a). By first choosing one point in each of these regions, and then determining the direction of the vector field at that point, we can decide the direction of the vector field at all points in the basic region. For example, the point $(0, 1)$ lies in region A and the vector field is $(1, -2)$ at this point, which points toward the southeast. Hence the vector field points southeast at all points in this region. Of course, the vector field may be nearly horizontal or nearly vertical in this region; when we say southeast we mean that the angle θ of the vector field lies in the sector $-\pi/2 < \theta < 0$. Continuing in this fashion we get the direction of the vector field in all four regions, as in Figure 9.1(b). This also determines the horizontal and vertical directions of the vector field on the nullclines.

Just from the direction field alone, it appears that the equilibrium point is a saddle. Indeed, this is the case because the linearized system at $(2, 4)$ is

$$X' = \begin{pmatrix} -4 & 1 \\ 1 & 0 \end{pmatrix} X,$$

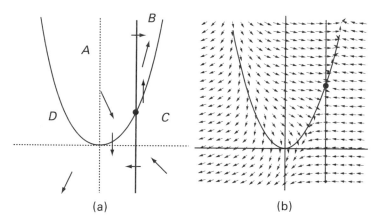

Figure 9.1 The (a) nullclines and (b) direction field.

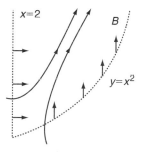

Figure 9.2 Solutions
enter the basic region
B and then tend to ∞.

which has eigenvalues $-2 \pm \sqrt{5}$, one of which is positive, the other negative.

More importantly, we can fill in the approximate behavior of solutions everywhere in the plane. For example, note that the vector field points into the basic region marked B at all points along its boundary, and then it points northeasterly at all points inside B. Thus any solution in region B must stay in region B for all time and tend toward ∞ in the northeast direction. See Figure 9.2. Similarly, solutions in the basic region D stay in that region and head toward ∞ in the southwest direction. Solutions starting in the basic regions A and C have a choice: They must eventually cross one of the nullclines and enter regions B and D (and therefore we know their ultimate behavior) or else they tend to the equilibrium point. However, there is only one curve of such solutions in each region, the stable curve at $(2, 4)$. Thus we completely understand the phase portrait for this system, at least from a qualitative point of view. See Figure 9.3. ∎

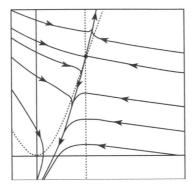

Figure 9.3   The nullclines and phase portrait for $x' = y - x^2$, $y' = x - 2$.

Example. (Heteroclinic Bifurcation) Next consider the system that depends on a parameter a:

$$x' = x^2 - 1$$
$$y' = -xy + a(x^2 - 1).$$

The x-nullclines are given by $x = \pm 1$ while the y-nullclines are $xy = a(x^2 - 1)$. The equilibrium points are $(\pm 1, 0)$. Since $x' = 0$ on $x = \pm 1$, the vector field is actually tangent to these nullclines. Moreover, we have $y' = -y$ on $x = 1$ and $y' = y$ on $x = -1$. So solutions tend to $(1, 0)$ along the vertical line $x = 1$ and tend away from $(-1, 0)$ along $x = -1$. This happens for all values of a.

Now, let's look at the case $a = 0$. Here the system simplifies to

$$x' = x^2 - 1$$
$$y' = -xy,$$

so $y' = 0$ along the axes. In particular, the vector field is tangent to the x-axis and is given by $x' = x^2 - 1$ on this line. So we have $x' > 0$ if $|x| > 1$ and $x' < 0$ if $|x| < 1$. Thus, at each equilibrium point, we have one straight-line solution tending to the equilibrium and one tending away. So it appears that each equilibrium is a saddle. This is indeed the case, as is easily checked by linearization.

There is a second y-nullcline along $x = 0$, but the vector field is not tangent to this nullcline. Computing the direction of the vector field in each of the basic regions determined by the nullclines yields Figure 9.4, from which we can deduce immediately the qualitative behavior of all solutions.

Note that, when $a = 0$, one branch of the unstable curve through $(1, 0)$ matches up exactly with a branch of the stable curve at $(-1, 0)$. All solutions

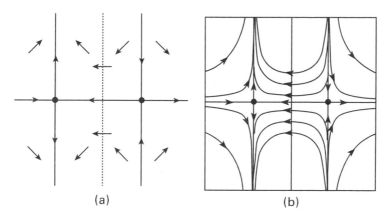

Figure 9.4   The (a) nullclines and (b) phase portrait for $x' = x^2 - 1$, $y' = -xy$.

on this curve simply travel from one saddle to the other. Such solutions are called *heteroclinic solutions* or *saddle connections*. Typically, for planar systems, stable and unstable curves rarely meet to form such heteroclinic "connections." When they do, however, one can expect a bifurcation.

Now consider the case where $a \neq 0$. The x-nullclines remain the same, at $x = \pm 1$. But the y-nullclines change drastically as shown in Figure 9.5. They are given by $y = a(x^2 - 1)/x$.

When $a > 0$, consider the basic region denoted by A. Here the vector field points southwesterly. In particular, the vector field points in this direction along the x-axis between $x = -1$ and $x = 1$. This breaks the heteroclinic connection: The right portion of the stable curve associated to $(-1, 0)$ must

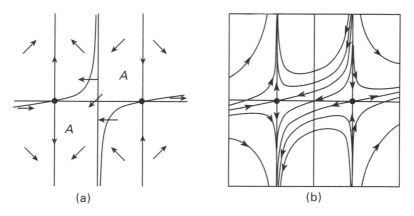

Figure 9.5   The (a) nullclines and (b) phase plane when $a > 0$ after the heteroclinic bifurcation.

now come from $y = \infty$ in the upper half plane, while the left portion of the unstable curve associated to $(1, 0)$ now descends to $y = -\infty$ in the lower half plane. This opens an "avenue" for certain solutions to travel from $y = +\infty$ to $y = -\infty$ between the two lines $x = \pm 1$. Whereas when $a = 0$ all solutions remain for all time confined to either the upper or lower half-plane, the *heteroclinic bifurcation* at $a = 0$ opens the door for certain solutions to make this transit.

A similar situation occurs when $a < 0$ (see Exercise 2 at the end of this chapter).   ■

9.2 Stability of Equilibria

Determining the stability of an equilibrium point is straightforward if the equilibrium is hyperbolic. When this is not the case, this determination becomes more problematic. In this section we develop an alternative method for showing that an equilibrium is asymptotically stable. Due to the Russian mathematician Liapunov, this method generalizes the notion that, for a linear system in canonical form, the radial component r decreases along solution curves. Liapunov noted that other functions besides r could be used for this purpose. Perhaps more importantly, Liapunov's method gives us a grasp on the size of the *basin of attraction* of an asymptotically stable equilibrium point. By definition, the basin of attraction is the set of all initial conditions whose solutions tend to the equilibrium point.

Let $L: \mathcal{O} \to \mathbb{R}$ be a differentiable function defined on an open set $\mathcal{O}$ in $\mathbb{R}^n$ that contains an equilibrium point X^* of the system $X' = F(X)$. Consider the function

$$\dot{L}(X) = DL_X(F(X)).$$

As we have seen, if $\phi_t(X)$ is the solution of the system passing through X when $t = 0$, then we have

$$\dot{L}(X) = \left.\frac{d}{dt}\right|_{t=0} L \circ \phi_t(X)$$

by the chain rule. Consequently, if $\dot{L}(X)$ is negative, then L decreases along the solution curve through X.

We can now state Liapunov's stability theorem:

Theorem. (Liapunov Stability) *Let X^* be an equilibrium point for $X' = F(X)$. Let $L: \mathcal{O} \to \mathbb{R}$ be a differentiable function defined on an open set $\mathcal{O}$*

containing X^*. Suppose further that

(a) $L(X^*) = 0$ and $L(X) > 0$ if $X \neq X^*$;
(b) $\dot{L} \leq 0$ in $\mathcal{O} - X^*$.

Then X^* is stable. Furthermore, if L also satisfies

(c) $\dot{L} < 0$ in $\mathcal{O} - X^*$,

then X^* is asymptotically stable.   ■

A function L satisfying (a) and (b) is called a *Liapunov function* for X^*. If (c) also holds, we call L a *strict* Liapunov function.

Note that Liapunov's theorem can be applied without solving the differential equation; all we need to compute is $DL_X(F(X))$. This is a real plus! On the other hand, there is no cut-and-dried method of finding Liapunov functions; it is usually a matter of ingenuity or trial and error in each case. Sometimes there are natural functions to try. For example, in the case of mechanical or electrical systems, energy is often a Liapunov function, as we shall see in Chapter 13.

Example. Consider the system of differential equations in $\mathbb{R}^3$ given by

$$x' = (\epsilon x + 2y)(z + 1)$$
$$y' = (-x + \epsilon y)(z + 1)$$
$$z' = -z^3$$

where ϵ is a parameter. The origin is the only equilibrium point for this system. The linearization of the system at $(0, 0, 0)$ is

$$Y' = \begin{pmatrix} \epsilon & 2 & 0 \\ -1 & \epsilon & 0 \\ 0 & 0 & 0 \end{pmatrix} Y.$$

The eigenvalues are 0 and $\epsilon \pm \sqrt{2}i$. Hence, from the linearization, we can only conclude that the origin is unstable if $\epsilon > 0$. When $\epsilon \leq 0$, all we can conclude is that the origin is not hyperbolic.

When $\epsilon \leq 0$ we search for a Liapunov function for $(0, 0, 0)$ of the form $L(x, y, z) = ax^2 + by^2 + cz^2$, with $a, b, c > 0$. For such an L, we have

$$\dot{L} = 2(axx' + byy' + czz'),$$

so that

$$\dot{L}/2 = ax(\epsilon x + 2y)(z + 1) + by(-x + \epsilon y)(z + 1) - cz^4$$
$$= \epsilon(ax^2 + by^2)(z + 1) + (2a - b)(xy)(z + 1) - cz^4.$$

For stability, we want $\dot{L} \leq 0$; this can be arranged, for example, by setting $a = 1$, $b = 2$, and $c = 1$. If $\epsilon = 0$, we then have $\dot{L} = -z^4 \leq 0$, so the origin is stable. It can be shown (see Exercise 4) that the origin is not asymptotically stable in this case.

If $\epsilon < 0$, then we find

$$\dot{L} = \epsilon(x^2 + 2y^2)(z + 1) - z^4$$

so that $\dot{L} < 0$ in the region $\mathcal{O}$ given by $z > -1$ (minus the origin). We conclude that the origin is asymptotically stable in this case, and, indeed, from Exercise 4, that all solutions that start in the region $\mathcal{O}$ tend to the origin.   ■

Example. (The Nonlinear Pendulum) Consider a pendulum consisting of a light rod of length ℓ to which is attached a ball of mass m. The other end of the rod is attached to a wall at a point so that the ball of the pendulum moves on a circle centered at this point. The position of the mass at time t is completely described by the angle $\theta(t)$ of the mass from the straight down position and measured in the counterclockwise direction. Thus the position of the mass at time t is given by $(\ell \sin \theta(t), -\ell \cos \theta(t))$.

The speed of the mass is the length of the velocity vector, which is $\ell \, d\theta/dt$, and the acceleration is $\ell \, d^2\theta/dt^2$. We assume that the only two forces acting on the pendulum are the force of gravity and a force due to friction. The gravitational force is a constant force equal to mg acting in the downward direction; the component of this force tangent to the circle of motion is given by $-mg \sin \theta$. We take the force due to friction to be proportional to velocity and so this force is given by $-b\ell \, d\theta/dt$ for some constant $b > 0$. When there is no force due to friction ($b = 0$), we have an *ideal pendulum*. Newton's law then gives the second-order differential equation for the pendulum

$$m\ell \frac{d^2\theta}{dt^2} = -b\ell \frac{d\theta}{dt} - mg \sin \theta.$$

For simplicity, we assume that units have been chosen so that $m = \ell = g = 1$. Rewriting this equation as a system, we introduce $v = d\theta/dt$ and get

$$\theta' = v$$
$$v' = -bv - \sin \theta.$$

Clearly, we have two equilibrium points (mod 2π): the downward rest position at $\theta = 0$, $v = 0$, and the straight-up position $\theta = \pi$, $v = 0$. This upward postion is an unstable equilibrium, both from a mathematical (check the linearization) and physical point of view.

For the downward equilibrium point, the linearized system is

$$Y' = \begin{pmatrix} 0 & 1 \\ -1 & -b \end{pmatrix} Y.$$

The eigenvalues here are either pure imaginary (when $b = 0$) or else have negative real parts (when $b > 0$). So the downward equilibrium is asymptotically stable if $b > 0$ as everyone on earth who has watched a real-life pendulum knows.

To investigate this equilibrium point further, consider the function $E(\theta, v) = \frac{1}{2}v^2 + 1 - \cos\theta$. For readers with a background in elementary mechanics, this is the well-known *total energy* function, which we will describe further in Chapter 13. We compute

$$\dot{E} = vv' + \sin\theta\,\theta' = -bv^2,$$

so that $\dot{E} \le 0$. Hence E is a Liapunov function. Thus the origin is a stable equilibrium. If $b = 0$ (that is, there is no friction), then $\dot{E} \equiv 0$. That is, E is constant along all solutions of the system. Hence we may simply plot the level curves of E to see where the solution curves reside. We find the phase portrait shown in Figure 9.6. Note that we do not have to solve the differential equation to paint this picture; knowing the level curves of E (and the direction of the vector field) tells us everything. We will encounter many such (very special) functions that are constant along solution curves later in this chapter.

The solutions encircling the origin have the property that $-\pi < \theta(t) < \pi$ for all t. Therefore these solutions correspond to the pendulum oscillating about the downward rest position without ever crossing the upward position $\theta = \pi$. The special solutions connecting the equilibrium points at $(\pm\pi, 0)$

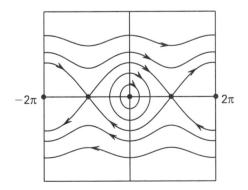

Figure 9.6   The phase portrait for the ideal pendulum.

correspond to the pendulum tending to the upward-pointing equilibrium in both the forward and backward time directions. (You don't often see such motions!) Beyond these special solutions we find solutions for which $\theta(t)$ either increases or decreases for all time; in these cases the pendulum spins forever in the counterclockwise or clockwise direction. ∎

We will return to the pendulum example for the case $b > 0$ later, but first we prove Liapunov's theorem.

Proof: Let $\delta > 0$ be so small that the closed ball $B_\delta(X^*)$ around the equilibrium point X^* of radius δ lies entirely in $\mathcal{O}$. Let α be the minimum value of L on the boundary of $B_\delta(X^*)$, that is, on the sphere $S_\delta(X^*)$ of radius δ and center X^*. Then $\alpha > 0$ by assumption. Let $\mathcal{U} = \{X \in B_\delta(X^*) \mid L(X) < \alpha\}$ and note that X^* lies in $\mathcal{U}$. Then no solution starting in $\mathcal{U}$ can meet $S_\delta(X^*)$ since L is nonincreasing on solution curves. Hence every solution starting in $\mathcal{U}$ never leaves $B_\delta(X^*)$. This proves that X^* is stable.

Now suppose that assumption (c) in the Liapunov stability theorem holds as well, so that L is strictly decreasing on solutions in $\mathcal{U} - X^*$. Let $X(t)$ be a solution starting in $\mathcal{U} - X^*$ and suppose that $X(t_n) \to Z_0 \in B_\delta(X^*)$ for some sequence $t_n \to \infty$. We claim that $Z_0 = X^*$. To see this, observe that $L(X(t)) > L(Z_0)$ for all $t \geq 0$ since $L(X(t))$ decreases and $L(X(t_n)) \to L(Z_0)$ by continuity of L. If $Z_0 \neq X^*$, let $Z(t)$ be the solution starting at Z_0. For any $s > 0$, we have $L(Z(s)) < L(Z_0)$. Hence for any solution $Y(s)$ starting sufficiently near Z_0 we have

$$L(Y(s)) < L(Z_0).$$

Setting $Y(0) = X(t_n)$ for sufficiently large n yields the contradiction

$$L(X(t_n + s)) < L(Z_0).$$

Therefore $Z_0 = X^*$. This proves that X^* is the only possible limit point of the set $\{X(t) \mid t \geq 0\}$ and completes the proof of Liapunov's theorem. ∎

Figure 9.7 makes the theorem intuitively obvious. The condition $\dot{L} < 0$ means that when a solution crosses a "level surface" $L^{-1}(c)$, it moves inside the set where $L \leq c$ and can never come out again. Unfortunately, it is sometimes difficult to justify the diagram shown in this figure; why should the sets $L^{-1}(c)$ shrink down to X^*? Of course, in many cases, Figure 9.7 is indeed correct, as, for example, if L is a quadratic function such as $ax^2 + by^2$ with $a, b > 0$. But what if the level surfaces look like those shown in Figure 9.8? It is hard to imagine such an L that fulfills all the requirements of a Liapunov function; but rather than trying to rule out that possibility, it is simpler to give the analytic proof as above.

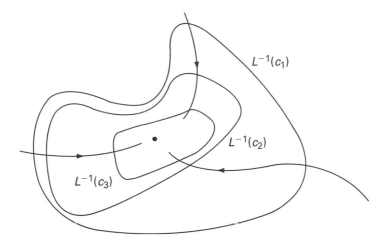

Figure 9.7   Solutions decrease through the level sets $L^{-1}(c_j)$ of a strict Liapunov function.

Example. Now consider the system

$$x' = -x^3$$
$$y' = -y(x^2 + z^2 + 1)$$
$$z' = -\sin z.$$

The origin is again an equilibrium point. It is not the only one, however, since $(0, 0, n\pi)$ is also an equilibrium point for each $n \in \mathbb{Z}$. Hence the origin cannot be globally asymptotically stable. Moreover, the planes $z = n\pi$ for $n \in \mathbb{Z}$ are *invariant* in the sense that any solution that starts on one of these planes remains there for all time. This occurs since $z' = 0$ when $z = n\pi$. In particular, any solution that begins in the region $|z| < \pi$ must remain trapped in this region for all time.

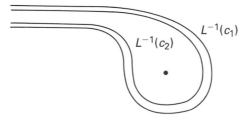

Figure 9.8   Level sets of a Liapunov function may look like this.

Linearization at the origin yields the system

$$Y' = \begin{pmatrix} 0 & 0 & 0 \\ 0 & -1 & 0 \\ 0 & 0 & -1 \end{pmatrix} Y$$

which tells us nothing about the stability of this equilibrium point.

However, consider the function

$$L(x, y, z) = x^2 + y^2 + z^2.$$

Clearly, $L > 0$ except at the origin. We compute

$$\dot{L} = -2x^4 - 2y^2(x^2 + z^2 + 1) - 2z \sin z.$$

Then $\dot{L} < 0$ at all points in the set $|z| < \pi$ (except the origin) since $z \sin z > 0$ when $z \neq 0$. Hence the origin is asymptotically stable.

Moreover, we can conclude that the basin of attraction of the origin is the entire region $|z| < \pi$. From the proof of the Liapunov stability theorem, it follows immediately that any solution that starts inside a sphere of radius $r < \pi$ must tend to the origin. Outside of the sphere of radius π and between the planes $z = \pm \pi$, the function L is still strictly decreasing. Since solutions are trapped between these two planes, it follows that they too must tend to the origin. ∎

Liapunov functions not only detect stable equilibria; they can also be used to estimate the size of the basin of attraction of an asymptotically stable equilibrium, as the above example shows. The following theorem gives a criterion for asymptotic stability and the size of the basin even when the Liapunov function is not strict. To state it, we need several definitions.

Recall that a set $\mathcal{P}$ is called *invariant* if for each $X \in \mathcal{P}$, $\phi_t(X)$ is defined and in $\mathcal{P}$ for all $t \in \mathbb{R}$. For example, the region $|z| < \pi$ in the above example is an invariant set. The set $\mathcal{P}$ is *positively invariant* if for each $X \in \mathcal{P}$, $\phi_t(X)$ is defined and in $\mathcal{P}$ for all $t \geq 0$. The portion of the region $|z| < \pi$ inside a sphere centered at the origin in the previous example is positively invariant but not invariant. Finally, an *entire solution* of a system is a set of the form $\{\phi_t(X) \mid t \in \mathbb{R}\}$.

Theorem. (Lasalle's Invariance Principle) *Let X^* be an equilibrium point for $X' = F(X)$ and let $L \colon \mathcal{U} \to \mathbb{R}$ be a Liapunov function for X^*, where $\mathcal{U}$ is an open set containing X^*. Let $\mathcal{P} \subset \mathcal{U}$ be a neighborhood of X^* that is closed and bounded. Suppose that $\mathcal{P}$ is positively invariant, and that there is no entire solution in $\mathcal{P} - X^*$ on which L is constant. Then X^* is asymptotically stable, and $\mathcal{P}$ is contained in the basin of attraction of X^*.* ∎

Before proving this theorem we apply it to the equilibrium $X^* = (0,0)$ of the damped pendulum discussed previously. Recall that a Liapunov function is given by $E(\theta, v) = \frac{1}{2}v^2 + 1 - \cos\theta$ and that $\dot{E} = -bv^2$. Since $\dot{E} = 0$ on $v = 0$, this Liapunov function is not strict.

To estimate the basin of $(0,0)$, fix a number c with $0 < c < 2$, and define

$$\mathcal{P}_c = \{(\theta, v) \mid E(\theta, v) \leq c \text{ and } |\theta| < \pi\}.$$

Clearly, $(0,0) \in \mathcal{P}_c$. We shall prove that $\mathcal{P}_c$ lies in the basin of attraction of $(0,0)$.

Note first that $\mathcal{P}_c$ is positively invariant. To see this suppose that $(\theta(t), v(t))$ is a solution with $(\theta(0), v(0)) \in \mathcal{P}_c$. We claim that $(\theta(t), v(t)) \in \mathcal{P}_c$ for all $t \geq 0$. We clearly have $E(\theta(t), v(t)) \leq c$ since $\dot{E} \leq 0$. If $|\theta(t)| \geq \pi$, then there must exist a smallest t_0 such that $\theta(t_0) = \pm\pi$. But then

$$E(\theta(t_0), v(t_0)) = E(\pm\pi, v(t_0))$$

$$= \frac{1}{2}v(t_0)^2 + 2$$

$$\geq 2.$$

However,

$$E(\theta(t_0), v(t_0)) \leq c < 2.$$

This contradiction shows that $\theta(t_0) < \pi$, and so $\mathcal{P}_c$ is positively invariant.

We now show that there is no entire solution in $\mathcal{P}_c$ on which E is constant (except the equilibrium solution). Suppose there is such a solution. Then, along that solution, $\dot{E} \equiv 0$ and so $v = 0$. Hence, $\theta' = 0$ so θ is constant on the solution. We also have $v' = -\sin\theta = 0$ on the solution. Since $|\theta| < \pi$, it follows that $\theta \equiv 0$. Thus the only entire solution in $\mathcal{P}_c$ on which E is constant is the equilibrium point $(0,0)$.

Finally, $\mathcal{P}_c$ is a closed set. Because if (θ_0, v_0) is a limit point of $\mathcal{P}_c$, then $|\theta_0| \leq \pi$, and $E(\theta_0, v_0) \leq c$ by continuity of E. But $|\theta_0| = \pi$ implies $E(\theta_0, v_0) > c$, as shown above. Hence $|\theta_0| < \pi$ and so (θ_0, v_0) does belong to $\mathcal{P}_c$, so $\mathcal{P}_c$ is closed.

From the theorem we conclude that $\mathcal{P}_c$ belongs to the basin of attraction of $(0,0)$ for each $c < 2$; hence the set

$$\mathcal{P} = \cup\{\mathcal{P}_c \mid 0 < c < 2\}$$

is also contained in this basin. Note that we may write

$$\mathcal{P} = \{(\theta, v) \mid E(\theta, v) < 2 \text{ and } |\theta| < \pi\}.$$

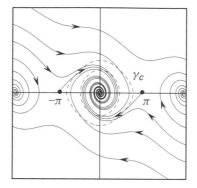

Figure 9.9   The curve γ_c
bounds the region $\mathcal{P}_c$.

Figure 9.9 displays the phase portrait for the damped pendulum. The curves marked γ_c are the level sets $E(\theta, v) = c$. Note that solutions cross each of these curves exactly once and eventually tend to the origin.

This result is quite natural on physical grounds. For if $\theta \neq \pm\pi$, then $E(\theta, 0) < 2$ and so the solution through $(\theta, 0)$ tends to $(0, 0)$. That is, if we start the pendulum from rest at any angle θ except the vertical position, the pendulum will eventually wind down to rest at its stable equilibrium postion.

There are other initial positions in the basin of $(0, 0)$ that are not in the set $\mathcal{P}$. For example, consider the solution through $(-\pi, u)$, where u is very small but not zero. Then $(-\pi, u) \notin \mathcal{P}$, but the solution through this point moves quickly into $\mathcal{P}$, and therefore eventually approaches $(0, 0)$. Hence $(-\pi, u)$ also lies in the basin of attraction of $(0, 0)$. This can be seen in Figure 9.9, where the solutions that begin just above the equilibrium point at $(-\pi, 0)$ and just below $(\pi, 0)$ quickly cross γ_c and then enter $\mathcal{P}_c$. See Exercise 5 for further examples of this.

We now prove the theorem.

Proof: Imagine a solution $X(t)$ that lies in the positively invariant set $\mathcal{P}$ for $0 \leq t \leq \infty$, but suppose that $X(t)$ does *not* tend to X^* as $t \to \infty$. Since $\mathcal{P}$ is closed and bounded, there must be a point $Z \neq X^*$ in $\mathcal{P}$ and a sequence $t_n \to \infty$ such that

$$\lim_{n \to \infty} X(t_n) = Z.$$

We may assume that the sequence $\{t_n\}$ is an increasing sequence.

We claim that the entire solution through Z lies in $\mathcal{P}$. That is, $\phi_t(Z)$ is defined and in $\mathcal{P}$ for all $t \in \mathbb{R}$, not just $t \geq 0$. This can be seen as follows.

First, $\phi_t(Z)$ is certainly defined for all $t \geq 0$ since $\mathcal{P}$ is positively invariant. On the other hand, $\phi_t(X(t_n))$ is defined and in $\mathcal{P}$ for all t in the interval $[-t_n, 0]$. Since $\{t_n\}$ is an increasing sequence, we have that $\phi_t(X(t_{n+k}))$ is also defined and in $\mathcal{P}$ for all $t \in [-t_n, 0]$ and all $k \geq 0$. Since the points $X(t_{n+k}) \to Z$ as $k \to \infty$, it follows from continuous dependence of solutions on initial conditions that $\phi_t(Z)$ is defined and in $\mathcal{P}$ for all $t \in [-t_n, 0]$. Since this holds for any t_n, we see that the solution through Z is an entire solution lying in $\mathcal{P}$.

Finally, we show that L is constant on the entire solution through Z. If $L(Z) = \alpha$, then we have $L(X(t_n)) \geq \alpha$ and moreover

$$\lim_{n \to \infty} L(X(t_n)) = \alpha.$$

More generally, if $\{s_n\}$ is any sequence of times for which $s_n \to \infty$ as $n \to \infty$, then $L(X(s_n)) \to \alpha$ as well. This follows from the fact that L is nonincreasing along solutions. Now the sequence $X(t_n + s)$ converges to $\phi_s(Z)$, and so $L(\phi_s(Z)) = \alpha$. This contradicts our assumption that there are no entire solutions lying in $\mathcal{P}$ on which L is constant and proves the theorem. ∎

In this proof we encountered certain points that were limits of a sequence of points on the solution through X. The set of all points that are limit points of a given solution is called the set of ω-*limit points*, or the ω-*limit set*, of the solution $X(t)$. Similarly, we define the set of α-*limit points*, or the α-*limit set*, of a solution $X(t)$ to be the set of all points Z such that $\lim_{n \to \infty} X(t_n) = Z$ for some sequence $t_n \to -\infty$. (The reason, such as it is, for this terminology is that α is the first letter and ω the last letter of the Greek alphabet.) The following facts, essentially proved above, will be used in the following chapter.

Proposition. *The α-limit set and the ω-limit set of a solution that is defined for all $t \in \mathbb{R}$ are closed, invariant sets.* ∎

9.3 Gradient Systems

Now we turn to a particular type of system for which the previous material on Liapunov functions is particularly germane. A *gradient system* on $\mathbb{R}^n$ is a system of differential equations of the form

$$X' = -\text{grad } V(X)$$

where $V: \mathbb{R}^n \to \mathbb{R}$ is a C^∞ function, and

$$\operatorname{grad} V = \left(\frac{\partial V}{\partial x_1}, \ldots, \frac{\partial V}{\partial x_n} \right).$$

(The negative sign in this system is traditional.) The vector field grad V is called the *gradient* of V. Note that $-\operatorname{grad} V(X) = \operatorname{grad}(-V(X))$.

Gradient systems have special properties that make their flows rather simple. The following equality is fundamental:

$$DV_X(Y) = \operatorname{grad} V(X) \cdot Y.$$

This says that the derivative of V at X evaluated at $Y = (y_1, \ldots, y_n) \in \mathbb{R}^n$ is given by the dot product of the vectors grad $V(X)$ and Y. This follows immediately from the formula

$$DV_X(Y) = \sum_{j=1}^{n} \frac{\partial V}{\partial x_j}(X) \, y_j.$$

Let $X(t)$ be a solution of the gradient system with $X(0) = X_0$, and let $\dot{V}: \mathbb{R}^n \to \mathbb{R}$ be the derivative of V along this solution. That is,

$$\dot{V}(X) = \frac{d}{dt} V(X(t)).$$

Proposition.   *The function V is a Liapunov function for the system $X' = -\operatorname{grad} V(X)$. Moreover, $\dot{V}(X) = 0$ if and only if X is an equilibrium point.*

Proof: By the chain rule we have

$$\dot{V}(X) = DV_X(X')$$
$$= \operatorname{grad} V(X) \cdot (-\operatorname{grad} V(X))$$
$$= -|\operatorname{grad} V(X)|^2 \leq 0.$$

In particular, $\dot{V}(X) = 0$ if and only if grad $V(X) = 0$.   ∎

An immediate consequence of this is the fact that if X^* is an isolated minimum of V, then X^* is an asymptotically stable equilibrium of the gradient system. Indeed, the fact that X^* is isolated guarantees that $\dot{V} < 0$ in a neighborhood of X^* (not including X^*).

To understand a gradient flow geometrically we look at the *level surfaces* of the function $V: \mathbb{R}^n \to \mathbb{R}$. These are the subsets $V^{-1}(c)$ with $c \in \mathbb{R}$. If

$X \in V^{-1}(c)$ is a *regular point*, that is, grad $V(X) \neq 0$, then $V^{-1}(c)$ looks like a "surface" of dimension $n - 1$ near X. To see this, assume (by renumbering the coordinates) that $\partial V/\partial x_n(X) \neq 0$. Using the implicit function theorem, we find a C^∞ function $g: \mathbb{R}^{n-1} \to \mathbb{R}$ such that, near X, the level set $V^{-1}(c)$ is given by

$$V\left(x_1, \ldots, x_{n-1}, g\left(x_1, \ldots, x_{n-1}\right)\right) = c.$$

That is, near X, $V^{-1}(c)$ looks like the graph of the function g. In the special case where $n = 2$, $V^{-1}(c)$ is a simple curve through X when X is a regular point. If all points in $V^{-1}(c)$ are regular points, then we say that c is a *regular value* for V. In the case $n = 2$, if c is a regular value, then the level set $V^{-1}(c)$ is a union of simple (or nonintersecting) curves. If X is a nonregular point for V, then grad $V(X) = 0$, so X is a *critical point* for the function V, since all partial derivatives of V vanish at X.

Now suppose that Y is a vector that is tangent to the level surface $V^{-1}(c)$ at X. Then we can find a curve $\gamma(t)$ in this level set for which $\gamma'(0) = Y$. Since V is constant along γ, it follows that

$$DV_X(Y) = \left.\frac{d}{dt}\right|_{t=0} V \circ \gamma(t) = 0.$$

We thus have, by our observations above, that grad $V(X) \cdot Y = 0$, or, in other words, grad $V(X)$ is perpendicular to every tangent vector to the level set $V^{-1}(c)$ at X. That is, the vector field grad $V(X)$ is perpendicular to the level surfaces $V^{-1}(c)$ at all regular points of V. We may summarize all of this in the following theorem.

Theorem. (Properties of Gradient Systems) *For the system $X' = -\text{grad } V(X)$:*

1. *If c is a regular value of V, then the vector field is perpendicular to the level set $V^{-1}(c)$.*
2. *The critical points of V are the equilibrium points of the system.*
3. *If a critical point is an isolated minimum of V, then this point is an asymptotically stable equilibrium point.* ∎

Example. Let $V: \mathbb{R}^2 \to \mathbb{R}$ be the function $V(x, y) = x^2(x-1)^2 + y^2$. Then the gradient system

$$X' = F(X) = -\text{grad } V(X)$$

is given by

$$x' = -2x(x-1)(2x-1)$$
$$y' = -2y.$$

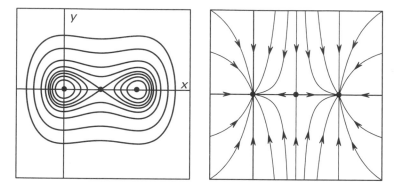

Figure 9.10 The level sets and phase portrait for the gradient system determined by $V(x, y) = x^2 (x - 1)^2 + y^2$.

There are three equilibrium points: $(0,0)$, $(1/2, 0)$, and $(1, 0)$. The linearizations at these three points yield the following matrices:

$$DF(0, 0) = \begin{pmatrix} -2 & 0 \\ 0 & -2 \end{pmatrix}, \quad DF(1/2, 0) = \begin{pmatrix} 1 & 0 \\ 0 & -2 \end{pmatrix},$$

$$DF(1, 0) = \begin{pmatrix} -2 & 0 \\ 0 & -2 \end{pmatrix}.$$

Hence $(0, 0)$ and $(1, 0)$ are sinks, while $(1/2, 0)$ is a saddle. Both the x- and y-axes are invariant, as are the lines $x = 1/2$ and $x = 1$. Since $y' = -2y$ on these vertical lines, it follows that the stable curve at $(1/2, 0)$ is the line $x = 1/2$, while the unstable curve at $(1/2, 0)$ is the interval $(0, 1)$ on the x-axis. ■

The level sets of V and the phase portrait are shown in Figure 9.10. Note that it appears that all solutions tend to one of the three equilibria. This is no accident, for we have:

Proposition. *Let Z be an α-limit point or an ω-limit point of a solution of a gradient flow. Then Z is an equilibrium point.*

Proof: Suppose Z is an ω-limit point. As in the proof of Lasalle's invariance principle from Section 9.2, one shows that V is constant along the solution through Z. Thus $\dot{V}(Z) = 0$, and so Z must be an equilibrium point. The case of an α-limit point is similar. In fact, an α-limit point Z of $X' = -\text{grad } V(X)$ is an ω-limit point of $X' = \text{grad } V(X)$, so that $\text{grad } V(Z) = 0$. ■

If a gradient system has only isolated equilibrium points, this result implies that every solution of the system must tend either to infinity or to an equilibrium point. In the previous example we see that the sets $V^{-1}([0, c])$ are closed, bounded, and positively invariant under the gradient flow. Therefore each solution entering such a set is defined for all $t \geq 0$, and tends to one of the three equilibria $(0, 0)$, $(1, 0)$, or $(1/2, 0)$. The solution through every point *does* enter such a set, since the solution through (x, y) enters the set $V^{-1}([0, c_0])$ where $V(x, y) = c_0$.

There is one final property that gradient systems share. Note that, in the previous example, all of the eigenvalues of the linearizations at the equilibria have real eigenvalues. Again, this is no accident, for the linearization of a gradient system at an equilibrium point X^* is a matrix $[a_{ij}]$ where

$$a_{ij} = -\left(\frac{\partial^2 V}{\partial x_i \partial x_j}\right)(X^*).$$

Since mixed partial derivatives are equal, we have

$$\left(\frac{\partial^2 V}{\partial x_i \partial x_j}\right)(X^*) = \left(\frac{\partial^2 V}{\partial x_j \partial x_i}\right)(X^*),$$

and so $a_{ij} = a_{ji}$. It follows that the matrix corresponding to the linearized system is a *symmetric matrix*. It is known that such matrices have only real eigenvalues. For example, in the 2×2 case, a symmetric matrix assumes the form

$$\begin{pmatrix} a & b \\ b & c \end{pmatrix}$$

and the eigenvalues are easily seen to be

$$\frac{a + c}{2} \pm \frac{\sqrt{(a - c)^2 + 4b^2}}{2},$$

both of which are real numbers. A more general case is relegated to Exercise 15. We therefore have:

Proposition. *For a gradient system* $X' = -\mathrm{grad}\, V(X)$, *the linearized system at any equilibrium point has only real eigenvalues.* ∎

9.4 Hamiltonian Systems

In this section we deal with another special type of system, a Hamiltonian system. As we shall see in Chapter 13, these are the types of systems that arise in classical mechanics.

We shall restrict attention in this section to Hamiltonian systems in $\mathbb{R}^2$. A *Hamiltonian system* on $\mathbb{R}^2$ is a system of the form

$$x' = \frac{\partial H}{\partial y}(x, y)$$

$$y' = -\frac{\partial H}{\partial x}(x, y)$$

where $H: \mathbb{R}^2 \to \mathbb{R}$ is a C^∞ function called the *Hamiltonian function*.

Example. (Undamped Harmonic Oscillator) Recall that this system is given by

$$x' = y$$
$$y' = -kx$$

where $k > 0$. A Hamiltonian function for this system is

$$H(x, y) = \frac{1}{2}y^2 + \frac{k}{2}x^2.$$ ∎

Example. (Ideal Pendulum) The equation for this system, as we saw in Section 9.2, is

$$\theta' = v$$
$$v' = -\sin\theta.$$

The total energy function

$$E(\theta, v) = \frac{1}{2}v^2 + 1 - \cos\theta$$

serves as a Hamiltonian function in this case. Note that we say a Hamiltonian function, since we can always add a constant to any Hamiltonian function without changing the equations.

What makes Hamiltonian systems so important is the fact that the Hamiltonian function is a *first integral* or *constant of the motion*. That is, H is constant along every solution of the system, or, in the language of the previous sections, $\dot{H} \equiv 0$. This follows immediately from

$$\dot{H} = \frac{\partial H}{\partial x}x' + \frac{\partial H}{\partial y}y'$$

$$= \frac{\partial H}{\partial x}\frac{\partial H}{\partial y} + \frac{\partial H}{\partial y}\left(-\frac{\partial H}{\partial x}\right) = 0.$$ ∎

Thus we have:

Proposition. *For a Hamiltonian system on $\mathbb{R}^2$, H is constant along every solution curve.* ∎

The importance of knowing that a given system is Hamiltonian is the fact that we can essentially draw the phase portrait without solving the system. Assuming that H is not constant on any open set, we simply plot the level curves $H(x, y) =$ constant. The solutions of the system lie on these level sets; all we need to do is figure out the directions of the solution curves on these level sets. But this is easy since we have the vector field. Note also that the equilibrium points for a Hamiltonian system occur at the critical points of H, that is, at points where both partial derivatives of H vanish.

Example. Consider the system

$$x' = y$$
$$y' = -x^3 + x.$$

A Hamiltonian function is

$$H(x, y) = \frac{x^4}{4} - \frac{x^2}{2} + \frac{y^2}{2} + \frac{1}{4}.$$

The constant value $1/4$ is irrelevant here; we choose it so that H has minimum value 0, which occurs at $(\pm 1, 0)$, as is easily checked. The only other equilibrium point lies at the origin. The linearized system is

$$X' = \begin{pmatrix} 0 & 1 \\ 1 - 3x^2 & 0 \end{pmatrix} X.$$

At $(0, 0)$, this system has eigenvalues ± 1, so we have a saddle. At $(\pm 1, 0)$, the eigenvalues are $\pm\sqrt{2}i$, so we have a center, at least for the linearized system.

Plotting the level curves of H and adding the directions at nonequilibrium points yields the phase portrait depicted in Figure 9.11. Note that the equilibrium points at $(\pm 1, 0)$ remain centers for the nonlinear system. Also note that the stable and unstable curves at the origin match up exactly. That is, we have solutions that tend to $(0, 0)$ in both forward and backward time. Such solutions are known as *homoclinic solutions* or *homoclinic orbits*. ∎

The fact that the eigenvalues of this system assume the special forms ± 1 and $\pm\sqrt{2}i$ is again no accident.

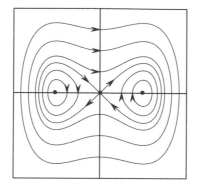

Figure 9.11   The phase
portrait for $x' = y$,
$y' = -x^3 + x$.

Proposition.   *Suppose (x_0, y_0) is an equilibrium point for a planar Hamiltonian system. Then the eigenvalues of the linearized system are either $\pm\lambda$ or $\pm i\lambda$ where $\lambda \in \mathbb{R}$.*   ■

The proof of the proposition is straightforward (see Exercise 11).

9.5 Exploration: The Pendulum with Constant Forcing

Recall from Section 9.2 that the equations for a nonlinear pendulum are

$$\theta' = v$$
$$v' = -bv - \sin\theta.$$

Here θ gives the angular position of the pendulum (which we assume to be measured in the counterclockwise direction) and v is its angular velocity. The parameter $b > 0$ measures the damping.

Now we apply a constant torque to the pendulum in the counterclockwise direction. This amounts to adding a constant to the equation for v', so the system becomes

$$\theta' = v$$
$$v' = -bv - \sin\theta + k$$

where we assume that $k \geq 0$. Since θ is measured mod 2π, we may think of this system as being defined on the cylinder $S^1 \times \mathbb{R}$, where S^1 denotes the unit circle.

1. Find all equilibrium points for this system and determine their stability.
2. Determine the regions in the bk–parameter plane for which there are different numbers of equilibrium points. Describe the motion of the pendulum in each different case.
3. Suppose $k > 1$. Prove that there exists a periodic solution for this system. *Hint:* What can you say about the vector field in a strip of the form $0 < v_1 < (k - \sin\theta)/b < v_2$?
4. Describe the qualitative features of a Poincaré map defined on the line $\theta = 0$ for this system.
5. Prove that when $k > 1$ there is a unique periodic solution for this system. *Hint:* Recall the energy function

$$E(\theta, y) = \frac{1}{2}y^2 - \cos\theta + 1$$

   and use the fact that the total change of E along any periodic solution must be 0.
6. Prove that there are parameter values for which a stable equilibrium and a periodic solution coexist.
7. Describe the bifurcation that must occur when the periodic solution ceases to exist.

EXERCISES

1. For each of the following systems, sketch the x- and y-nullclines and use this information to determine the nature of the phase portrait. You may assume that these systems are defined only for $x, y \geq 0$.

   (a) $x' = x(y + 2x - 2),\ y' = y(y - 1)$
   (b) $x' = x(y + 2x - 2),\ y' = y(y + x - 3)$
   (c) $x' = x(2 - y - 2x),\ y' = y(3 - 3y - x)$
   (d) $x' = x(2 - y - 2x),\ y' = y(3 - y - 4x)$
   (e) $x' = x(2500 - x^2 - y^2),\ y' = y(70 - y - x)$

2. Describe the phase portrait for

$$x' = x^2 - 1$$
$$y' = -xy + a(x^2 - 1)$$

when $a < 0$. What qualitative features of this flow change as a passes
from negative to positive?

3. Consider the system of differential equations

$$x' = x(-x - y + 1)$$
$$y' = y(-ax - y + b)$$

where a and b are parameters with $a, b > 0$. Suppose that this system is
only defined for $x, y \geq 0$.

(a) Use the nullclines to sketch the phase portrait for this system for
various a- and b-values.

(b) Determine the values of a and b at which a bifurcation occurs.

(c) Sketch the regions in the ab–plane where this system has qualitatively
similar phase portraits, and describe the bifurcations that occur as
the parameters cross the boundaries of these regions.

4. Consider the system

$$x' = (\epsilon x + 2y)(z + 1)$$
$$y' = (-x + \epsilon y)(z + 1)$$
$$z' = -z^3.$$

(a) Show that the origin is not asymptotically stable when $\epsilon = 0$.

(b) Show that when $\epsilon < 0$, the basin of attraction of the origin contains
the region $z > -1$.

5. For the nonlinear damped pendulum, show that for every integer n
and every angle θ_0 there is an initial condition (θ_0, v_0) whose solution
corresponds to the pendulum moving around the circle at least n times,
but not $n + 1$ times, before settling down to the rest position.

6. Find a strict Liapunov function for the equilibrium point $(0, 0)$ of

$$x' = -2x - y^2$$
$$y' = -y - x^2.$$

Find $\delta > 0$ as large as possible so that the open disk of radius δ and center
$(0, 0)$ is contained in the basin of $(0, 0)$.

7. For each of the following functions $V(X)$, sketch the phase portrait of
the gradient flow $X' = -\text{grad } V(X)$. Sketch the level surfaces of V on
the same diagram. Find all of the equilibrium points and determine their
type.

(a) $x^2 + 2y^2$

(b) $x^2 - y^2 - 2x + 4y + 5$

(c) $y \sin x$

(d) $2x^2 - 2xy + 5y^2 + 4x + 4y + 4$

(e) $x^2 + y^2 - z$

(f) $x^2(x - 1) + y^2(y - 2) + z^2$

8. Sketch the phase portraits for the following systems. Determine if the system is Hamiltonian or gradient along the way. (That's a little hint, by the way.)

(a) $x' = x + 2y$, $y' = -y$

(b) $x' = y^2 + 2xy$, $y' = x^2 + 2xy$

(c) $x' = x^2 - 2xy$, $y' = y^2 - 2xy$

(d) $x' = x^2 - 2xy$, $y' = y^2 - x^2$

(e) $x' = -\sin^2 x \sin y$, $y' = -2 \sin x \cos x \cos y$

9. Let $X' = AX$ be a linear system where

$$A = \begin{pmatrix} a & b \\ c & d \end{pmatrix}.$$

(a) Determine conditions on a, b, c, and d that guarantee that this system is a gradient system. Give a gradient function explicitly.

(b) Repeat the previous question for a Hamiltonian system.

10. Consider the planar system

$$x' = f(x, y)$$
$$y' = g(x, y).$$

Determine explicit conditions on f and g that guarantee that this system is a gradient system or a Hamiltonian system.

11. Prove that the linearization at an equilibrium point of a planar Hamiltonian system has eigenvalues that are either $\pm\lambda$ or $\pm i\lambda$ where $\lambda \in \mathbb{R}$.

12. Let T be the torus defined as the square $0 \leq \theta_1, \theta_2 \leq 2\pi$ with opposite sides identified. Let $F(\theta_1, \theta_2) = \cos\theta_1 + \cos\theta_2$. Sketch the phase portrait for the system $-\text{grad } F$ in T. Sketch a three-dimensional representation of this phase portrait with T represented as the surface of a doughnut.

13. Repeat the previous exercise, but assume now that F is a Hamiltonian function.

14. On the torus T above, let $F(\theta_1, \theta_2) = \cos\theta_1(2 - \cos\theta_2)$. Sketch the phase portrait for the system $-\text{grad } F$ in T. Sketch a three-dimensional representation of this phase portrait with T represented as the surface of a doughnut.

15. Prove that a 3 × 3 symmetric matrix has only real eigenvalues.

16. A solution $X(t)$ of a system is called *recurrent* if $X(t_n) \rightarrow X(0)$ for some sequence $t_n \rightarrow \infty$. Prove that a gradient dynamical system has no nonconstant recurrent solutions.

17. Show that a closed bounded omega limit set is connected. Give an example of a planar system having an unbounded omega limit set consisting of two parallel lines.

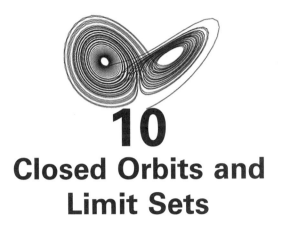

10

Closed Orbits and Limit Sets

In the past few chapters we have concentrated on equilibrium solutions of systems of differential equations. These are undoubtedly among the most important solutions, but there are other types of solutions that are important as well. In this chapter we will investigate another important type of solution, the *periodic solution* or *closed orbit*. Recall that a periodic solution occurs for $X' = F(X)$ if we have a nonequilibrium point X and a time $\tau > 0$ for which $\phi_\tau(X) = X$. It follows that $\phi_{t+\tau}(X) = \phi_t(X)$ for all t, so ϕ_t is a periodic function. The least such $\tau > 0$ is called the *period* of the solution. As an example, all nonzero solutions of the undamped harmonic oscillator equation are periodic solutions. Like equilibrium points that are asymptotically stable, periodic solutions may also attract other solutions. That is, solutions may limit on periodic solutions just as they can approach equilibria.

In the plane, the limiting behavior of solutions is essentially restricted to equilibria and closed orbits, although there are a few exceptional cases. We will investigate this phenomenon in this chapter in the guise of the important Poincaré-Bendixson theorem. We will see later that, in dimensions greater than two, the limiting behavior of solutions can be quite a bit more complicated.

10.1 Limit Sets

We begin by describing the limiting behavior of solutions of systems of differential equations. Recall that $Y \in \mathbb{R}^n$ is an ω-limit point for the solution through X if there is a sequence $t_n \to \infty$ such that $\lim_{n \to \infty} \phi_{t_n}(X) = Y$. That is, the solution curve through X accumulates on the point Y as time moves forward. The set of all ω-limit points of the solution through X is the ω-limit set of X and is denoted by $\omega(X)$. The α-limit points and the α-limit set $\alpha(X)$ are defined by replacing $t_n \to \infty$ with $t_n \to -\infty$ in the above definition. By a *limit set* we mean a set of the form $\omega(X)$ or $\alpha(X)$.

Here are some examples of limit sets. If X^* is an asymptotically stable equilibrium, it is the ω-limit set of every point in its basin of attraction. Any equilibrium is its own α- and ω-limit set. A periodic solution is the α-limit and ω-limit set of every point on it. Such a solution may also be the ω-limit set of many other points.

Example. Consider the planar system given in polar coordinates by

$$r' = \frac{1}{2}(r - r^3)$$

$$\theta' = 1.$$

As we saw in Section 8.1, all nonzero solutions of this equation tend to the periodic solution that resides on the unit circle in the plane. See Figure 10.1. Consequently, the ω-limit set of any nonzero point is this closed orbit.   ■

Example. Consider the system

$$x' = \sin x(-.1 \cos x - \cos y)$$

$$y' = \sin y(\cos x - .1 \cos y).$$

There are equilibria which are saddles at the corners of the square $(0, 0)$, $(0, \pi)$, (π, π), and $(\pi, 0)$, as well as at many other points. There are heteroclinic

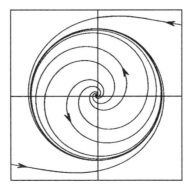

Figure 10.1   The phase plane for $r' = \frac{1}{2}(r - r^3)$, $\theta' = 1$.

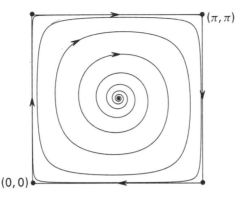

Figure 10.2   The ω-limit set of any
solution emanating from the source at
$(\pi/2, \pi/2)$ is the square bounded by the
four equilibria and the heteroclinic
solutions.

solutions connecting these equilibria in the order listed. See Figure 10.2. There
is also a spiral source at $(\pi/2, \pi/2)$. All solutions emanating from this source
accumulate on the four heteroclinic solutions connecting the equilibria (see
Exercise 4 at the end of this chapter). Hence the ω-limit set of any point on
these solutions is the square bounded by $x = 0, \pi$ and $y = 0, \pi$.   ∎

   In three dimensions there are extremely complicated examples of limit
sets, which are not very easy to describe. In the plane, however, limit sets
are fairly simple. In fact, Figure 10.2 is typical in that one can show that
a closed and bounded limit set other than a closed orbit or equilibrium
point is made up of equilibria and solutions joining them. The Poincaré-
Bendixson theorem discussed in Section 10.5 states that if a closed and
bounded limit set in the plane contains no equilibria, then it must be a closed
orbit.
   Recall from Section 9.2 that a limit set is closed in $\mathbb{R}^n$ and is invariant under
the flow. We shall also need the following result:

Proposition.

1. *If X and Z lie on the same solution curve, then $\omega(X) = \omega(Z)$ and $\alpha(X) = \alpha(Z)$;*
2. *If D is a closed, positively invariant set and $Z \in D$, then $\omega(Z) \subset D$, and similarly for negatively invariant sets and α-limits;*
3. *A closed invariant set, in particular, a limit set, contains the α-limit and ω-limit sets of every point in it.*

Proof: For (1), suppose that $Y \in \omega(X)$, and $\phi_s(X) = Z$. If $\phi_{t_n}(X) \to Y$, then we have

$$\phi_{t_n - s}(Z) = \phi_{t_n}(X) \to Y.$$

Hence $Y \in \omega(Z)$ as well. For (2), if $\phi_{t_n}(Z) \to Y \in \omega(Z)$ as $t_n \to \infty$, then we have $t_n \geq 0$ for sufficiently large n so that $\phi_{t_n}(Z) \in D$. Hence $Y \in D$ since D is a closed set. Finally, part (3) follows immediately from part (2).   ∎

10.2 Local Sections and Flow Boxes

For the rest of this chapter, we restrict the discussion to planar systems. In this section we describe the local behavior of the flow associated to $X' = F(X)$ near a given point X_0, which is not an equilibrium point. Our goal is to construct first a local section at X_0, and then a flow box neighborhood of X_0. In this flow box, solutions of the system behave particularly simply.

Suppose $F(X_0) \neq 0$. The *transverse line* at X_0, denoted by $\ell(X_0)$, is the straight line through X_0, which is perpendicular to the vector $F(X_0)$ based at X_0. We parametrize $\ell(X_0)$ as follows. Let V_0 be a unit vector based at X_0 and perpendicular to $F(X_0)$. Then define $h : \mathbb{R} \to \ell(X_0)$ by $h(u) = X_0 + uV_0$.

Since $F(X)$ is continuous, the vector field is not tangent to $\ell(X_0)$, at least in some open interval in $\ell(X_0)$ surrounding X_0. We call such an open subinterval containing X_0 a *local section* at X_0. At each point of a local section S, the vector field points "away from" S, so solutions must cut across a local section. In particular $F(X) \neq 0$ for $X \in S$. See Figure 10.3.

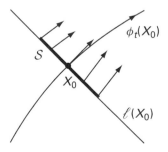

Figure 10.3   A local section S at X_0 and several representative vectors from the vector field along S.

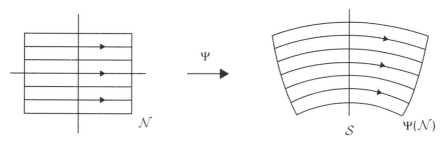

Figure 10.4   The flow box associated to $\mathcal{S}$.

Our first use of a local section at X_0 will be to construct an associated *flow box* in a neighborhood of X_0. A flow box gives a complete description of the behavior of the flow in a neighborhood of a nonequilibrium point by means of a special set of coordinates. An intuitive description of the flow in a flow box is simple: points move in parallel straight lines at constant speed.

Given a local section $\mathcal{S}$ at X_0, we may construct a map Ψ from a neighborhood $\mathcal{N}$ of the origin in $\mathbb{R}^2$ to a neighborhood of X_0 as follows. Given $(s, u) \in \mathbb{R}^2$, we define

$$\Psi(s, u) = \phi_s(h(u))$$

where h is the parameterization of the transverse line described above. Note that Ψ maps the vertical line $(0, u)$ in $\mathcal{N}$ to the local section $\mathcal{S}$; Ψ also maps horizontal lines in $\mathcal{N}$ to pieces of solution curves of the system. Provided that we choose $\mathcal{N}$ sufficiently small, the map Ψ is then one to one on $\mathcal{N}$. Also note that $D\Psi$ takes the constant vector field $(1, 0)$ in $\mathcal{N}$ to vector field $F(X)$. Using the language of Chapter 4, Ψ is a local conjugacy between the flow of this constant vector field and the flow of the nonlinear system.

We usually take $\mathcal{N}$ in the form $\{(s, u) \mid |s| < \sigma\}$ where $\sigma > 0$. In this case we sometimes write $\mathcal{V}_\sigma = \Psi(\mathcal{N})$ and call $\mathcal{V}_\sigma$ the *flow box* at (or about) X_0. See Figure 10.4. An important property of a flow box is that if $X \in \mathcal{V}_\sigma$, then $\phi_t(X) \in \mathcal{S}$ for a unique $t \in (-\sigma, \sigma)$.

If $\mathcal{S}$ is a local section, the solution through a point Z_0 (perhaps far from $\mathcal{S}$) may reach $X_0 \in \mathcal{S}$ at a certain time t_0; see Figure 10.5. We show that in a certain local sense, this "time of first arrival" at $\mathcal{S}$ is a continuous function of Z_0. More precisely:

Proposition.   *Let $\mathcal{S}$ be a local section at X_0 and suppose $\phi_{t_0}(Z_0) = X_0$. Let $\mathcal{W}$ be a neighborhood of Z_0. Then there is an open set $\mathcal{U} \subset \mathcal{W}$ containing Z_0 and a differentiable function $\tau : \mathcal{U} \to \mathbb{R}$ such that $\tau(Z_0) = t_0$ and*

$$\phi_{\tau(X)}(X) \in \mathcal{S}$$

for each $X \in \mathcal{U}$.

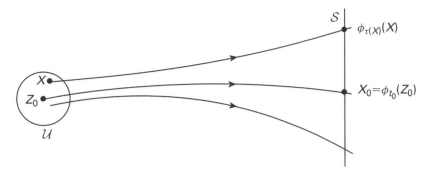

Figure 10.5   Solutions crossing the local section S.

Proof: Suppose $F(X_0)$ is the vector (α, β) and recall that $(\alpha, \beta) \neq (0, 0)$. For $Y = (y_1, y_2) \in \mathbb{R}^2$, define $\eta : \mathbb{R}^2 \to \mathbb{R}$ by

$$\eta(Y) = Y \cdot F(X_0) = \alpha y_1 + \beta y_2.$$

Recall that Y belongs to the transverse line $\ell(X_0)$ if and only if $Y = X_0 + V$ where $V \cdot F(X_0) = 0$. Hence $Y \in \ell(X_0)$ if and only if $\eta(Y) = Y \cdot F(X_0) = X_0 \cdot F(X_0)$.

Now define $G : \mathbb{R}^2 \times \mathbb{R} \to \mathbb{R}$ by

$$G(X, t) = \eta(\phi_t(X)) = \phi_t(X) \cdot F(X_0).$$

We have $G(Z_0, t_0) = X_0 \cdot F(X_0)$ since $\phi_{t_0}(Z_0) = X_0$. Furthermore

$$\frac{\partial G}{\partial t}(Z_0, t_0) = |F(X_0)|^2 \neq 0.$$

We may thus apply the implicit function theorem to find a smooth function $\tau : \mathbb{R}^2 \to \mathbb{R}$ defined on a neighborhood $\mathcal{U}_1$ of (Z_0, t_0) such that $\tau(Z_0) = t_0$ and

$$G(X, \tau(X)) \equiv G(Z_0, t_0) = X_0 \cdot F(X_0).$$

Hence $\phi_{\tau(X)}(X)$ belongs to the transverse line $\ell(X_0)$. If $\mathcal{U} \subset \mathcal{U}_1$ is a sufficiently small neighborhood of Z_0, then $\phi_{\tau(X)}(X) \in \mathcal{S}$, as required. ∎

10.3 The Poincaré Map

As in the case of equilibrium points, closed orbits may also be stable, asymptotically stable, or unstable. The definitions of these concepts for closed orbits are entirely analogous to those for equilibria as in Section 8.4. However, determining the stability of closed orbits is much more difficult than the corresponding problem for equilibria. While we do have a tool that resembles the linearization technique that is used to determine the stability of (most) equilibria, generally this tool is much more difficult to use in practice. Here is the tool.

Given a closed orbit γ, there is an associated *Poincaré map* for γ, some examples of which we previously encountered in Sections 1.4 and 6.2. Near a closed orbit, this map is defined as follows. Choose $X_0 \in \gamma$ and let $\mathcal{S}$ be a local section at X_0. We consider the first return map on $\mathcal{S}$. This is the function P that associates to $X \in \mathcal{S}$ the point $P(X) = \phi_t(X) \in \mathcal{S}$ where t is the smallest positive time for which $\phi_t(X) \in \mathcal{S}$. Now P may not be defined at all points on $\mathcal{S}$ as the solutions through certain points in $\mathcal{S}$ may never return to $\mathcal{S}$. But we certainly have $P(X_0) = X_0$, and the previous proposition guarantees that P is defined and continuously differentiable in a neighborhood of X_0.

In the case of planar systems, a local section is a subset of a straight line through X_0, so we may regard this local section as a subset of $\mathbb{R}$ and take $X_0 = 0 \in \mathbb{R}$. Hence the Poincaré map is a real function taking 0 to 0. If $|P'(0)| < 1$, it follows that P assumes the form $P(x) = ax +$ higher order terms, where $|a| < 1$. Hence, for x near 0, $P(x)$ is closer to 0 than x. This means that the solution through the corresponding point in $\mathcal{S}$ moves closer to γ after one passage through the local section. Continuing, we see that each passage through $\mathcal{S}$ brings the solution closer to γ, and so we see that γ is asymptotically stable. We have:

Proposition. Let $X' = \Gamma(X)$ be a planar system and suppose that X_0 lies on a closed orbit γ. Let P be a Poincaré map defined on a neighborhood of X_0 in some local section. If $|P'(X_0)| < 1$, then γ is asymptotically stable. ∎

Example. Consider the planar system given in polar coordinates by

$$r' = r(1 - r)$$
$$\theta' = 1.$$

Clearly, there is a closed orbit lying on the unit circle $r = 1$. This solution in rectangular coordinates is given by $(\cos t, \sin t)$ when the initial condition is $(1, 0)$. Also, there is a local section lying along the positive real axis since $\theta' = 1$. Furthermore, given any $x \in (0, \infty)$, we have $\phi_{2\pi}(x, 0)$, which also lies on the positive real axis $\mathbb{R}^+$. Thus we have a Poincaré map $P: \mathbb{R}^+ \to \mathbb{R}^+$. Moreover, $P(1) = 1$ since the point $x = 1$, $y = 0$ is the initial condition giving the periodic solution. To check the stability of this solution, we need to compute $P'(1)$.

To do this, we compute the solution starting at $(x, 0)$. We have $\theta(t) = t$, so we need to find $r(2\pi)$. To compute $r(t)$, we separate variables to find

$$\int \frac{dr}{r(1 - r)} = t + \text{constant}.$$

Evaluating this integral yields

$$r(t) = \frac{xe^t}{1 - x + xe^t}.$$

Hence

$$P(x) = r(2\pi) = \frac{xe^{2\pi}}{1 - x + xe^{2\pi}}.$$

Differentiating, we find $P'(1) = 1/e^{2\pi}$ so that $0 < P'(1) < 1$. Thus the periodic solution is asymptotically stable.   ∎

The astute reader may have noticed a little scam here. To determine the Poincaré map, we actually first found formulas for all of the solutions starting at $(x, 0)$. So why on earth would we need to compute a Poincaré map? Well, good question. Actually, it is usually very difficult to compute the exact form of a Poincaré map or even its derivative along a closed orbit, since in practice we rarely have a closed-form expression for the closed orbit, never mind the nearby solutions. As we shall see, the Poincaré map is usually more useful when setting up a geometric model of a specific system (see the Lorenz system in Chapter 14). There are some cases where we can circumvent this problem and gain insight into the Poincaré map, as we shall see when we investigate the Van der Pol equation in Section 12.3.

10.4 Monotone Sequences in Planar Dynamical Systems

Let $X_0, X_1, \ldots \in \mathbb{R}^2$ be a finite or infinite sequence of distinct points on the solution curve through X_0. We say that the sequence is *monotone along the solution* if $\phi_{t_n}(X_0) = X_n$ with $0 \leq t_1 < t_2 \cdots$.

Let $Y_0, Y_1, \ldots$ be a finite or infinite sequence of points on a line segment I in $\mathbb{R}^2$. We say that this sequence is *monotone along I* if Y_n is between Y_{n-1} and Y_{n+1} in the natural order along I for all $n \geq 1$.

A sequence of points may be on the intersection of a solution curve and a segment I; they may be monotone along the solution curve but not along

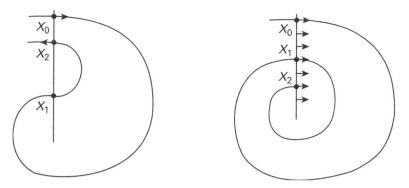

Figure 10.6 Two solutions crossing a straight line. On the left, X_0, X_1, X_2 is monotone along the solution but not along the straight line. On the right, X_0, X_1, X_2 is monotone along both the solution and the line.

the segment, or vice versa; see Figure 10.6. However, this is impossible if the segment is a local section in the plane.

Proposition. *Let S be a local section for a planar system of differential equations and let Y_0, Y_1, Y_2, ... be a sequence of distinct points in S that lie on the same solution curve. If this sequence is monotone along the solution, then it is also monotone along S.*

Proof: It suffices to consider three points Y_0, Y_1, and Y_2 in S. Let Σ be the simple closed curve made up of the part of the solution between Y_0 and Y_1 and the segment $T \subset S$ between Y_0 and Y_1. Let D be the region bounded by Σ. We suppose that the solution through Y_1 leaves D at Y_1 (see Figure 10.7; if the solution enters D, the argument is similar). Hence the solution leaves D at every point in T since T is part of the local section.

It follows that the complement of D is positively invariant. For no solution can enter D at a point of T; nor can it cross the solution connecting Y_0 and Y_1, by uniqueness of solutions.

Therefore $\phi_t(Y_1) \in \mathbb{R}^2 - D$ for all $t > 0$. In particular, $Y_2 \in S - T$. The set $S - T$ is the union of two half open intervals I_0 and I_1 with Y_j an endpoint of I_j for $j = 0, 1$. One can draw an arc from a point $\phi_\epsilon(Y_1)$ (with $\epsilon > 0$ very small) to a point of I_1, without crossing Σ. Therefore I_1 is outside D. Similarly I_0 is inside D. It follows that $Y_2 \in I_1$ since it must be outside D. This shows that Y_1 is between Y_0 and Y_2 in I, proving the proposition. ∎

We now come to an important property of limit points.

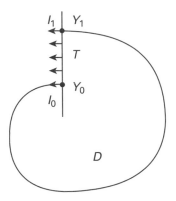

Figure 10.7   Solutions exit
the region D through T.

Proposition.   *For a planar system, suppose that $Y \in \omega(X)$. Then the solution through Y crosses any local section at no more than one point. The same is true if $Y \in \alpha(X)$.*

Proof: Suppose for the sake of contradiction that Y_1 and Y_2 are distinct points on the solution through Y and S is a local section containing Y_1 and Y_2. Suppose $Y \in \omega(X)$ (the argument for $\alpha(X)$ is similar). Then $Y_k \in \omega(X)$ for $k = 1, 2$. Let $\mathcal{V}_k$ be flow boxes at Y_k defined by some intervals $J_k \subset S$; we assume that J_1 and J_2 are disjoint as depicted in Figure 10.8. The solution through X enters each $\mathcal{V}_k$ infinitely often; hence it crosses J_k infinitely often.

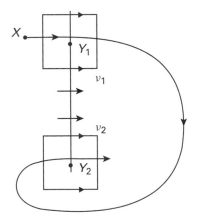

Figure 10.8   The solution
through X cannot cross $\mathcal{V}_1$ and
$\mathcal{V}_2$ infinitely often.

Hence there is a sequence

$$a_1, b_1, a_2, b_2, a_3, b_3, \ldots,$$

which is monotone along the solution through X, with $a_n \in J_1, b_n \in J_2$ for $n = 1, 2, \ldots$. But such a sequence cannot be monotone along S since J_1 and J_2 are disjoint, contradicting the previous proposition. ∎

10.5 The Poincaré-Bendixson Theorem

In this section we prove a celebrated result concerning planar systems:

Theorem. (Poincaré-Bendixson) *Suppose that Ω is a nonempty, closed and bounded limit set of a planar system of differential equations that contains no equilibrium point. Then Ω is a closed orbit.*

Proof: Suppose that $\omega(X)$ is closed and bounded and that $Y \in \omega(X)$. (The case of α-limit sets is similar.) We show first that Y lies on a closed orbit and later that this closed orbit actually is $\omega(X)$.

Since Y belongs to $\omega(X)$ we know from Section 10.1 that $\omega(Y)$ is a nonempty subset of $\omega(X)$. Let $Z \in \omega(Y)$ and let S be a local section at Z. Let V be a flow box associated to S. By the results of the previous section, the solution through Y meets S in exactly one point. On the other hand, there is a sequence $t_n \to \infty$ such that $\phi_{t_n}(Y) \to Z$; hence infinitely many $\phi_{t_n}(Y)$ belong to V. Therefore we can find $r, s \in \mathbb{R}$ such that $r > s$ and $\phi_r(Y), \phi_s(Y) \in S$. It follows that $\phi_r(Y) = \phi_s(Y)$; hence $\phi_{r-s}(Y) = Y$ and $r - s > 0$. Since $\omega(X)$ contains no equilibria, Y must lie on a closed orbit.

It remains to prove that if γ is a closed orbit in $\omega(X)$, then $\gamma = \omega(X)$. For this, it is enough to show that

$$\lim_{t \to \infty} d(\phi_t(X), \gamma) = 0,$$

where $d(\phi_t(x), \gamma)$ is the distance from $\phi_t(X)$ to the set γ (that is, the distance from $\phi_t(X)$ to the nearest point of γ).

Let S be a local section at $Y \in \gamma$. Let $\epsilon > 0$ and consider a flow box V_ϵ associated to S. Then there is a sequence $t_0 < t_1 < \cdots$ such that

1. $\phi_{t_n}(X) \in S$;
2. $\phi_{t_n}(X) \to Y$;
3. $\phi_t(X) \notin S$ for $t_{n-1} < t < t_n$, $n = 1, 2, \ldots$.

Let $X_n = \phi_{t_n}(X)$. By the first proposition in the previous section, X_n is a monotone sequence in S that converges to Y.

We claim that there exists an upper bound for the set of positive numbers $t_{n+1} - t_n$ for n sufficiently large. To see this, suppose $\phi_\tau(Y) = Y$ where $\tau > 0$. Then for X_n sufficiently near Y, $\phi_\tau(X_n) \in V_\epsilon$ and hence

$$\phi_{\tau+t}(X_n) \in S$$

for some $t \in [-\epsilon, \epsilon]$. Thus

$$t_{n+1} - t_n \leq \tau + \epsilon.$$

This provides the upper bound for $t_{n+1} - t_n$. Also, $t_{n+1} - t_n$ is clearly at least 2ϵ, so $t_n \to \infty$ as $n \to \infty$.

Let $\beta > 0$ be small. By continuity of solutions with respect to initial conditions, there exists $\delta > 0$ such that, if $|Z - Y| < \delta$ and $|t| \leq \tau + \epsilon$ then $|\phi_t(Z) - \phi_t(Y)| < \beta$. That is, the distance from the solution $\phi_t(Z)$ to γ is less than β for all t satisfying $|t| \leq \tau + \epsilon$. Let n_0 be so large that $|X_n - Y| < \delta$ for all $n \geq n_0$. Then

$$|\phi_t(X_n) - \phi_t(Y)| < \beta$$

if $|t| \leq \tau + \epsilon$ and $n \geq n_0$. Now let $t \geq t_{n_0}$. Let $n \geq n_0$ be such that

$$t_n \leq t \leq t_{n+1}.$$

Then

$$
\begin{aligned}
d(\phi_t(X), \gamma) &\leq |\phi_t(X) - \phi_{t-t_n}(Y)| \\
&= |\phi_{t-t_n}(X_n) - \phi_{t-t_n}(Y)| \\
&< \beta
\end{aligned}
$$

since $|t - t_n| \leq \tau + \epsilon$. This shows that the distance from $\phi_t(X)$ to γ is less than β for all sufficiently large t. This completes the proof of the Poincaré-Bendixson theorem. ∎

Example. Another example of an ω-limit set that is neither a closed orbit nor an equilibrium is provided by a *homoclinic solution*. Consider the system

$$x' = -y - \left(\frac{x^4}{4} - \frac{x^2}{2} + \frac{y^2}{2}\right)(x^3 - x)$$

$$y' = x^3 - x - \left(\frac{x^4}{4} - \frac{x^2}{2} + \frac{y^2}{2}\right)y.$$

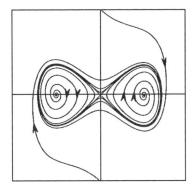

Figure 10.9   A pair of
homoclinic solutions in the
ω-limit set.

A computation shows that there are three equilibria: at $(0,0)$, $(-1,0)$, and $(1,0)$. The origin is a saddle, while the other two equilibria are sources. The phase portrait of this system is shown in Figure 10.9. Note that solutions far from the origin tend to accumulate on the origin and a pair of homoclinic solutions, each of which leaves and then returns to the origin. Solutions emanating from either source have ω-limit set that consists of just one homoclinic solution and $(0,0)$. See Exercise 6 for proofs of these facts.   ■

10.6 Applications of Poincaré-Bendixson

The Poincaré-Bendixson theorem essentially determines all of the possible limiting behaviors of a planar flow. We give a number of corollaries of this important theorem in this section.

A *limit cycle* is a closed orbit γ such that $\gamma \subset \omega(X)$ or $\gamma \subset \alpha(X)$ for some $X \notin \gamma$. In the first case γ is called an ω-limit cycle; in the second case, an α-limit cycle. We deal only with ω-limit sets in this section; the case of α-limit sets is handled by simply reversing time.

In the proof of the Poincaré-Bendixson theorem it was shown that limit cycles have the following property: If γ is an ω-limit cycle, there exists $X \notin \gamma$ such that

$$\lim_{t \to \infty} d(\phi_t(X), \gamma) = 0.$$

Geometrically this means that some solution spirals toward γ as $t \to \infty$. See Figure 10.10. Not all closed orbits have this property. For example, in the

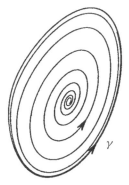

Figure 10.10   A
solution spiraling
toward a limit cycle.

case of a linear system with a center at the origin in $\mathbb{R}^2$, the closed orbits that surround the origin have no solutions approaching them, and so they are not limit cycles.

Limit cycles possess a kind of (one-sided, at least) stability. Let γ be an ω-limit cycle and suppose $\phi_t(X)$ spirals toward γ as $t \to \infty$. Let S be a local section at $Z \in \gamma$. Then there is an interval $T \subset S$ disjoint from γ, bounded by $\phi_{t_0}(X)$ and $\phi_{t_1}(X)$ with $t_0 < t_1$, and not meeting the solution through X for $t_0 < t < t_1$. See Figure 10.11. The annular region A that is bounded on one side by γ and on the other side by the union of T and the curve

$$\{\phi_t(X) \mid t_0 \leq t \leq t_1\}$$

is positively invariant, as is the set $B = A - \gamma$. It is easy to see that $\phi_t(Y)$ spirals toward γ for all $Y \in B$. Hence we have:

Corollary 1.   *Let γ be an ω-limit cycle. If $\gamma = \omega(X)$ where $X \notin \gamma$, then X has a neighborhood $\mathcal{O}$ such that $\gamma = \omega(Y)$ for all $Y \in \mathcal{O}$. In other words, the set*

$$\{Y \mid \omega(Y) = \gamma\} - \gamma$$

is open.   ■

Recall that a subset of R^n that is closed and bounded is said to be *compact*. As another consequence of the Poincaré-Bendixson theorem, suppose that K is a positively invariant set that is compact. If $X \in K$, then $\omega(X)$ must also lie in K. Hence K must contain either an equilibrium point or a limit cycle.

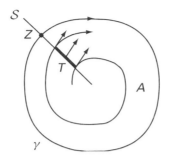

Figure 10.11   The region
A is positively invariant.

Corollary 2.   *A compact set K that is positively or negatively invariant contains either a limit cycle or an equilibrium point.*   ■

The next result exploits the spiraling property of limit cycles.

Corollary 3.   *Let γ be a closed orbit and let U be the open region in the interior of γ. Then U contains either an equilibrium point or a limit cycle.*

Proof: Let D be the compact set $U \cup \gamma$. Then D is invariant since no solution in U can cross γ. If U contains no limit cycle and no equilibrium, then, for any $X \in U$,

$$\omega(X) = \alpha(X) = \gamma$$

by Poincaré-Bendixson. If S is a local section at a point $Z \in \gamma$, there are sequences $t_n \to \infty$, $s_n \to -\infty$ such that $\phi_{t_n}(X), \phi_{s_n}(X) \in S$ and both $\phi_{t_n}(X)$ and $\phi_{s_n}(X)$ tend to Z as $n \to \infty$. But this leads to a contradiction of the proposition in Section 10.4 on monotone sequences.   ■

Actually this last result can be considerably sharpened:

Corollary 4.   *Let γ be a closed orbit that forms the boundary of an open set U. Then U contains an equilibrium point.*

Proof: Suppose U contains no equilibrium point. Consider first the case that there are only finitely many closed orbits in U. We may choose the closed orbit that bounds the region with smallest area. There are then no closed orbits or equilibrium points inside this region, and this contradicts corollary 3.

Now suppose that there are infinitely many closed orbits in U. If $X_n \to X$ in U and each X_n lies on a closed orbit, then X must lie on a closed orbit. Otherwise, the solution through X would spiral toward a limit cycle since there are no equilibria in U. By corollary 1, so would the solution through some nearby X_n, which is impossible.

Let $v \geq 0$ be the greatest lower bound of the areas of regions enclosed by closed orbits in U. Let $\{\gamma_n\}$ be a sequence of closed orbits enclosing regions of areas v_n such that $\lim_{n \to \infty} v_n = v$. Let $X_n \in \gamma_n$. Since $\gamma \cup U$ is compact, we may assume that $X_n \to X \in U$. Then if U contains no equilibrium, X lies on a closed orbit β bounding a region of area v. The usual section argument shows that as $n \to \infty$, γ_n gets arbitrarily close to β and hence the area $v_n - v$ of the region between γ_n and β goes to 0. Then the argument above shows that there can be no closed orbits or equilibrium points inside γ, and this provides a contradiction to corollary 3.   ∎

The following result uses the spiraling properties of limit cycles in a subtle way.

Corollary 5.   *Let H be a first integral of a planar system. If H is not constant on any open set, then there are no limit cycles.*

Proof: Suppose there is a limit cycle γ; let $c \in \mathbb{R}$ be the constant value of H on γ. If $X(t)$ is a solution that spirals toward γ, then $H(X(t)) \equiv c$ by continuity of H. In corollary 1 we found an open set whose solutions spiral toward γ; thus H is constant on an open set.   ∎

Finally, the following result is implicit in our development of the theory of Liapunov functions in Section 9.2.

Corollary 6.   *If L is a strict Liapunov function for a planar system, then there are no limit cycles.*   ∎

10.7 Exploration: Chemical Reactions That Oscillate

For much of the 20th century, chemists believed that all chemical reactions tended monotonically to equilibrium. This belief was shattered in the 1950s when the Russian biochemist Belousov discovered that a certain reaction involving citric acid, bromate ions, and sulfuric acid, when combined with a cerium catalyst, could oscillate for long periods of time before settling to equilibrium. The concoction would turn yellow for a while, then fade, then turn yellow again, then fade, and on and on like this for over an hour. This reaction, now called the Belousov-Zhabotinsky reaction (the BZ reaction, for short), was a major turning point in the history of chemical reactions. Now, many systems are known to oscillate. Some have even been shown to behave chaotically.

One particularly simple chemical reaction is given by a chlorine dioxide–iodine–malonic acid interaction. The exact differential equations modeling this reaction are extremely complicated. However, there is a planar nonlinear system that closely approximates the concentrations of two of the reactants. The system is

$$x' = a - x - \frac{4xy}{1 + x^2}$$

$$y' = bx\left(1 - \frac{y}{1 + x^2}\right)$$

where x and y represent the concentrations of I^- and ClO_2^-, respectively, and a and b are positive parameters.

1. Begin the exploration by investigating these reaction equations numerically. What qualitatively different types of phase portraits do you find?
2. Find all equilibrium points for this system.
3. Linearize the system at your equilibria and determine the type of each equilibrium.
4. In the ab–plane, sketch the regions where you find asymptotically stable or unstable equilibria.
5. Identify the a, b-values where the system undergoes bifurcations.
6. Using the nullclines for the system together with the Poincaré-Bendixson theorem, find the a, b-values for which a stable limit cycle exists. Why do these values correspond to oscillating chemical reactions?

For more details on this reaction, see [27]. The very interesting history of the BZ-reaction is described in [47]. The original paper by Belousov is reprinted in [17].

EXERCISES

1. For each of the following systems, identify all points that lie in either an ω- or an α-limit set

    (a) $r' = r - r^2$, $\theta' = 1$
    (b) $r' = r^3 - 3r^2 + 2r$, $\theta' = 1$

(c) $r' = \sin r, \ \theta' = -1$

(d) $x' = \sin x \sin y, \ y' = - \cos x \cos y$

2. Consider the three-dimensional system

$$r' = r(1 - r)$$
$$\theta' = 1$$
$$z' = -z.$$

Compute the Poincaré map along the closed orbit lying on the unit circle given by $r = 1$ and show that this closed orbit is asymptotically stable.

3. Consider the three-dimensional system

$$r' = r(1 - r)$$
$$\theta' = 1$$
$$z' = z.$$

Again compute the Poincaré map for this system. What can you now say about the behavior of solutions near the closed orbit? Sketch the phase portrait for this system.

4. Consider the system

$$x' = \sin x(-0.1 \cos x - \cos y)$$
$$y' = \sin y(\cos x - 0.1 \cos y).$$

Show that all solutions emanating from the source at $(\pi/2, \pi/2)$ have ω-limit sets equal to the square bounded by $x = 0, \pi$ and $y = 0, \pi$.

5. The system

$$r' = ar + r^3 - r^5$$
$$\theta' = 1$$

depends on a parameter a. Determine the phase plane for representative a values and describe all bifurcations for the system.

6. Consider the system

$$x' = -y - \left(\frac{x^4}{4} - \frac{x^2}{2} + \frac{y^2}{2} \right)(x^3 - x)$$

$$y' = x^3 - x - \left(\frac{x^4}{4} - \frac{x^2}{2} + \frac{y^2}{2} \right)y.$$

(a) Find all equilibrium points.

(b) Determine the types of these equilibria.

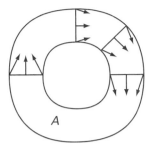

Figure 10.12 The
region A is positively
invariant.

(c) Prove that all nonequilibrium solutions have ω-limit sets consisting of either one or two homoclinic solutions plus a saddle point.

7. Let A be an annular region in $\mathbb{R}^2$. Let F be a planar vector field that points inward along the two boundary curves of A. Suppose also that every radial segment of A is local section. See Figure 10.12. Prove there is a periodic solution in A.

8. Let F be a planar vector field and again consider an annular region A as in the previous problem. Suppose that F has no equilibria and that F points inward along the boundary of the annulus, as before.

(a) Prove there is a closed orbit in A. (Notice that the hypothesis is weaker than in the previous problem.)

(b) If there are exactly seven closed orbits in A, show that one of them has orbits spiraling toward it from both sides.

9. Let F be a planar vector field on a neighborhood of the annular region A above. Suppose that for every boundary point X of A, $F(X)$ is a nonzero vector tangent to the boundary.

(a) Sketch the possible phase portraits in A under the further assumption that there are no equilibria and no closed orbits besides the boundary circles. Include the case where the solutions on the boundary travel in opposite directions.

(b) Suppose the boundary solutions are oppositely oriented and that the flow preserves area. Show that A contains an equilibrium.

10. Show that a closed orbit of a planar system meets a local section in at most one point.

11. Show that a closed and bounded limit set is connected (that is, not the union of two disjoint nonempty closed sets).

12. Let $X' = F(X)$ be a planar system with no equilibrium points. Suppose the flow ϕ_t generated by F preserves area (that is, if U is any open set, the area of $\phi_t(U)$ is independent of t). Show that every solution curve is a closed set.

13. Let γ be a closed orbit of a planar system. Let λ be the period of γ. Let $\{\gamma_n\}$ be a sequence of closed orbits. Suppose the period of γ_n is λ_n. If there are points $X_n \in \gamma_n$ such that $X_n \to X \in \gamma$, prove that $\lambda_n \to \lambda$. (This result can be false for higher dimensional systems. It is true, however, that if $\lambda_n \to \mu$, then μ is an integer multiple of λ.)

14. Consider a system in $\mathbb{R}^2$ having only a finite number of equilibria.

   (a) Show that every limit set is either a closed orbit or the union of equilibrium points and solutions $\phi_t(X)$ such that $\lim_{t\to\infty} \phi_t(X)$ and $\lim_{t\to-\infty} \phi_t(X)$ are these equilibria.

   (b) Show by example (draw a picture) that the number of distinct solution curves in $\omega(X)$ may be infinite.

15. Let X be a *recurrent* point of a planar system, that is, there is a sequence $t_n \to \pm\infty$ such that

$$\phi_{t_n}(X) \to X.$$

   (a) Prove that either X is an equilibrium or X lies on a closed orbit.

   (b) Show by example that there can be a recurrent point for a nonplanar system that is not an equilibrium and does not lie on a closed orbit.

16. Let $X' = F(X)$ and $X' = G(X)$ be planar systems. Suppose that

$$F(X) \cdot G(X) = 0$$

for all $X \in \mathbb{R}^2$. If F has a closed orbit, prove that G has an equilibrium point.

17. Let γ be a closed orbit for a planar system, and let $\mathcal{U}$ be the bounded, open region inside γ. Show that γ is not simultaneously the omega and alpha limit set of points of $\mathcal{U}$. Use this fact and the Poincaré-Bendixson theorem to prove that $\mathcal{U}$ contains an equilibrium that is not a saddle. (*Hint*: Consider the limit sets of points on the stable and unstable curves of saddles.)

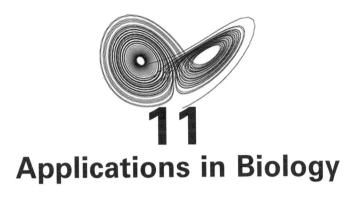

11
Applications in Biology

In this chapter we make use of the techniques developed in the previous few chapters to examine some nonlinear systems that have been used as mathematical models for a variety of biological systems. In Section 11.1 we utilize the preceding results involving nullclines and linearization to describe several biological models involving the spread of communicable diseases. In Section 11.2 we investigate the simplest types of equations that model a predator/prey ecology. A more sophisticated approach is used in Section 11.3 to study the populations of a pair of competing species. Instead of developing explicit formulas for these differential equations, we instead make only qualitative assumptions about the form of the equations. We then derive geometric information about the behavior of solutions of such systems based on these assumptions.

11.1 Infectious Diseases

The spread of infectious diseases such as measles or malaria may be modeled as a nonlinear system of differential equations. The simplest model of this type is the SIR model. Here we divide a given population into three disjoint groups. The population of susceptible individuals is denoted by S, the infected population by I, and the recovered population by R. As usual, each of these is a function of time. We assume for simplicity that the total population is constant, so that $(S + I + R)' = 0$.

In the most basic case we make the assumption that, once an individual has been infected and subsequently has recovered, that individual cannot be reinfected. This is the situation that occurs for such diseases as measles, mumps, and smallpox, among many others. We also assume that the rate of transmission of the disease is proportional to the number of encounters between susceptible and infected individuals. The easiest way to characterize this assumption mathematically is to put $S' = -\beta SI$ for some constant $\beta > 0$. We finally assume that the rate at which infected individuals recover is proportional to the number of infected. The SIR model is then

$$S' = -\beta SI$$
$$I' = \beta SI - \nu I$$
$$R' = \nu I$$

where β and ν are positive parameters.

As stipulated, we have $(S + I + R)' = 0$, so that $S + I + R$ is a constant. This simplifies the system, for if we determine $S(t)$ and $I(t)$, we then derive $R(t)$ for free. Hence it suffices to consider the two-dimensional system

$$S' = -\beta SI$$
$$I' = \beta SI - \nu I.$$

The equilibria for this system are given by the S-axis ($I = 0$). Linearization at $(S, 0)$ yields the matrix

$$\begin{pmatrix} 0 & -\beta S \\ 0 & \beta S - \nu \end{pmatrix},$$

so the eigenvalues are 0 and $\beta S - \nu$. This second eigenvalue is negative if $0 < S < \nu/\beta$ and positive if $S > \nu/\beta$.

The S-nullclines are given by the S and I axes. On the I-axis, we have $I' = -\nu I$, so solutions simply tend to the origin along this line. The I-nullclines are $I = 0$ and the vertical line $S = \nu/\beta$. Hence we have the nullcline diagram depicted in Figure 11.1. From this it appears that, given any initial population (S_0, I_0) with $S_0 > \nu/\beta$ and $I_0 > 0$, the susceptible population decreases monotonically, while the infected population at first rises, but eventually reaches a maximum and then declines to 0.

We can actually prove this analytically, because we can explicitly compute a function that is constant along solution curves. Note that the slope of the vector field is a function of S alone:

$$\frac{I'}{S'} = \frac{\beta SI - \nu I}{-\beta SI} = -1 + \frac{\nu}{\beta S}.$$

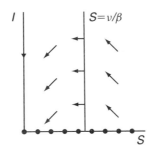

Figure 11.1   The
nullclines and direction
field for the SIR model.

Hence we have

$$\frac{dI}{dS} = \frac{dI/dt}{dS/dt} = -1 + \frac{\nu}{\beta S},$$

which we can immediately integrate to find

$$I = I(S) = -S + \frac{\nu}{\beta} \log S + \text{ constant.}$$

Hence the function $I + S - (\nu/\beta) \log S$ is constant along solution curves. It then follows that there is a unique solution curve connecting each equilibrium point in the interval $\nu/\beta < S < \infty$ to one in the interval $0 < S < \nu/\beta$ as shown in Figure 11.2.

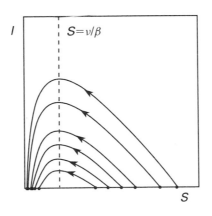

Figure 11.2   The phase portrait
for the SIR system.

A slightly more complicated model for infectious diseases arises when we assume that recovered individuals may lose their immunity and become reinfected with the disease. Examples of this type of disease include malaria and tuberculosis. We assume that the return of recovered individuals to the class S occurs at a rate proportional to the population of recovered individuals. This leads to the SIRS model (where the extra S indicates that recovered individuals may reenter the susceptible group). The system becomes

$$S' = -\beta SI + \mu R$$
$$I' = \beta SI - \nu I$$
$$R' = \nu I - \mu R.$$

Again we see that the total population $S + I + R$ is a constant, which we denote by τ. We may eliminate R from this system by setting $R = \tau - S - I$:

$$S' = -\beta SI + \mu(\tau - S - I)$$
$$I' = \beta SI - \nu I.$$

Here β, μ, ν, and τ are all positive parameters.

Unlike the SIR model, we now have at most two equilibria, one at $(\tau, 0)$ and the other at

$$(S^*, I^*) = \left(\frac{\nu}{\beta}, \frac{\mu(\tau - \frac{\nu}{\beta})}{\nu + \mu} \right).$$

The first equilibrium point corresponds to no disease whatsoever in the population. The second equilibrium point only exists when $\tau \geq \nu/\beta$. When $\tau = \nu/\beta$, we have a bifurcation as the two equilibria coalesce at $(\tau, 0)$. The quantity ν/β is called the *threshold level* for the disease.

The linearized system is given by

$$Y' = \begin{pmatrix} -\beta I - \mu & -\beta S - \mu \\ \beta I & \beta S - \nu \end{pmatrix} Y.$$

At the equilibrium point $(\tau, 0)$, the eigenvalues are $-\mu$ and $\beta\tau - \nu$, so this equilibrium point is a saddle provided that the total population exceeds the threshold level. At the second equilibrium point, a straightforward computation shows that the trace of the matrix is negative, while the determinant is positive. It then follows from the results in Chapter 4 that both eigenvalues have negative real parts, and so this equilibrium point is asymptotically stable.

Biologically, this means that the disease may become established in the community only when the total population exceeds the threshold level. We will only consider this case in what follows.

Note that the SIRS system is only of interest in the region given by $I, S \geq 0$ and $S + I \leq \tau$. Denote this triangular region by Δ (of course!). Note that the I-axis is no longer invariant, while on the S-axis, solutions increase up to the equilibrium at $(\tau, 0)$.

Proposition.   *The region Δ is positively invariant.*

Proof: We check the direction of the vector field along the boundary of Δ. The field is tangent to the boundary along the lower edge $I = 0$ as well as at $(0, \tau)$. Along $S = 0$ we have $S' = \mu(\tau - I) > 0$, so the vector field points inward for $0 < I < \tau$. Along the hypoteneuse, if $0 < S \leq \nu/\beta$, we have $S' = -\beta SI < 0$ and $I' = I(\beta S - \nu) \leq 0$ so the vector field points inward. When $\nu/\beta < S < \tau$ we have

$$-1 < \frac{I'}{S'} = -1 + \frac{\nu}{\beta S} \leq 0$$

so again the vector field points inward. This completes the proof.   ∎

The I-nullclines are given as in the SIR model by $I = 0$ and $S = \nu/\beta$. The S-nullcline is given by the graph of the function

$$I = I(S) = \frac{\mu(\tau - S)}{\beta S + \mu}.$$

A calculus student will compute that $I'(S) < 0$ and $I''(S) > 0$ when $0 \leq S < \tau$. So this nullcline is the graph of a decreasing and concave up function that passes through both $(\tau, 0)$ and $(0, \tau)$, as displayed in Figure 11.3. Note that in this phase portrait, all solutions appear to tend to the equilibrium point (S^*, I^*); the proportion of infected to susceptible individuals tends to a "steady state." To prove this, however, one would need to eliminate the possibility of closed orbits encircling the equilibrium point for a given set of parameters β, μ, ν, and τ.

11.2 Predator/Prey Systems

We next consider a pair of species, one of which consists of predators whose population is denoted by y and the other its prey with population x. We assume that the prey population is the total food supply for the predators. We also

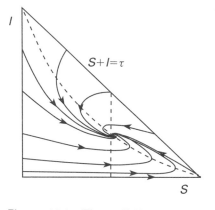

Figure 11.3	The nullclines and phase portrait in Δ for the SIRS system. Here $\beta = \nu = \mu = 1$ and $\tau = 2$.

assume that, in the absence of predators, the prey population grows at a rate proportional to the current population. That is, as in Chapter 1, when $y = 0$ we have $x' = ax$ where $a > 0$. So in this case $x(t) = x_0 \exp(at)$. When predators are present, we assume that the prey population decreases at a rate proportional to the number of predator/prey encounters. As in the previous section, one simple model for this is bxy where $b > 0$. So the differential equation for the prey population is $x' = ax - bxy$.

For the predator population, we make more or less the opposite assumptions. In the absence of prey, the predator population declines at a rate proportional to the current population. So when $x = 0$ we have $y' = -cy$ with $c > 0$, and thus $y(t) = y_0 \exp(-ct)$. The predator species becomes extinct in this case. When there are prey in the environment, we assume that the predator population increases at a rate proportional to the predator/prey meetings, or dxy. We do not at this stage assume anything about overcrowding. Thus our simplified predator/prey system (also called the Volterra-Lotka system) is

$$x' = ax - bxy = x(a - by)$$
$$y' = -cy + dxy = y(-c + dx)$$

where the parameters a, b, c, and d are all assumed to be positive. Since we are dealing with populations, we only consider $x, y \geq 0$.

As usual, our first job is to locate the equilibrium points. These occur at the origin and at $(x, y) = (c/d, a/b)$. The linearized system is

$$X' = \begin{pmatrix} a - by & -bx \\ dy & -c + dx \end{pmatrix} X,$$

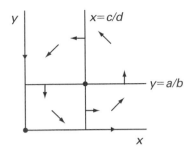

Figure 11.4   The nullclines
and direction field for the
predator/prey system.

so when $x = y = 0$ we have a saddle with eigenvalues a and $-c$. We know the
stable and unstable curves: They are the y- and x-axes, respectively.

At the other equilibrium point $(c/d, a/b)$, the eigenvalues are pure imaginary
$\pm i\sqrt{ac}$, and so we cannot conclude anything at this stage about stability of
this equilibrium point.

We next sketch the nullclines for this system. The x-nullclines are given by
the straight lines $x = 0$ and $y = a/b$, whereas the y-nullclines are $y = 0$ and
$x = c/d$. The nonzero nullcline lines separate the region $x, y > 0$ into four basic
regions in which the vector field points as indicated in Figure 11.4. Hence the
solutions wind in the counterclockwise direction about the equilibrium point.

From this, we cannot determine the precise behavior of solutions: They
could possibly spiral in toward the equilibrium point, spiral toward a limit
cycle, spiral out toward "infinity" and the coordinate axes, or else lie on closed
orbits. To make this determination, we search for a Liapunov function L.
Employing the trick of *separation of variables*, we look for a function of the
form

$$L(x, y) = F(x) + G(y).$$

Recall that $\dot{L}$ denotes the time derivative of L along solutions. We compute

$$\dot{L}(x, y) = \frac{d}{dt} L(x(t), y(t))$$

$$= \frac{dF}{dx} x' + \frac{dG}{dy} y'.$$

Hence

$$\dot{L}(x, y) = x \frac{dF}{dx}(a - by) + y \frac{dG}{dy}(-c + dx).$$

We obtain $\dot{L} \equiv 0$ provided

$$\frac{x \, dF/dx}{dx - c} \equiv \frac{y \, dG/dy}{by - a}.$$

Since x and y are independent variables, this is possible if and only if

$$\frac{x \, dF/dx}{dx - c} = \frac{y \, dG/dy}{by - a} = \text{constant}.$$

Setting the constant equal to 1, we obtain

$$\frac{dF}{dx} = d - \frac{c}{x},$$
$$\frac{dG}{dy} = b - \frac{a}{y}.$$

Integrating, we find

$$F(x) = dx - c \log x,$$
$$G(y) = by - a \log y.$$

Thus the function

$$L(x, y) = dx - c \log x + by - a \log y$$

is constant on solution curves of the system when $x, y > 0$.

By considering the signs of $\partial L/\partial x$ and $\partial L/\partial y$ it is easy to see that the equilibrium point $Z = (c/d, a/b)$ is an absolute minimum for L. It follows that L [or, more precisely, $L - L(Z)$] is a Liapunov function for the system. Therefore Z is a stable equilibrium.

We note next that there are no limit cycles; this follows from Corollary 5 in Section 10.6 because L is not constant on any open set. We now prove the following theorem.

Theorem. *Every solution of the predator/prey system is a closed orbit (except the equilibrium point Z and the coordinate axes).*

Proof: Consider the solution through $W \neq Z$, where W does not lie on the x- or y-axis. This solution spirals around Z, crossing each nullcline infinitely often. Thus there is a doubly infinite sequence $\cdots < t_{-1} < t_0 < t_1 < \cdots$ such that $\phi_{t_n}(W)$ is on the line $x = c/d$, and $t_n \to \pm\infty$ as $n \to \pm\infty$. If W is not on a closed orbit, the points $\phi_{t_n}(W)$ are monotone along the line

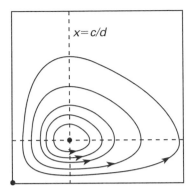

Figure 11.5 The nullclines and phase portrait for the predator/prey system.

$x = c/d$, as discussed in the previous chapter. Since there are no limit cycles, either $\phi_{t_n}(W) \to Z$ as $n \to \infty$ or $\phi_{t_n}(W) \to Z$ as $n \to -\infty$. Since L is constant along the solution through W, this implies that $L(W) = L(Z)$. But this contradicts minimality of $L(Z)$. This completes the proof. ∎

The phase portrait for this predator/prey system is displayed in Figure 11.5. We conclude that, for any given initial populations $(x(0), y(0))$ with $x(0) \neq 0$ and $y(0) \neq 0$, other than Z, the populations of predator and prey oscillate cyclically. No matter what the populations of prey and predator are, neither species will die out, nor will its population grow indefinitely.

Now let us introduce overcrowding into the prey equation. As in the logistic model in Chapter 1, the equations for prey, in the absence of predators, may be written in the form

$$x' = ax - \lambda x^2.$$

We also assume that the predator population obeys a similar equation

$$y' = -cy - \mu y^2$$

when $x = 0$. Incorporating the assumptions above yields the *predator/prey equations for species with limited growth*:

$$x' = x(a - by - \lambda x)$$
$$y' = y(-c + dx - \mu y).$$

As before, the parameters a, b, c, d as well as λ and μ are all positive. When $y = 0$, we have the logistic equation $x' = x(a - \lambda x)$, which yields equilibria at the origin and at $(a/\lambda, 0)$. As we saw in Chapter 1, all nonzero solutions on the x-axis tend to a/λ.

When $x = 0$, the equation for y is $y' = -cy - \mu y^2$. Since $y' < 0$ when $y > 0$, it follows that all solutions on this axis tend to the origin. Thus we confine attention to the upper-right quadrant $\mathcal{Q}$ where $x, y > 0$.

The nullclines are given by the x- and y-axes, together with the lines

$$L : \quad a - by - \lambda x = 0$$
$$M : \quad -c + dx - \mu y = 0.$$

Along the lines L and M, we have $x' = 0$ and $y' = 0$, respectively. There are two possibilities, according to whether these lines intersect in $\mathcal{Q}$ or not.

We first consider the case where the two lines do not meet in $\mathcal{Q}$. In this case we have the nullcline configuration depicted in Figure 11.6. All solutions to the right of M head upward and to the left until they meet M; between the lines L and M solutions now head downward and to the left. Thus they either meet L or tend directly to the equilibrium point at $(a/\lambda, 0)$. If solutions cross L, they then head right and downward, but they cannot cross L again. Thus they too tend to $(a/\lambda, 0)$. Thus all solutions in $\mathcal{Q}$ tend to this equilibrium point. We conclude that, in this case, the predator population becomes extinct and the prey population approaches its limiting value of a/λ.

We may interpret the behavior of solutions near the nullclines as follows. Since both x' and y' are never both positive, it is impossible for both prey

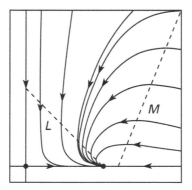

Figure 11.6 The nullclines and phase portrait for a predator/prey system with limited growth when the nullclines do not meet in $\mathcal{Q}$.

and predators to increase at the same time. If the prey population is above its limiting value, it must decrease. After a while the lack of prey causes the predator population to begin to decrease (when the solution crosses M). After that point the prey population can never increase past a/λ, and so the predator population continues to decrease. If the solution crosses L, the prey population increases again (but not past a/λ), while the predators continue to die off. In the limit the predators disappear and the prey population stabilizes at a/λ.

Suppose now that L and M cross at a point $Z = (x_0, y_0)$ in the quadrant $\mathcal{Q}$; of course, Z is an equilibrium. The linearization of the vector field at Z is

$$X' = \begin{pmatrix} -\lambda x_0 & -bx_0 \\ dy_0 & -\mu y_0 \end{pmatrix} X.$$

The characteristic polynomial has trace given by $-\lambda x_0 - \mu y_0 < 0$ and determinant $(bd + \lambda\mu)x_0 y_0 > 0$. From the trace-determinant plane of Chapter 4, we see that Z has eigenvalues that are either both negative or both complex with negative real parts. Hence Z is asymptotically stable.

Note that, in addition to the equilibria at Z and $(0,0)$, there is still an equilibrium at $(a/\lambda, 0)$. Linearization shows that this equilibrium is a saddle; its stable curve lies on the x-axis. See Figure 11.7.

It is not easy to determine the basin of Z, nor do we know whether there are any limit cycles. Nevertheless we can obtain some information. The line L meets the x-axis at $(a/\lambda, 0)$ and the y-axis at $(0, a/b)$. Let Γ be a rectangle whose corners are $(0,0)$, $(p,0)$, $(0,q)$, and (p,q) with $p > a/\lambda$, $q > a/b$, and the point (p, q) lying in M. Every solution at a boundary point of Γ either

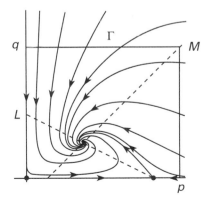

Figure 11.7 The nullclines and phase portrait for a predator/prey system with limited growth when the nullclines do meet in $\mathcal{Q}$.

enters Γ or is part of the boundary. Therefore Γ is positively invariant. Every point in Q is contained in such a rectangle.

By the Poincaré-Bendixson theorem, the ω-limit set of any point (x, y) in Γ, with $x, y > 0$, must be a limit cycle or contain one of the three equilibria $(0, 0)$, Z, or $(a/\lambda, 0)$. We rule out $(0, 0)$ and $(a/\lambda, 0)$ by noting that these equilibria are saddles whose stable curves lie on the x- or y-axes. Therefore $\omega(x, y)$ is either Z or a limit cycle in Γ. By Corollary 4 of the Poincaré-Bendixson theorem any limit cycle must surround Z.

We observe further that any such rectangle Γ contains *all* limit cycles, because a limit cycle (like any solution) must enter Γ, and Γ is positively invariant. Fixing (p, q) as above, it follows that for any initial values $(x(0), y(0))$, there exists $t_0 > 0$ such that $x(t) < p$, $y(t) < q$ if $t \geq t_0$. We conclude that in the long run, a solution either approaches Z or else spirals down to a limit cycle.

From a practical standpoint a solution that tends toward Z is indistinguishable from Z after a certain time. Likewise, a solution that approaches a limit cycle γ can be identified with γ after it is sufficiently close. We conclude that any population of predators and prey that obeys these equations eventually settles down to either a constant or periodic population. Furthermore, there are absolute upper bounds that no population can exceed in the long run, no matter what the initial populations are.

11.3 Competitive Species

We consider now two species that compete for a common food supply. Instead of analyzing specific equations, we follow a different procedure: We consider a large class of equations about which we assume only a few qualitative features. In this way considerable generality is gained, and little is lost because specific equations can be very difficult to analyze.

Let x and y denote the populations of the two species. The equations of growth of the two populations may be written in the form

$$x' = M(x, y)x$$
$$y' = N(x, y)y$$

where the growth rates M and N are functions of both variables. As usual, we assume that x and y are nonnegative. So the x-nullclines are given by $x = 0$ and $M(x, y) = 0$ and the y-nullclines are $y = 0$ and $N(x, y) = 0$. We make the following assumptions on M and N:

1. Because the species compete for the same resources, if the population of either species increases, then the growth rate of the other goes down.

Hence

$$\frac{\partial M}{\partial y} < 0 \quad \text{and} \quad \frac{\partial N}{\partial x} < 0.$$

2. If either population is very large, both populations decrease. Hence there exists $K > 0$ such that

$$M(x, y) < 0 \quad \text{and} \quad N(x, y) < 0 \quad \text{if } x \geq K \text{ or } y \geq K.$$

3. In the absence of either species, the other has a positive growth rate up to a certain population and a negative growth rate beyond it. Therefore there are constants $a, b > 0$ such that

$$M(x, 0) > 0 \text{ for } x < a \quad \text{and} \quad M(x, 0) < 0 \text{ for } x > a,$$
$$N(0, y) > 0 \text{ for } y < b \quad \text{and} \quad N(0, y) < 0 \text{ for } y > b.$$

By conditions (1) and (3) each vertical line $\{x\} \times \mathbb{R}$ meets the set $\mu = M^{-1}(0)$ exactly once if $0 \leq x \leq a$ and not at all if $x > a$. By condition (1) and the implicit function theorem, μ is the graph of a nonnegative function $f: [0, a] \to \mathbb{R}$ such that $f^{-1}(0) = a$. Below the curve μ, M is positive and above it, M is negative. In the same way the set $\nu = N^{-1}(0)$ is a smooth curve of the form

$$\{(x, y) \mid x = g(y)\},$$

where $g: [0, b] \to \mathbb{R}$ is a nonnegative function with $g^{-1}(0) = b$. The function N is positive to the left of ν and negative to the right.

Suppose first that μ and ν do not intersect and that μ is below ν. Then the phase portrait can be determined immediately from the nullclines. The equilibria are $(0, 0)$, $(a, 0)$, and $(0, b)$. The origin is a source, while $(a, 0)$ is a saddle (assuming that $(\partial M/\partial x)(a, 0) < 0$). The equilibrium at $(0, b)$ is a sink [again assuming that $(\partial N/\partial y)(0, b) < 0$]. All solutions with $y_0 > 0$ tend to the asymptotically stable equilibrium $(0, b)$ with the exception of solutions on the x-axis. See Figure 11.8. In the case where μ lies above ν, the situation is reversed, and all solutions with $x_0 > 0$ tend to the sink that now appears at $(a, 0)$.

Suppose now that μ and ν intersect. We make the assumption that $\mu \cap \nu$ is a finite set, and at each intersection point, μ and ν *cross transversely*, that is, they have distinct tangent lines at the intersection points. This assumption may be eliminated; we make it only to simplify the process of determining the flow.

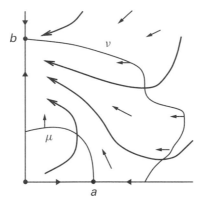

Figure 11.8   The phase portrait
when μ and ν do not meet.

The nullclines μ and ν and the coordinate axes bound a finite number of connected open sets in the upper-right quadrant: These are the basic regions where $x' \neq 0$ and $y' \neq 0$. They are of four types:

$$ A: \quad x' > 0, \; y' > 0, \qquad B: \quad x' < 0, \; y' > 0; $$
$$ C: \quad x' < 0, \; y' < 0, \qquad D: \quad x' > 0, \; y' < 0. $$

Equivalently, these are the regions where the vector field points northeast, northwest, southwest, or southeast, respectively. Some of these regions are indicated in Figure 11.9. The boundary $\partial \mathcal{R}$ of a basic region $\mathcal{R}$ is made up of points of the following types: points of $\mu \cap \nu$, called *vertices*; points on μ or ν but not on both nor on the coordinate axes, called *ordinary* boundary points; and points on the axes.

A vertex is an equilibrium; the other equilibria lie on the axes at $(0,0)$, $(a,0)$, and $(0,b)$. At an ordinary boundary point $Z \in \partial \mathcal{R}$, the vector field is either vertical (if $Z \in \mu$) or horizontal (if $Z \in \nu$). This vector points either into or out of $\mathcal{R}$ since μ has no vertical tangents and ν has no horizontal tangents. We call Z an *inward* or *outward* point of $\partial \mathcal{R}$, accordingly. Note that, in Figure 11.9, the vector field either points inward at all ordinary points on the boundary of a basic region, or else it points outward at all such points. This is no accident, for we have:

Proposition.   *Let $\mathcal{R}$ be a basic region for the competitive species model. Then the ordinary boundary points of $\mathcal{R}$ are either all inward or all outward.*

Proof: There are only two ways in which the curves μ and ν can intersect at a vertex P. As y increases along ν, the curve ν may either pass from below μ

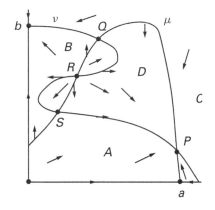

Figure 11.9   The basic regions
when the nullclines μ and ν
intersect.

to above μ, or from above to below μ. These two scenarios are illustrated in
Figures 11.10a and b. There are no other possibilities since we have assumed
that these curves cross transversely.

Since $x' > 0$ below μ and $x' < 0$ above μ, and since $y' > 0$ to the left of ν
and $y' < 0$ to the right, we therefore have the following configurations for the
vector field in these two cases. See Figure 11.11.

In each case we see that the vector field points inward in two opposite basic
regions abutting P, and outward in the other two basic regions.

If we now move along μ or ν to the next vertex along this curve, we see that
adjacent basic regions must maintain their inward or outward configuration.
Therefore, at all ordinary boundary points on each basic region, the vector
field either points outward or points inward, as required.   ∎

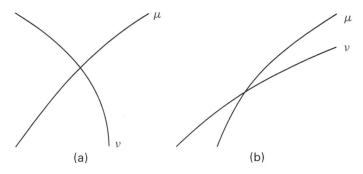

Figure 11.10   In (a), ν passes from below μ to above μ as y
increases. The situation is reversed in (b).

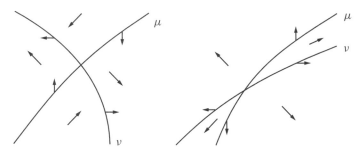

Figure 11.11   Configurations of the vector field near vertices.

As a consequence of the proposition, it follows that each basic region and its closure is either positively or negatively invariant. What are the possible ω-limit points of this system? There are no closed orbits. A closed orbit must be contained in a basic region, but this is impossible since $x(t)$ and $y(t)$ are monotone along any solution curve in a basic region. Therefore all ω-limit points are equilibria.

We note also that each solution is defined for all $t \geq 0$, because any point lies in a large rectangle Γ with corners at $(0, 0)$, $(x_0, 0)$, $(0, y_0)$, and (x_0, y_0) with $x_0 > a$ and $y_0 > b$; such a rectangle is positively invariant. See Figure 11.12. Thus we have shown:

Theorem. *The flow ϕ_t of the competitive species system has the following property: For all points (x, y), with $x \geq 0, y \geq 0$, the limit*

$$\lim_{t \to \infty} \phi_t(x, y)$$

exists and is one of a finite number of equilibria.   ■

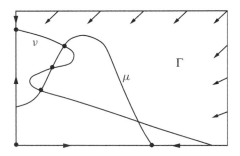

Figure 11.12   All solutions must enter and then remain in Γ.

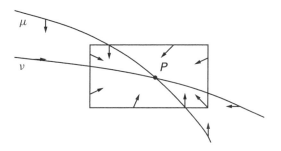

Figure 11.13 This configuration of μ and ν
leads to an asymptotically stable equilibrium
point.

We conclude that the populations of two competing species always tend to one of a finite number of limiting populations.

Examining the equilibria for stability, one finds the following results. A vertex where μ and ν each have negative slope, but μ is steeper, is asymptotically stable. See Figure 11.13. One sees this by drawing a small rectangle with sides parallel to the axes around the equilibrium, putting one corner in each of the four adjacent basic regions. Such a rectangle is positively invariant; since it can be arbitrarily small, the equilibrium is asymptotically stable.

This may also be seen as follows. We have

$$\text{slope of } \mu = -\frac{M_x}{M_y} < \text{slope of } \nu = -\frac{N_x}{N_y} < 0,$$

where $M_x = \partial M / \partial x$, $M_y = \partial M / \partial y$, and so on, at the equilibrium. Now recall that $M_y < 0$ and $N_x < 0$. Therefore, at the equilibrium point, we also have $M_x < 0$ and $N_y < 0$. Linearization at the equilibrium point yields the matrix

$$\begin{pmatrix} xM_x & xM_y \\ yN_x & yN_y \end{pmatrix}.$$

The trace of this matrix is $xM_x + yN_y < 0$ while the determinant is $xy(M_x N_y - M_y N_x) > 0$. Thus the eigenvalues have negative real parts, and so we have a sink.

A case-by-case study of the different ways μ and ν can cross shows that the only other asymptotically stable equilibrium in this model is $(0, b)$ when $(0, b)$ is above μ, or $(a, 0)$ when $(a, 0)$ is to the right of ν. All other equilibria are unstable. There must be at least one asymptotically stable equilibrium. If $(0, b)$ is not one, then it lies under μ; and if $(a, 0)$ is not one, it lies over μ. In that case μ and ν cross, and the first crossing to the left of $(a, 0)$ is asymptotically stable.

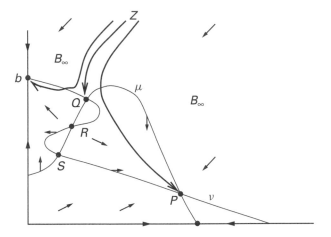

Figure 11.14   Note that solutions on either side of the point Z in the stable curve of Q have very different fates.

For example, this analysis tells us that, in Figure 11.14, only P and $(0, b)$ are asymptotically stable; all other equilibria are unstable. In particular, assuming that the equilibrium Q in Figure 11.14 is hyperbolic, then it must be a saddle because certain nearby solutions tend toward it, while others tend away. The point Z lies on one branch of the stable curve through Q. All points in the region denoted B_∞ to the left of Z tend to the equilibrium at $(0, b)$, while points to the right go to P. Thus as we move across the branch of the stable curve containing Z, the limiting behavior of solutions changes radically. Since solutions just to the right of Z tend to the equilibrium point P, it follows that the populations in this case tend to stabilize. On the other hand, just to the left of Z, solutions tend to an equilibrium point where $x = 0$. Thus in this case, one of the species becomes extinct. A small change in initial conditions has led to a dramatic change in the fate of populations. Ecologically, this small change could have been caused by the introduction of a new pesticide, the importation of additional members of one of the species, a forest fire, or the like. Mathematically, this event is a jump from the basin of P to that of $(0, b)$.

11.4 Exploration: Competition and Harvesting

In this exploration we will investigate the competitive species model where we allow either harvesting (emigration) or immigration of one of the species. We consider the system

$$x' = x(1 - ax - y)$$
$$y' = y(b - x - y) + h.$$

Here a, b, and h are parameters. We assume that $a, b > 0$. If $h < 0$, then we are harvesting species y at a constant rate, whereas if $h > 0$, we add to the population y at a constant rate. The goal is to understand this system completely for all possible values of these parameters. As usual, we only consider the regime where $x, y \geq 0$. If $y(t) < 0$ for any $t > 0$, then we consider this species to have become extinct.

1. First assume that $h = 0$. Give a complete synopsis of the behavior of this system by plotting the different behaviors you find in the a, b parameter plane.
2. Identify the points or curves in the ab–plane where bifurcations occur when $h = 0$ and describe them.
3. Now let $h < 0$. Describe the ab–parameter plane for various (fixed) h-values.
4. Repeat the previous exploration for $h > 0$.
5. Describe the full three-dimensional parameter space using pictures, flip books, 3D models, movies, or whatever you find most appropriate.

EXERCISES

1. For the SIRS model, prove that all solutions in the triangular region Δ tend to the equilibrium point $(\tau, 0)$ when the total population does not exceed the threshold level for the disease.
2. Sketch the phase plane for the following variant of the predator/prey system:

$$x' = x(1 - x) - xy$$
$$y' = y\left(1 - \frac{y}{x}\right).$$

3. A modification of the predator/prey equations is given by

$$x' = x(1 - x) - \frac{axy}{x + 1}$$
$$y' = y(1 - y)$$

where $a > 0$ is a parameter.

(a) Find all equilibrium points and classify them.

(b) Sketch the nullclines and the phase portraits for different values of a.

(c) Describe any bifurcations that occur as a varies.

4. Another modification of the predator/prey equations is given by

$$x' = x(1 - x) - \frac{xy}{x + b}$$

$$y' = y(1 - y)$$

where $b > 0$ is a parameter.

(a) Find all equilibrium points and classify them.

(b) Sketch the nullclines and the phase portraits for different values of b.

(c) Describe any bifurcations that occur as b varies.

5. The equations

$$x' = x(2 - x - y),$$

$$y' = y(3 - 2x - y)$$

satisfy conditions (1) through (3) in Section 11.3 for competing species. Determine the phase portrait for this system. Explain why these equations make it mathematically possible, but extremely unlikely, for both species to survive.

6. Consider the competing species model

$$x' = x(a - x - ay)$$

$$y' = y(b - bx - y)$$

where the parameters a and b are positive.

(a) Find all equilibrium points for this system and determine their stability type. These types will, of course, depend on a and b.

(b) Use the nullclines to determine the various phase portraits that arise for different choices of a and b.

(c) Determine the values of a and b for which there is a bifurcation in this system and describe the bifurcation that occurs.

(d) Record your findings by drawing a picture of the ab–plane and indicating in each open region of this plane the qualitative structure of the corresponding phase portraits.

7. Two species x, y are in *symbiosis* if an increase of either population leads to an increase in the growth rate of the other. Thus we assume

$$x' = M(x, y)x$$
$$y' = N(x, y)y$$

with

$$\frac{\partial M}{\partial y} > 0 \quad \text{and} \quad \frac{\partial N}{\partial x} > 0$$

and $x, y \geq 0$. We also suppose that the total food supply is limited; hence for some $A > 0, B > 0$ we have

$$M(x, y) < 0 \quad \text{if } x > A,$$
$$N(x, y) < 0 \quad \text{if } y > B.$$

If both populations are very small, they both increase; hence

$$M(0, 0) > 0 \quad \text{and} \quad N(0, 0) > 0.$$

Assuming that the intersections of the curves $M^{-1}(0), N^{-1}(0)$ are finite, and that all are transverse, show the following:

(a) Every solution tends to an equilibrium in the region $0 < x < A$, $0 < y < B$.
(b) There are no sources.
(c) There is at least one sink.
(d) If $\partial M/\partial x < 0$ and $\partial N/\partial y < 0$, there is a unique sink Z, and Z is the ω-limit set for all (x, y) with $x > 0, y > 0$.

8. Give a system of differential equations for a pair of mutually destructive species. Then prove that, under plausible hypotheses, two mutually destructive species cannot coexist in the long run.

9. Let y and x denote predator and prey populations. Let

$$x' = M(x, y)x$$
$$y' = N(x, y)y$$

where M and N satisfy the following conditions.

(a) If there are not enough prey, the predators decrease. Hence for some $b > 0$

$$N(x, y) < 0 \quad \text{if } x < b.$$

(b) An increase in the prey improves the predator growth rate; hence $\partial N/\partial x > 0$.

(c) In the absence of predators a small prey population will increase; hence $M(0,0) > 0$.

(d) Beyond a certain size, the prey population must decrease; hence there exists $A > 0$ with $M(x, y) < 0$ if $x > A$.

(e) Any increase in predators decreases the rate of growth of prey; hence $\partial M/\partial y < 0$.

(f) The two curves $M^{-1}(0), N^{-1}(0)$ intersect transversely and at only a finite number of points.

Show that if there is some (u, v) with $M(u, v) > 0$ and $N(u, v) > 0$ then there is either an asymptotically stable equilibrium or an ω-limit cycle. Moreover, show that, if the number of limit cycles is finite and positive, one of them must have orbits spiraling toward it from both sides.

10. Consider the following modification of the predator/prey equations:

$$x' = x(1 - x) - \frac{axy}{x + c}$$

$$y' = by\left(1 - \frac{y}{x}\right)$$

where $a, b,$ and c are positive constants. Determine the region in the parameter space for which this system has a stable equilibrium with both $x, y \neq 0$. Prove that, if the equilibrium point is unstable, this system has a stable limit cycle.

12

Applications in Circuit Theory

In this chapter we first present a simple but very basic example of an electrical circuit and then derive the differential equations governing this circuit. Certain special cases of these equations are analyzed using the techniques developed in Chapters 8 through 10 in the next two sections; these are the classical equations of Lienard and van der Pol. In particular, the van der Pol equation could perhaps be regarded as one of the fundamental examples of a nonlinear ordinary differential equation. It possesses an oscillation or periodic solution that is a periodic attractor. Every nontrivial solution tends to this periodic solution; no linear system has this property. Whereas asymptotically stable equilibria sometimes imply death in a system, attracting oscillators imply life. We give an example in Section 12.4 of a continuous transition from one such situation to the other.

12.1 An *RLC* Circuit

In this section, we present our first example of an electrical circuit. This circuit is the simple but fundamental series *RLC* circuit displayed in Figure 12.1. We begin by explaining what this diagram means in mathematical terms. The circuit has three *branches*, one resistor marked by R, one inductor marked by L, and one capacitor marked by C. We think of a branch of this circuit as a

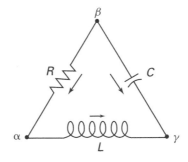

Figure 12.1   An *RLC* circuit.

certain electrical device with two terminals. For example, in this circuit, the branch *R* has terminals α and β and all of the terminals are wired together to form the points or *nodes* α, β, and γ.

In the circuit there is a current flowing through each branch that is measured by a real number. More precisely, the currents in the circuit are given by the three real numbers i_R, i_L, and i_C where i_R measures the current through the resistor, and so on. Current in a branch is analogous to water flowing in a pipe; the corresponding measure for water would be the amount flowing in unit time, or better, the rate at which water passes by a fixed point in the pipe. The arrows in the diagram that orient the branches tell us how to read which way the current (read water!) is flowing; for example, if i_R is positive, then according to the arrow, current flows through the resistor from β to α (the choice of the arrows is made once and for all at the start).

The state of the currents at a given time in the circuit is thus represented by a point $i = (i_R, i_L, i_C) \in \mathbb{R}^3$. But *Kirchhoff's current law* (KCL) says that in reality there is a strong restriction on which i can occur. KCL asserts that the total current flowing into a node is equal to the total current flowing out of that node. (Think of the water analogy to make this plausible.) For our circuit this is equivalent to

$$\text{KCL:} \quad i_R = i_L = -i_C.$$

This defines the one-dimensional subspace K_1 of $\mathbb{R}^3$ of *physical current states*.

Our choice of orientation of the capacitor branch may seem unnatural. In fact, these orientations are arbitrary; in this example they were chosen so that the equations eventually obtained relate most directly to the history of the subject.

The *state* of the circuit is characterized by the current $i = (i_R, i_L, i_C)$ together with the voltage (or, more precisely, the voltage drop) across each branch. These voltages are denoted by v_R, v_L, and v_C for the resistor branch, inductor branch, and capacitor branch, respectively. In the water analogy one thinks of the voltage drop as the difference in pressures at the two ends of a pipe. To measure voltage, one places a voltmeter (imagine a water pressure meter) at each of the nodes α, β, and γ that reads $V(\alpha)$ at α, and so on. Then v_R is the difference in the reading at α and β

$$V(\beta) - V(\alpha) = v_R.$$

The direction of the arrow tells us that $v_R = V(\beta) - V(\alpha)$ rather than $V(\alpha) - V(\beta)$.

An *unrestricted voltage state* of the circuit is then a point $v = (v_R, v_L, v_C) \in \mathbb{R}^3$. This time the *Kirchhoff voltage law* puts a physical restriction on v:

$$\text{KVL}: \quad v_R + v_L - v_C = 0.$$

This defines a two-dimensional subspace K_2 of $\mathbb{R}^3$. KVL follows immediately from our definition of the v_R, v_L, and v_C in terms of voltmeters; that is,

$$v_R + v_L - v_C = (V(\beta) - V(\alpha)) + (V(\alpha) - V(\gamma)) - (V(\beta) - V(\gamma)) = 0.$$

The product space $\mathbb{R}^3 \times \mathbb{R}^3$ is called the *state space* for the circuit. Those states $(i, v) \in \mathbb{R}^3 \times \mathbb{R}^3$ satisfying Kirchhoff's laws form a three-dimensional subspace of the state space.

Now we give a mathematical definition of the three kinds of electrical devices in the circuit. First consider the resistor element. A resistor in the R branch imposes a "functional relationship" on i_R and v_R. We take in our example this relationship to be defined by a function $f: \mathbb{R} \to \mathbb{R}$, so that $v_R = f(i_R)$. If R is a conventional linear resistor, then f is linear and so $f(i_R) = k i_R$. This relation is known as *Ohm's law*. Nonlinear functions yield a generalized Ohm's law. The graph of f is called the *characteristic* of the resistor. A couple of examples of characteristics are given in Figure 12.2. (A characteristic like that in Figure 12.2b occurs in a "tunnel diode.")

A *physical state* $(i, v) \in \mathbb{R}^3 \times \mathbb{R}^3$ is a point that satisfies KCL, KVL, and also $f(i_R) = v_R$. These conditions define the *set of physical states* $\Sigma \subset \mathbb{R}^3 \times \mathbb{R}^3$. Thus Σ is the set of points $(i_R, i_L, i_C, v_R, v_L, v_C)$ in $\mathbb{R}^3 \times \mathbb{R}^3$ that satisfy:

1. $i_R = i_L = -i_C$ (KCL);
2. $v_R + v_L - v_C = 0$ (KVL);
3. $f(i_R) = v_R$ (generalized Ohm's law).

Now we turn to the differential equations governing the circuit. The inductor (which we think of as a coil; it is hard to find a water analogy) specifies that

$$L\frac{di_L(t)}{dt} = v_L(t) \quad \text{(Faraday's law)},$$

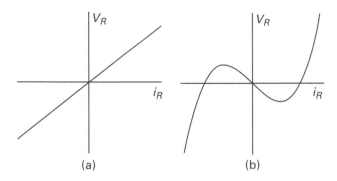

Figure 12.2   Several possible characteristics for a
resistor.

where L is a positive constant called the *inductance*. On the other hand, the
capacitor (which may be thought of as two metal plates separated by some
insulator; in the water model it is a tank) imposes the condition

$$C\frac{dv_C(t)}{dt} = i_C(t)$$

where C is a positive constant called the *capacitance*.

Let's summarize the development so far: a state of the circuit is given by the
six numbers $(i_R, i_L, i_C, v_R, v_L, v_C)$, that is, a point in the state space $\mathbb{R}^3 \times \mathbb{R}^3$.
These numbers are subject to three restrictions: Kirchhoff's current law,
Kirchhoff's voltage law, and the resistor characteristic or generalized Ohm's
law. Therefore the set of physical states is a certain subset $\Sigma \subset \mathbb{R}^3 \times \mathbb{R}^3$.
The way a state changes in time is determined by the two previous differential
equations.

Next, we simplify the set of physical states Σ by observing that i_L and v_C
determine the other four coordinates. This follows since $i_R = i_L$ and $i_C = -i_L$
by KCL, $v_R = f(i_R) = f(i_L)$ by the generalized Ohm's law, and $v_L = v_C - v_R =$
$v_C - f(i_L)$ by KVL. Therefore we can use $\mathbb{R}^2$ as the state space, with coordinates
given by (i_L, v_C). Formally, we define a map $\pi : \mathbb{R}^3 \times \mathbb{R}^3 \to \mathbb{R}^2$, which sends
$(i, v) \in \mathbb{R}^3 \times \mathbb{R}^3$ to (i_L, v_C). Then we set $\pi_0 = \pi \mid \Sigma$, the restriction of π to
Σ. The map $\pi_0 : \Sigma \to \mathbb{R}^2$ is one to one and onto; its inverse is given by the
map $\psi : \mathbb{R}^2 \to \Sigma$, where

$$\psi(i_L, v_C) = (i_L, i_L, -i_L, f(i_L), v_C - f(i_L), v_C).$$

It is easy to check that $\psi(i_L, v_C)$ satisfies KCL, KVL, and the generalized Ohm's
law, so ψ does map $\mathbb{R}^2$ into Σ. It is also easy to see that π_0 and ψ are inverse
to each other.

We therefore adopt $\mathbb{R}^2$ as our state space. The differential equations
governing the change of state must be rewritten in terms of our new

coordinates (i_L, v_C):

$$L\frac{di_L}{dt} = v_L = v_C - f(i_L)$$

$$C\frac{dv_C}{dt} = i_C = -i_L.$$

For simplicity, and since this is only an example, we set $L = C = 1$. If we write $x = i_L$ and $y = v_C$, we then have a system of differential equations in the plane of the form

$$\frac{dx}{dt} = y - f(x)$$

$$\frac{dy}{dt} = -x.$$

This is one form of the equation known as the *Lienard equation*. We analyze this system in the following section.

12.2 The Lienard Equation

In this section we begin the study of the phase portrait of the Lienard system from the circuit of the previous section, namely:

$$\frac{dx}{dt} = y - f(x)$$

$$\frac{dy}{dt} = -x.$$

In the special case where $f(x) = x^3 - x$, this system is called the *van der Pol equation*.

First consider the simplest case where f is linear. Suppose $f(x) = kx$, where $k > 0$. Then the Lienard system takes the form $Y' = AY$ where

$$A = \begin{pmatrix} -k & 1 \\ -1 & 0 \end{pmatrix}.$$

The eigenvalues of A are given by $\lambda_\pm = (-k \pm (k^2 - 4)^{1/2})/2$. Since $\lambda_\pm$ is either negative or else has a negative real part, the equilibrium point at the origin is a sink. It is a spiral sink if $k < 2$. For any $k > 0$, all solutions of the system tend to the origin; physically, this is the dissipative effect of the resistor.

Note that we have

$$y'' = -x' = -y + kx = -y - ky',$$

so that the system is equivalent to the second-order equation $y'' + ky' + y = 0$, which is often encountered in elementary differential equations courses.

Next we consider the case of a general characteristic f. There is a unique equilibrium point for the Lienard system that is given by $(0, f(0))$. Linearization yields the matrix

$$\begin{pmatrix} -f'(0) & 1 \\ -1 & 0 \end{pmatrix}$$

whose eigenvalues are given by

$$\lambda_\pm = \frac{1}{2}\left(-f'(0) \pm \sqrt{(f'(0))^2 - 4}\right).$$

We conclude that this equilibrium point is a sink if $f'(0) > 0$ and a source if $f'(0) < 0$. In particular, for the van der Pol equation where $f(x) = x^3 - x$, the unique equilibrium point is a source.

To analyze the system further, we define the function $W : \mathbb{R}^2 \to \mathbb{R}^2$ by $W(x, y) = \frac{1}{2}(x^2 + y^2)$. Then we have

$$\dot{W} = x(y - f(x)) + y(-x) = -xf(x).$$

In particular, if f satisfies $f(x) > 0$ if $x > 0$, $f(x) < 0$ if $x < 0$, and $f(0) = 0$, then W is a strict Liapunov function on all of $\mathbb{R}^2$. It follows that, in this case, all solutions tend to the unique equilibrium point lying at the origin.

In circuit theory, a resistor is called *passive* if its characteristic is contained in the set consisting of $(0, 0)$ and the interior of the first and third quadrant. Therefore in the case of a passive resistor, $-xf(x)$ is negative except when $x = 0$, and so all solutions tend to the origin. Thus the word *passive* correctly describes the dynamics of such a circuit.

12.3 The van der Pol Equation

In this section we continue the study of the Lienard equation in the special case where $f(x) = x^3 - x$. This is the *van der Pol equation*:

$$\frac{dx}{dt} = y - x^3 + x$$

$$\frac{dy}{dt} = -x.$$

Let ϕ_t denote the flow of this system. In this case we can give a fairly complete phase portrait analysis.

Theorem. *There is one nontrivial periodic solution of the van der Pol equation and every other solution (except the equilibrium point at the origin) tends to this periodic solution. "The system oscillates."* ■

We know from the previous section that this system has a unique equilibrium point at the origin, and that this equilibrium is a source, since $f'(0) < 0$. The next step is to show that every nonequilibrium solution "rotates" in a certain sense around the equilibrium in a clockwise direction. To see this, note that the x-nullcline is given by $y = x^3 - x$ and the y-nullcline is the y-axis. We subdivide each of these nullclines into two pieces given by

$$v^+ = \{ (x, y) \mid y > 0, x = 0 \}$$
$$v^- = \{ (x, y) \mid y < 0, x = 0 \}$$
$$g^+ = \{ (x, y) \mid x > 0, y = x^3 - x \}$$
$$g^- = \{ (x, y) \mid x < 0, y = x^3 - x \}.$$

These curves are disjoint; together with the origin they form the boundaries of the four basic regions A, B, C, and D depicted in Figure 12.3.

From the configuration of the vector field in the basic regions, it appears that all nonequilibrium solutions wind around the origin in the clockwise direction. This is indeed the case.

Proposition. *Solution curves starting on v^+ cross successively through g^+, v^-, and g^- before returning to v^+.*

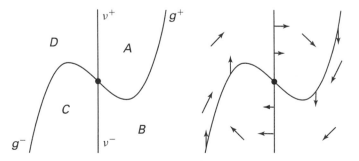

Figure 12.3 The basic regions and nullclines for the van der Pol system.

Proof: Any solution starting on v^+ immediately enters the region A since $x'(0) > 0$. In A we have $y' < 0$, so this solution must decrease in the y direction. Since the solution cannot tend to the source, it follows that this solution must eventually meet g^+. On g^+ we have $x' = 0$ and $y' < 0$. Consequently, the solution crosses g^+, then enters the region B. Once inside B, the solution heads southwest. Note that the solution cannot reenter A since the vector field points straight downward on g^+. There are thus two possibilities: Either the solution crosses v^-, or else the solution tends to $-\infty$ in the y direction and never crosses v^-.

We claim that the latter cannot happen. Suppose that it does. Let (x_0, y_0) be a point on this solution in region B and consider $\phi_t(x_0, y_0) = (x(t), y(t))$. Since $x(t)$ is never 0, it follows that this solution curve lies for all time in the strip S given by $0 < x \le x_0$, $y \le y_0$, and we have $y(t) \to -\infty$ as $t \to t_0$ for some t_0. We first observe that, in fact, $t_0 = \infty$. To see this, note that

$$y(t) - y_0 = \int_0^t y'(s)\, ds = \int_0^t -x(s)\, ds.$$

But $0 < x(s) \le x_0$, so we may only have $y(t) \to -\infty$ if $t \to \infty$.

Now consider $x(t)$ for $0 \le t < \infty$. We have $x' = y - x^3 + x$. Since the quantity $-x^3 + x$ is bounded in the strip S and $y(t) \to -\infty$ as $t \to \infty$, it follows that

$$x(t) - x_0 = \int_0^t x'(s)\, ds \to -\infty$$

as $t \to \infty$ as well. But this contradicts our assumption that $x(t) > 0$.

Hence this solution must cross v^-. Now the vector field is skew-symmetric about the origin. That is, if $G(x, y)$ is the van der Pol vector field, then $G(-x, -y) = -G(x, y)$. Exploiting this symmetry, it follows that solutions must then pass through regions C and D in similar fashion. ∎

As a consequence of this result, we may define a Poincaré map P on the half-line v^+. Given $(0, y_0) \in v^+$, we define $P(y_0)$ to be the y coordinate of the first return of $\phi_t(0, y_0)$ to v^+ with $t > 0$. See Figure 12.4. As in Section 10.3, P is a one to one C^∞ function. The Poincaré map is also onto. To see this, simply follow solutions starting on v^+ backward in time until they reintersect v^+, as they must by the proposition. Let $P^n = P \circ P^{n-1}$ denote the n-fold composition of P with itself.

Our goal now is to prove the following theorem:

Theorem. *The Poincaré map has a unique fixed point in v^+. Furthermore, the sequence $P^n(y_0)$ tends to this fixed point as $n \to \infty$ for any nonzero $y_0 \in v^+$.* ∎

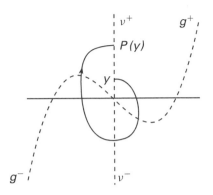

Figure 12.4   The Poincaré map
on v^+.

Clearly, any fixed point of P lies on a periodic solution. On the other hand, if $P(y_0) \neq y_0$, then the solution through $(0, y_0)$ can never be periodic. Indeed, if $P(y_0) > y_0$, then the successive intersection of $\phi_t(0, y_0)$ with v^+ is a monotone sequence as in Section 10.4. Hence the solution crosses v^+ in an increasing sequence of points and so the solution can never meet itself. The case $P(y_0) < y_0$ is analogous.

We may define a "semi-Poincaré map" $\alpha : v^+ \to v^-$ by letting $\alpha(y)$ be the y coordinate of the first point of intersection of $\phi_t(0, y)$ with v^- where $t > 0$. Also define

$$\delta(y) = \frac{1}{2}\left(\alpha(y)^2 - y^2\right).$$

Note for later use that there is a unique point $(0, y^*) \in v^+$ and time t^* such that

1. $\phi_t(0, y^*) \in A$ for $0 < t < t^*$;
2. $\phi_{t^*}(0, y^*) = (1, 0) \in g^+$.

See Figure 12.5.

The theorem will now follow directly from the following rather delicate result.

Proposition.   *The function $\delta(y)$ satisfies:*

1. *$\delta(y) > 0$ if $0 < y < y^*$;*
2. *$\delta(y)$ decreases monotonically to $-\infty$ as $y \to \infty$ for $y > y^*$.*   ■

We will prove the proposition shortly; first we use it to complete the proof of the theorem. We exploit the fact that the vector field is skew-symmetric

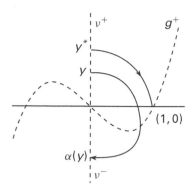

Figure 12.5   The
semi-Poincaré map.

about the origin. This implies that if $(x(t), y(t))$ is a solution curve, then so is $(-x(t), -y(t))$.

Part of the graph of $\delta(y)$ is shown schematically in Figure 12.6. The intermediate value theorem and the proposition imply that there is a unique $y_0 \in v^+$ with $\delta(y_0) = 0$. Consequently, $\alpha(y_0) = -y_0$ and it follows from the skew-symmetry that the solution through $(0, y_0)$ is periodic. Since $\delta(y) \neq 0$ except at y_0, we have $\alpha(y) \neq y$ for all other y-values, and so it follows that $\phi_t(0, y_0)$ is the unique periodic solution.

We next show that all other solutions (except the equilibrium point) tend to this periodic solution. Toward that end, we define a map $\beta : v^- \to v^+$, sending each point of v^- to the first intersection of the solution (for $t > 0$) with v^+. By symmetry we have

$$\beta(y) = -\alpha(-y).$$

Note also that $P(y) = \beta \circ \alpha(y)$.

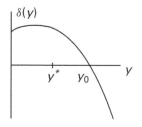

Figure 12.6   The
graph of $\delta(y)$.

We identify the y-axis with the real numbers in the y coordinate. Thus if $y_1, y_2 \in v^+ \cup v^-$ we write $y_1 > y_2$ if y_1 is above y_2. Note that α and β reverse this ordering while P preserves it.

Now choose $y \in v_+$ with $y > y_0$. Since $\alpha(y_0) = -y_0$ we have $\alpha(y) < -y_0$ and $P(y) > y_0$. On the other hand, $\delta(y) < 0$, which implies that $\alpha(y) > -y$. Therefore $P(y) = \beta(\alpha(y)) < y$. We have shown that $y > y_0$ implies $y > P(y) > y_0$. Similarly $P(y) > P(P(y)) > y_0$ and by induction $P^n(y) > P^{n+1}(y) > y_0$ for all $n > 0$.

The decreasing sequence $P^n(y)$ has a limit $y_1 \geq y_0$ in v^+. Note that y_1 is a fixed point of P, because, by continuity of P, we have

$$P(y_1) - y_1 = \lim_{n \to \infty} P(P^n(y)) - y_1$$
$$= y_1 - y_1 = 0.$$

Since P has only one fixed point, we have $y_1 = y_0$. This shows that the solution through y spirals toward the periodic solution as $t \to \infty$. See Figure 12.7. The same is true if $y < y_0$; the details are left to the reader. Since every solution except the equilibrium meets v^+, the proof of the theorem is complete. ∎

Finally, we turn to the proof of the proposition. We adopt the following notation. Let $\gamma : [a, b] \to \mathbb{R}^2$ be a smooth curve in the plane and let $F : \mathbb{R}^2 \to \mathbb{R}$. We write $\gamma(t) = (x(t), y(t))$ and define

$$\int_\gamma F(x, y) = \int_a^b F(x(t), y(t)) \, dt.$$

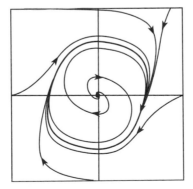

Figure 12.7   The phase portrait of the van der Pol equation.

If it happens that $x'(t) \neq 0$ for $a \leq t \leq b$, then along γ, y is a function of x, so we may write $y = y(x)$. In this case we can change variables:

$$\int_a^b F(x(t), y(t)) \, dt = \int_{x(a)}^{x(b)} F(x, y(x)) \frac{dt}{dx} \, dx.$$

Therefore

$$\int_\gamma F(x, y) = \int_{x(a)}^{x(b)} \frac{F(x, y(x))}{dx/dt} \, dx.$$

We have a similar expression if $y'(t) \neq 0$.

Now recall the function

$$W(x, y) = \frac{1}{2} \left(x^2 + y^2 \right)$$

introduced in the previous section. Let $p \in v^+$. Suppose $\alpha(p) = \phi_\tau(p)$. Let $\gamma(t) = (x(t), y(t))$ for $0 \leq t \leq \tau = \tau(p)$ be the solution curve joining $p \in v^+$ to $\alpha(p) \in v^-$. By definition

$$\delta(p) = \frac{1}{2} \left(y(\tau)^2 - y(0)^2 \right)$$
$$= W(x(\tau), y(\tau)) - W(x(0), y(0)).$$

Thus

$$\delta(p) = \int_0^\tau \frac{d}{dt} W(x(t), y(t)) \, dt.$$

Recall from Section 12.2 that we have

$$\dot{W} = -xf(x) = -x(x^3 - x).$$

Thus we have

$$\delta(p) = \int_0^\tau -x(t)(x(t)^3 - x(t)) \, dt$$
$$= \int_0^\tau x(t)^2 (1 - x(t)^2) \, dt.$$

This immediately proves part (1) of the proposition because the integrand is positive for $0 < x(t) < 1$.

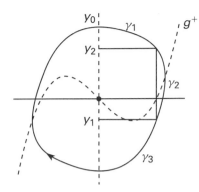

Figure 12.8 The curves $\gamma_1, \gamma_2,$ and γ_3 depicted on the closed orbit through y_0.

We may rewrite the last equality as

$$\delta(p) = \int_{\gamma} x^2 \left(1 - x^2\right).$$

We restrict attention to points $p \in v^+$ with $p > y^*$. We divide the corresponding solution curve γ into three curves $\gamma_1, \gamma_2, \gamma_3$ as displayed in Figure 12.8. The curves γ_1 and γ_3 are defined for $0 \le x \le 1$, while the curve γ_2 is defined for $y_1 \le y \le y_2$. Then

$$\delta(p) = \delta_1(p) + \delta_2(p) + \delta_3(p)$$

where

$$\delta_i(p) = \int_{\gamma_i} x^2 \left(1 - x^2\right), \quad i = 1, 2, 3.$$

Notice that, along γ_1, $y(t)$ may be regarded as a function of x. Hence we have

$$\delta_1(p) = \int_0^1 \frac{x^2 \left(1 - x^2\right)}{dx/dt} \, dx$$

$$= \int_0^1 \frac{x^2 \left(1 - x^2\right)}{y - f(x)} \, dx$$

where $f(x) = x^3 - x$. As p moves up the y-axis, $y - f(x)$ increases [for (x, y) on γ_1]. Hence $\delta_1(p)$ decreases as p increases. Similarly $\delta_3(p)$ decreases as p increases.

On γ_2, $x(t)$ may be regarded as a function of y, which is defined for $y \in [y_1, y_2]$ and $x \geq 1$. Therefore since $dy/dt = -x$, we have

$$\delta_2(p) = \int_{y_2}^{y_1} -x(y)\left(1 - x(y)^2\right) dy$$

$$= \int_{y_1}^{y_2} x(y)\left(1 - x(y)^2\right) dy$$

so that $\delta_2(p)$ is negative.

As p increases, the domain $[y_1, y_2]$ of integration becomes steadily larger. The function $y \to x(y)$ depends on p, so we write it as $x_p(y)$. As p increases, the curves γ_2 move to the right; hence $x_p(y)$ increases and so $x_p(y)(1 - x_p(y)^2)$ decreases. It follows that $\delta_2(p)$ decreases as p increases; and evidently $\lim_{p \to \infty} \delta_2(p) = -\infty$. Consequently, $\delta(p)$ also decreases and tends to $-\infty$ as $p \to \infty$. This completes the proof of the proposition.

12.4 A Hopf Bifurcation

We now describe a more general class of circuit equations where the resistor characteristic depends on a parameter μ and is denoted by f_μ. (Perhaps μ is the temperature of the resistor.) The physical behavior of the circuit is then described by the system of differential equations on $\mathbb{R}^2$:

$$\frac{dx}{dt} = y - f_\mu(x)$$

$$\frac{dy}{dt} = -x.$$

Consider as an example the special case where f_μ is described by

$$f_\mu(x) = x^3 - \mu x$$

and the parameter μ lies in the interval $[-1, 1]$. When $\mu = 1$ we have the van der Pol system from the previous section. As before, the only equilibrium point lies at the origin. The linearized system is

$$Y' = \begin{pmatrix} \mu & 1 \\ -1 & 0 \end{pmatrix} Y,$$

and the eigenvalues are

$$\lambda_{\pm} = \frac{1}{2}\left(\mu \pm \sqrt{\mu^2 - 4}\right).$$

Thus the origin is a spiral sink for $-1 \leq \mu < 0$ and a spiral source for $0 < \mu \leq 1$. Indeed, when $-1 \leq \mu \leq 0$, the resistor is passive as the graph of f_{μ} lies in the first and third quadrants. Therefore all solutions tend to the origin in this case. This holds even in the case where $\mu = 0$ and the linearization yields a center. Physically the circuit is dead in that, after a period of transition, all currents and voltages stay at 0 (or as close to 0 as we want).

However, as μ becomes positive, the circuit becomes alive. It begins to oscillate. This follows from the fact that the analysis of Section 12.3 applies to this system for all μ in the interval $(0, 1]$. We therefore see the birth of a (unique) periodic solution γ_{μ} as μ increases through 0 (see Exercise 4 at the end of this chapter). Just as above, this solution attracts all other nonzero solutions. As in Section 8.5, this is an example of a *Hopf bifurcation*. Further elaboration of the ideas in Section 12.3 can be used to show that $\gamma_{\mu} \to 0$ as $\mu \to 0$ with $\mu > 0$. Figure 12.9 shows some phase portraits associated to this bifurcation.

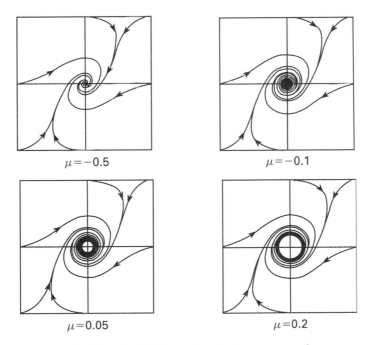

$\mu = -0.5$ $\mu = -0.1$

$\mu = 0.05$ $\mu = 0.2$

Figure 12.9   The Hopf bifurcation in the system $x' = y - x^3 + \mu x$, $y' = -x$.

12.5 Exploration: Neurodynamics

One of the most important developments in the study of the firing of nerve cells or neurons was the development of a model for this phenomenon in giant squid in the 1950s by Hodgkin and Huxley [23]. They developed a four-dimensional system of differential equations that described the electrochemical transmission of neuronal signals along the cell membrane, a work for which they later received the Nobel prize. Roughly speaking, this system is similar to systems that arise in electrical circuits. The neuron consists of a cell body, or *soma*, which receives electrical stimuli. This stimulus is then conducted along the *axon*, which can be thought of as an electrical cable that connects to other neurons via a collection of synapses. Of course, the motion is not really electrical, because the current is not really made up of electrons, but rather ions (predominantly sodium and potassium). See [15] or [34] for a primer on the neurobiology behind these systems.

The four-dimensional Hodgkin-Huxley system is difficult to deal with, primarily because of the highly nonlinear nature of the equations. An important breakthrough from a mathematical point of view was achieved by Fitzhugh [18] and Nagumo *et al.* [35], who produced a simpler model of the Hodgkin-Huxley model. Although this system is not as biologically accurate as the original system, it nevertheless does capture the essential behavior of nerve impulses, including the phenomenon of *excitability* alluded to below.

The Fitzhugh-Nagumo system of equations is given by

$$x' = y + x - \frac{x^3}{3} + I$$
$$y' = -x + a - by$$

where a and b are constants satisfying

$$0 < \frac{3}{2}(1 - a) < b < 1$$

and I is a parameter. In these equations x is similar to the voltage and represents the *excitability* of the system; the variable y represents a combination of other forces that tend to return the system to rest. The parameter I is a stimulus parameter that leads to excitation of the system; I is like an applied current. Note the similarity of these equations with the van der Pol equation of Section 12.3.

1. First assume that $I = 0$. Prove that this system has a unique equilibrium point (x_0, y_0). *Hint:* Use the geometry of the nullclines for this rather than

explicitly solving the equations. Also remember the restrictions placed on a and b.

2. Prove that this equilibrium point is always a sink.
3. Now suppose that $I \neq 0$. Prove that there is still a unique equilibrium point (x_I, y_I) and that x_I varies monotonically with I.
4. Determine values of x_I for which the equilibrium point is a source and show that there must be a stable limit cycle in this case.
5. When $I \neq 0$, the point (x_0, y_0) is no longer an equilibrium point. Nonetheless we can still consider the solution through this point. Describe the qualitative nature of this solution as I moves away from 0. Explain in mathematical terms why biologists consider this phenomenon the "excitement" of the neuron.
6. Consider the special case where $a = I = 0$. Describe the phase plane for each $b > 0$ (no longer restrict to $b < 1$) as completely as possible. Describe any bifurcations that occur.
7. Now let I vary as well and again describe any bifurcations that occur. Describe in as much detail as possible the phase portraits that occur in the I, b–plane, with $b > 0$.
8. Extend the analysis of the previous problem to the case $b \leq 0$.
9. Now fix $b = 0$ and let a and I vary. Sketch the bifurcation plane (the I, a–plane) in this case.

EXERCISES

1. Find the phase portrait for the differential equation

$$x' = y - f(x), \quad f(x) = x^2,$$
$$y' = -x.$$

   Hint: Exploit the symmetry about the y-axis.

2. Let

$$f(x) = \begin{cases} 2x - 3 & \text{if } x > 1 \\ -x & \text{if } -1 \leq x \leq 1 \\ 2x + 3 & \text{if } x < -1. \end{cases}$$

   Consider the system

$$x' = y - f(x)$$
$$y' = -x.$$

   (a) Sketch the phase plane for this system.
   (b) Prove that this system has a unique closed orbit.

3. Let

$$
f_a(x) = \begin{cases} 2x + a - 2 & \text{if } x > 1 \\ ax & \text{if } -1 \leq x \leq 1 \\ 2x - a + 2 & \text{if } x < -1. \end{cases}
$$

Consider the system

$$
x' = y - f_a(x)
$$
$$
y' = -x.
$$

(a) Sketch the phase plane for this system for various values of a.
(b) Describe the bifurcation that occurs when $a = 0$.

4. Consider the system described in Section 12.4

$$
x' = y - (x^3 - \mu x)
$$
$$
y' = -x
$$

where the parameter μ satisfies $0 \leq \mu < 1$. Fill in the details of the proof that a Hopf bifurcation occurs at $\mu = 0$.

5. Consider the system

$$
x' = \mu(y - (x^3 - x)), \quad \mu > 0
$$
$$
y' = -x.
$$

Prove that this system has a unique nontrivial periodic solution γ_μ. Show that as $\mu \to \infty$, γ_μ tends to the closed curve consisting of two horizontal line segments and two arcs on $y = x^3 - x$ as in Figure 12.10. This type of solution is called a *relaxation oscillation*. When μ is large, there are two quite different timescales along the periodic solution. When moving horizontally, we have x' very large, and so the solution makes this transit

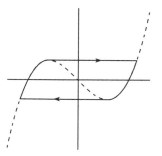

Figure 12.10

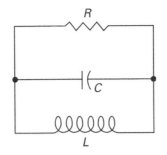

Figure 12.11

very quickly. On the other hand, near the cubic nullcline, $x' = 0$ while y' is bounded. Hence this transit is comparatively much slower.

6. Find the differential equations for the network in Figure 12.11, where the resistor is voltage controlled; that is, the resistor characteristic is the graph of a function $g: \mathbb{R} \to \mathbb{R}$, $i_R = g(v_R)$.

7. Show that the LC circuit consisting of one inductor and one capacitor wired in a closed loop oscillates.

8. Determine the phase portrait of the following differential equation and in particular show there is a unique nontrivial periodic solution:

$$x' = y - f(x)$$
$$y' = -g(x)$$

where all of the following are assumed:

(a) $g(-x) = -g(x)$ and $xg(x) > 0$ for all $x \neq 0$;
(b) $f(-x) = -f(x)$ and $f(x) < 0$ for $0 < x < a$;
(c) for $x > a, f(x)$ is positive and increasing;
(d) $f(x) \to \infty$ as $x \to \infty$.

9. Consider the system

$$x' = y$$
$$y' = a(1 - x^4)y - x$$

(a) Find all equilibrium points and classify them.
(b) Sketch the phase plane.
(c) Describe the bifurcation that occurs when a becomes positive.
(d) Prove that there exists a unique closed orbit for this system when $a > 0$.
(e) Show that all nonzero solutions of the system tend to this closed orbit when $a > 0$.

13

Applications in Mechanics

We turn our attention in this chapter to the earliest important examples of differential equations that, in fact, are connected with the origins of calculus. These equations were used by Newton to derive and unify the three laws of Kepler. In this chapter we give a brief derivation of two of Kepler's laws and then discuss more general problems in mechanics.

The equations of Newton, our starting point, have retained importance throughout the history of modern physics and lie at the root of that part of physics called *classical mechanics*. The examples here provide us with concrete examples of historical and scientific importance. Furthermore, the case we consider most thoroughly here, that of a particle moving in a central force gravitational field, is simple enough so that the differential equations can be solved explicitly using exact, classical methods (just calculus!). However, with an eye toward the more complicated mechanical systems that cannot be solved in this way, we also describe a more geometric approach to this problem.

13.1 Newton's Second Law

We will be working with a particle moving in a *force field* F. Mathematically, F is just a vector field on the (configuration) space of the particle, which in our case will be $\mathbb{R}^n$. From the physical point of view, $F(X)$ is the force exerted on a particle located at position $X \in \mathbb{R}^n$.

The example of a force field we will be most concerned with is the gravitational field of the sun: $F(X)$ is the force on a particle located at X that attracts the particle to the sun. We go into details of this system in Section 13.3.

The connection between the physical concept of a force field and the mathematical concept of a differential equation is *Newton's second law*: $F = ma$. This law asserts that a particle in a force field moves in such a way that the force vector at the location X of the particle, at any instant, equals the acceleration vector of the particle times the mass m. That is, Newton's law gives the second-order differential equation

$$mX'' = F(X).$$

As a system, this equation becomes

$$X' = V$$

$$V' = \frac{1}{m}F(X)$$

where $V = V(t)$ is the velocity of the particle. This is a system of equations on $\mathbb{R}^n \times \mathbb{R}^n$. This type of system is often called a *mechanical system with n degrees of freedom*.

A solution $X(t) \subset \mathbb{R}^n$ of the second-order equation is said to lie in *configuration space*. The solution of the system $(X(t), V(t)) \subset \mathbb{R}^n \times \mathbb{R}^n$ lies in *phase space* or *state space* of the system.

Example. Recall the simple undamped harmonic oscillator from Chapter 2. In this case the mass moves in one dimension and its position at time t is given by a function $x(t)$, where $x\colon \mathbb{R} \to \mathbb{R}$. As we saw, the differential equation governing this motion is

$$mx'' = -kx$$

for some constant $k > 0$. That is, the force field at the point $x \in \mathbb{R}$ is given by $-kx$. ∎

Example. The two-dimensional version of the harmonic oscillator allows the mass to move in the plane, so the position is now given by the vector $X(t) = (x_1(t), x_2(t)) \in \mathbb{R}^2$. As in the one-dimensional case, the force field is $F(X) = -kX$ so the equations of motion are the same

$$mX'' = -kX$$

with solutions in configuration space given by

$$x_1(t) = c_1 \cos(\sqrt{k/m}\, t) + c_2 \sin(\sqrt{k/m}\, t)$$

$$x_2(t) = c_3 \cos(\sqrt{k/m}\, t) + c_4 \sin(\sqrt{k/m}\, t)$$

for some choices of the c_j as is easily checked using the methods of Chapter 6. ∎

Before dealing with more complicated cases of Newton's law, we need to recall a few concepts from multivariable calculus. Recall that the *dot product* (or *inner product*) of two vectors $X, Y \in \mathbb{R}^n$ is denoted by $X \cdot Y$ and defined by

$$X \cdot Y = \sum_{i=1}^{n} x_i y_i$$

where $X = (x_1, \ldots, x_n)$ and $Y = (y_1, \ldots, y_n)$. Thus $X \cdot X = |X|^2$. If $X, Y : I \to \mathbb{R}^n$ are smooth functions, then a version of the product rule yields

$$(X \cdot Y)' = X' \cdot Y + X \cdot Y',$$

as can be easily checked using the coordinate functions x_i and y_i.

Recall also that if $g : \mathbb{R}^n \to \mathbb{R}$, the gradient of g, denoted grad g, is defined by

$$\operatorname{grad} g(X) = \left(\frac{\partial g}{\partial x_1}(X), \ldots, \frac{\partial g}{\partial x_n}(X) \right).$$

As we saw in Chapter 9, grad g is a vector field on $\mathbb{R}^n$.

Next, consider the composition of two smooth functions $g \circ F$ where $F : \mathbb{R} \to \mathbb{R}^n$ and $g : \mathbb{R}^n \to \mathbb{R}$. The chain rule applied to $g \circ F$ yields

$$\frac{d}{dt} g(F(t)) = \operatorname{grad} g(F(t)) \cdot F'(t)$$

$$= \sum_{i=1}^{n} \frac{\partial g}{\partial x_i}(F(t)) \frac{dF_i}{dt}(t).$$

We will also use the *cross product* (or vector product) $U \times V$ of vectors $U, V \in \mathbb{R}^3$. By definition,

$$U \times V = (u_2 v_3 - u_3 v_2, \; u_3 v_1 - u_1 v_3, \; u_1 v_2 - u_2 v_1) \in \mathbb{R}^3.$$

Recall from multivariable calculus that we have

$$U \times V = -V \times U = |U||V|N \sin \theta$$

where N is a unit vector perpendicular to U and V with the orientations of the vectors U, V, and N given by the "right-hand rule." Here θ is the angle between U and V.

Note that $U \times V = 0$ if and only if one vector is a scalar multiple of the other. Also, if $U \times V \neq 0$, then $U \times V$ is perpendicular to the plane containing U and V. If U and V are functions of t in $\mathbb{R}$, then another version of the product rule asserts that

$$\frac{d}{dt}(U \times V) = U' \times V + U \times V'$$

as one can again check by using coordinates.

13.2 Conservative Systems

Many force fields appearing in physics arise in the following way. There is a smooth function $U: \mathbb{R}^n \to \mathbb{R}$ such that

$$F(X) = -\left(\frac{\partial U}{\partial x_1}(X), \frac{\partial U}{\partial x_2}(X), \ldots, \frac{\partial U}{\partial x_n}(X)\right)$$
$$= -\operatorname{grad} U(X).$$

(The negative sign is traditional.) Such a force field is called *conservative*. The associated system of differential equations

$$X' = V$$

$$V' = -\frac{1}{m} \operatorname{grad} U(X)$$

is called a *conservative system*. The function U is called the *potential energy* of the system. [More properly, U should be called *a* potential energy since adding a constant to it does not change the force field $-\operatorname{grad} U(X)$.]

Example. The planar harmonic oscillator above corresponds to the force field $F(X) = -kX$. This field is conservative, with potential energy

$$U(X) = \frac{1}{2}k|X|^2.$$

For any moving particle $X(t)$ of mass m, the *kinetic energy* is defined to be

$$K = \frac{1}{2}m|V(t)|^2.$$

Note that the kinetic energy depends on velocity, while the potential energy is a function of position. The *total energy* (or sometimes simply *energy*) is defined on phase space by $E = K + U$. The total energy function is important in mechanics because it is constant along any solution curve of the system. That is, in the language of Section 9.4, E is a constant of the motion or a first integral for the flow. ■

Theorem. (Conservation of Energy) *Let* $(X(t), V(t))$ *be a solution curve of a conservative system. Then the total energy E is constant along this solution curve.*

Proof: To show that $E(X(t))$ is constant in t, we compute

$$\dot{E} = \frac{d}{dt}\left(\frac{1}{2}m|V(t)|^2 + U(X(t))\right)$$

$$= mV \cdot V' + (\text{grad } U) \cdot X'$$

$$= V \cdot (-\text{grad } U) + (\text{grad } U) \cdot V$$

$$= 0. \qquad\qquad ■$$

We remark that we may also write this type of system in *Hamiltonian form*. Recall from Chapter 9 that a Hamiltonian system on $\mathbb{R}^n \times \mathbb{R}^n$ is a system of the form

$$x_i' = \frac{\partial H}{\partial y_i}$$

$$y_i' = -\frac{\partial H}{\partial x_i}$$

where $H: \mathbb{R}^n \times \mathbb{R}^n \to \mathbb{R}$ is the *Hamiltonian function*. As we have seen, the function H is constant along solutions of such a system. To write the conservative system in Hamiltonian form, we make a simple change of variables. We introduce the *momentum vector* $Y = mV$ and then set

$$H = K + U = \frac{1}{2m}\sum y_i^2 + U(x_1, \ldots, x_n).$$

This puts the conservative system in Hamiltonian form, as is easily checked.

13.3 Central Force Fields

A force field F is called *central* if $F(X)$ points directly toward or away from the origin for every X. In other words, the vector $F(X)$ is always a scalar multiple of X:

$$F(X) = \lambda(X)X$$

where the coefficient $\lambda(X)$ depends on X. We often tacitly exclude from consideration a particle at the origin; many central force fields are not defined (or are "infinite") at the origin. We deal with these types of singularities in Section 13.7. Theoretically, the function $\lambda(X)$ could vary for different values of X on a sphere given by $|X| = $ constant. However, if the force field is conservative, this is not the case.

Proposition.   *Let F be a conservative force field. Then the following statements are equivalent:*

1. *F is central;*
2. *$F(X) = f(|X|)X$;*
3. *$F(X) = -\mathrm{grad}\ U(X)$ and $U(X) = g(|X|)$.*

Proof: Suppose (3) is true. To prove (2) we find, from the chain rule:

$$\frac{\partial U}{\partial x_j} = g'(|X|)\frac{\partial}{\partial x_j}\left(x_1^2 + x_2^2 + x_3^2\right)^{1/2}$$

$$= \frac{g'(|X|)}{|X|}x_j.$$

This proves (2) with $f(|X|) = -g'(|X|)/|X|$. It is clear that (2) implies (1). To show that (1) implies (3) we must prove that U is constant on each sphere

$$S_\alpha = \{X \in \mathbb{R}^n \mid |X| = \alpha > 0\}.$$

Because any two points in S_α can be connected by a curve in S_α, it suffices to show that U is constant on any curve in S_α. Hence if $J \subset \mathbb{R}$ is an interval and $\gamma : J \to S_\alpha$ is a smooth curve, we must show that the derivative of the composition $U \circ \gamma$ is identically 0. This derivative is

$$\frac{d}{dt}U(\gamma(t)) = \mathrm{grad}\ U(\gamma(t)) \cdot \gamma'(t)$$

as in Section 13.1. Now grad $U(X) = -F(X) = -\lambda(X)X$ since F is central. Thus we have

$$\frac{d}{dt}U(\gamma(t)) = -\lambda(\gamma(t))\gamma(t) \cdot \gamma'(t)$$

$$= -\frac{\lambda(\gamma(t))}{2}\frac{d}{dt}|\gamma(t)|^2$$

$$= 0$$

because $|\gamma(t)| \equiv \alpha$.   ∎

Consider now a central force field, not necessarily conservative, defined on $\mathbb{R}^3$. Suppose that, at some time t_0, $\mathcal{P} \subset \mathbb{R}^3$ denotes the plane containing the position vector $X(t_0)$, the velocity vector $V(t_0)$, and the origin (assuming, for the moment, that the position and velocity vectors are not collinear). Note that the force vector $F(X(t_0))$ also lies in $\mathcal{P}$. This makes it plausible that the particle stays in the plane $\mathcal{P}$ for all time. In fact, this is true:

Proposition.   *A particle moving in a central force field in $\mathbb{R}^3$ always moves in a fixed plane containing the origin.*

Proof: Suppose $X(t)$ is the path of a particle moving under the influence of a central force field. We have

$$\frac{d}{dt}(X \times V) = V \times V + X \times V'$$

$$= X \times X''$$

$$= 0$$

because X'' is a scalar multiple of X. Therefore $Y = X(t) \times V(t)$ is a constant vector. If $Y \neq 0$, this means that X and V always lie in the plane orthogonal to Y, as asserted. If $Y = 0$, then $X'(t) = g(t)X(t)$ for some real function $g(t)$. This means that the velocity vector of the moving particle is always directed along the line through the origin and the particle, as is the force on the particle. This implies that the particle always moves along the same line through the origin. To prove this, let $(x_1(t), x_2(t), x_3(t))$ be the coordinates of $X(t)$. Then we have three separable differential equations:

$$\frac{dx_k}{dt} = g(t)x_k(t), \quad \text{for } k = 1, 2, 3.$$

Integrating, we find

$$x_k(t) = e^{h(t)}x_k(0), \quad \text{where } h(t) = \int_0^t g(s)\, ds.$$

Therefore $X(t)$ is always a scalar multiple of $X(0)$ and so $X(t)$ moves in a fixed line and hence in a fixed plane.   ∎

The vector $m(X \times V)$ is called the *angular momentum* of the system, where m is the mass of the particle. By the proof of the preceding proposition, this vector is also conserved by the system.

Corollary. (Conservation of Angular Momentum) *Angular momentum is constant along any solution curve in a central force field.*   ∎

We now restrict attention to a conservative central force field. Because of the previous proposition, the particle remains for all time in a plane, which we may take to be $x_3 = 0$. In this case angular momentum is given by the vector $(0, 0, m(x_1 v_2 - x_2 v_1))$. Let

$$\ell = m(x_1 v_2 - x_2 v_1).$$

So the function ℓ is also constant along solutions. In the planar case we also call ℓ the angular momentum. Introducing polar coordinates $x_1 = r \cos \theta$ and $x_2 = r \sin \theta$, we find

$$v_1 = x_1' = r' \cos \theta - r \sin \theta \, \theta'$$
$$v_2 = x_2' = r' \sin \theta + r \cos \theta \, \theta'.$$

Then

$$
\begin{aligned}
x_1 v_2 - x_2 v_1 &= r \cos \theta (r' \sin \theta + r \cos \theta \, \theta') - r \sin \theta (r' \cos \theta - r \sin \theta \, \theta') \\
&= r^2 (\cos^2 \theta + \sin^2 \theta) \theta' \\
&= r^2 \theta'.
\end{aligned}
$$

Hence in polar coordinates, $\ell = mr^2 \theta'$.

We can now prove one of Kepler's laws. Let $A(t)$ denote the area swept out by the vector $X(t)$ in the time from t_0 to t. In polar coordinates we have $dA = \frac{1}{2} r^2 \, d\theta$. We define the *areal velocity* to be

$$A'(t) = \frac{1}{2} r^2(t) \theta'(t),$$

the rate at which the position vector sweeps out area. Kepler observed that the line segment joining a planet to the sun sweeps out equal areas in equal times, which we interpret to mean $A' = $ constant. We have therefore proved more generally that this is true for any particle moving in a conservative central force field.

We now have found two constants of the motion or first integrals for a conservative system generated by a central force field: total energy and angular momentum. In the 19th century, the idea of solving a differential equation was tied to the construction of a sufficient number of such constants of the motion. In the 20th century, it became apparent that first integrals do not exist for differential equations very generally; the culprit here is chaos, which we will discuss in the next two chapters. Basically, chaotic behavior of solutions of a differential equation in an open set precludes the existence of first integrals in that set.

13.4 The Newtonian Central Force System

We now specialize the discussion to the Newtonian central force system. This system deals with the motion of a single planet orbiting around the sun. We assume that the sun is fixed at the origin in $\mathbb{R}^3$ and that the relatively small planet exerts no force on the sun. The sun exerts a force on a planet given by *Newton's law of gravitation*, which is also called the *inverse square law*. This law states that the sun exerts a force on a planet located at $X \in \mathbb{R}^3$ whose magnitude is $gm_s m_p/r^2$, where m_s is the mass of the sun, m_p is the mass of the planet, and g is the gravitational constant. The direction of the force is toward the sun. Therefore Newton's law yields the differential equation

$$m_p X'' = -gm_s m_p \frac{X}{|X|^3}.$$

For clarity, we change units so that the constants are normalized to one and so the equation becomes more simply

$$X'' = F(X) = -\frac{X}{|X|^3}.$$

where F is now the force field. As a system of differential equations, we have

$$X' = V$$

$$V' = -\frac{X}{|X|^3}.$$

This system is called the *Newtonian central force system*. Our goal in this section is to describe the geometry of this system; in the next section we derive a complete analytic solution of this system.

Clearly, this is a central force field. Moreover, it is conservative, since

$$\frac{X}{|X|^3} = \operatorname{grad} U(X)$$

where the potential energy U is given by

$$U(X) = -\frac{1}{|X|}.$$

Observe that $F(X)$ is not defined at 0; indeed, the force field becomes infinite as the moving mass approaches collision with the stationary mass at the origin.

As in the previous section we may restrict attention to particles moving in the plane $\mathbb{R}^2$. Thus we look at solutions in the configuration space $\mathcal{C} = \mathbb{R}^2 - \{0\}$. We denote the phase space by $\mathcal{P} = (\mathbb{R}^2 - \{0\}) \times \mathbb{R}^2$.

We visualize phase space as the collection of all tangent vectors at each point $X \in C$. For a given $X \in \mathbb{R}^2 - \{0\}$, let $T_X = \{(X, V) \mid V \in \mathbb{R}^2\}$. T_X is the *tangent plane* to the configuration space at X. Then

$$\mathcal{P} = \bigcup_{X \in C} T_X$$

is the *tangent space* to the configuration space, which we may naturally identify with a subset of $\mathbb{R}^4$.

The dimension of phase space is four. However, we can cut this dimension in half by making use of the two known first integrals, total energy and angular momentum. Recall that energy is constant along solutions and is given by

$$E(X, V) = K(V) + U(X) = \frac{1}{2}|V|^2 - \frac{1}{|X|}.$$

Let Σ_h denote the subset of $\mathcal{P}$ consisting of all points (X, V) with $E(X, V) = h$. The set Σ_h is called an *energy surface* with total energy h. If $h \geq 0$, then Σ_h meets each T_X in a circle of tangent vectors satisfying

$$|V|^2 = 2\left(h + \frac{1}{|X|}\right).$$

The radius of these circles in the tangent planes at X tends to ∞ as $X \to 0$ and decreases to $2h$ as $|X|$ tends to ∞.

When $h < 0$, the structure of the energy surface Σ_h is different. If $|X| > -1/h$, then there are no vectors in $T_X \cap \Sigma_h$. When $|X| = -1/h$, only the zero vector in T_X lies in Σ_h. The circle $r = -1/h$ in configuration space is therefore known as the *zero velocity curve*. If X lies inside the zero velocity curve, then T_X meets the energy surface in a circle of tangent vectors as before. Figure 13.1 gives a caricature of Σ_h for the case where $h < 0$.

We now introduce polar coordinates in configuration space and new variables (v_r, v_θ) in the tangent planes via

$$V = v_r \begin{pmatrix} \cos\theta \\ \sin\theta \end{pmatrix} + v_\theta \begin{pmatrix} -\sin\theta \\ \cos\theta \end{pmatrix}.$$

We have

$$V = X' = r' \begin{pmatrix} \cos\theta \\ \sin\theta \end{pmatrix} + r\theta' \begin{pmatrix} -\sin\theta \\ \cos\theta \end{pmatrix}$$

so that $r' = v_r$ and $\theta' = v_\theta/r$. Differentiating once more, we find

$$\frac{-1}{r^2} \begin{pmatrix} \cos\theta \\ \sin\theta \end{pmatrix} = -\frac{X}{|X|^3} = V' = \left(v_r' - \frac{v_\theta^2}{r}\right) \begin{pmatrix} \cos\theta \\ \sin\theta \end{pmatrix} + \left(\frac{v_r v_\theta}{r} + v_\theta'\right) \begin{pmatrix} -\sin\theta \\ \cos\theta \end{pmatrix}.$$

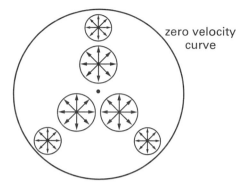

Figure 13.1   Over each nonzero point inside the zero velocity curve, T_X meets the energy surface Σ_h in a circle of tangent vectors.

Therefore in the new coordinates $(r, \theta, v_r, v_\theta)$, the system becomes

$$r' = v_r$$

$$\theta' = v_\theta / r$$

$$v_r' = -\frac{1}{r^2} + \frac{v_\theta^2}{r}$$

$$v_\theta' = -\frac{v_r v_\theta}{r}.$$

In these coordinates, total energy is given by

$$\frac{1}{2}\left(v_r^2 + v_\theta^2\right) - \frac{1}{r} = h$$

and angular momentum is given by $\ell = r v_\theta$. Let $\Sigma_{h,\ell}$ consist of all points in phase space with total energy h and angular momentum ℓ. For simplicity, we will restrict attention to the case where $h < 0$.

If $\ell = 0$, we must have $v_\theta = 0$. So if X lies inside the zero velocity curve, the tangent space at X meets $\Sigma_{h,0}$ in precisely two vectors of the form

$$\pm v_r \begin{pmatrix} \cos\theta \\ \sin\theta \end{pmatrix},$$

both of which lie on the line connecting 0 and X, one pointing toward 0, the other pointing away. On the zero velocity curve, only the zero vector lies in $\Sigma_{h,0}$. Hence we see immediately that each solution in $\Sigma_{h,0}$ lies on a straight line through the origin. The solution leaves the origin and travels along a straight line until reaching the zero velocity curve, after which time it recedes back to the origin. In fact, since the vectors in $\Sigma_{h,0}$ have magnitude tending to ∞ as $X \to 0$, these solutions reach the singularity in finite time in both directions. Solutions of this type are called *collision-ejection orbits*.

When $\ell \neq 0$, a different picture emerges. Given X inside the zero velocity curve, we have $v_\theta = \ell/r$, so that, from the total energy formula,

$$r^2 v_r^2 = 2hr^2 + 2r - \ell^2. \tag{A}$$

The quadratic polynomial in r on the right in Eq. (A) must therefore be nonnegative, so this puts restrictions on which r-values can occur for $X \in \Sigma_{h,\ell}$. The graph of this quadratic polynomial is concave down since $h < 0$. It has no real roots if $\ell^2 > -1/2h$. Therefore the space $\Sigma_{h,\ell}$ is empty in this case. If $\ell^2 = -1/2h$, we have a single root that occurs at $r = -1/2h$. Hence this is the only allowable r-value in $\Sigma_{h,\ell}$ in this case. In the tangent plane at (r, θ), we have $v_r = 0$, $v_\theta = -2h\ell$, so this represents a circular closed orbit (traversed clockwise if $\ell < 0$, counterclockwise if $\ell > 0$).

If $\ell^2 < -1/2h$, then this polynomial has a pair of distinct roots at α, β with $\alpha < -1/2h < \beta$. Note that $\alpha > 0$. Let $A_{\alpha,\beta}$ be the annular region $\alpha \leq r \leq \beta$ in configuration space. We therefore have that motion in configuration space is confined to $A_{\alpha,\beta}$.

Proposition. *Suppose $h < 0$ and $\ell^2 < -1/2h$. Then $\Sigma_{h,\ell} \subset \mathcal{P}$ is a two-dimensional torus.*

Proof: We compute the set of tangent vectors lying in $T_X \cap \Sigma_{h,\ell}$ for each $X \in A_{\alpha,\beta}$. If X lies on the boundary of the annulus, the quadratic term on the right of Eq. (A) vanishes, and so $v_r = 0$ while $v_\theta = \ell/r$. Hence there is a unique tangent vector in $T_X \cap \Sigma_{h,\ell}$ when X lies on the boundary of the annulus. When X is in the interior of $A_{\alpha,\beta}$, we have

$$v_r^\pm = \pm \frac{1}{r} \sqrt{2hr^2 + 2r - \ell^2}, \quad v_\theta = \ell/r$$

so that we have a pair of vectors in $T_X \cap \Sigma_{h,\ell}$ in this case. Note that these vectors all point either clockwise or counterclockwise in $A_{\alpha,\beta}$, since v_θ has the same sign for all X. See Figure 13.2. Thus we can think of $\Sigma_{h,\ell}$ as being given by a pair of graphs over $A_{\alpha,\beta}$: a positive graph given by v_r^+ and a negative graph given by v_r^- which are joined together along the boundary circles $r = \alpha$ and $r = \beta$. (Of course, the "real" picture is a subset of $\mathbb{R}^4$.) This yields the torus.   ∎

It is tempting to think that the two curves in the torus given by $r = \alpha$ and $r = \beta$ are closed orbits for the system, but this is not the case. This follows

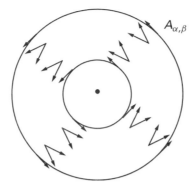

Figure 13.2   A selection of
vectors in $\Sigma_{h,\ell}$.

since, when $r = \alpha$, we have

$$v'_r = -\frac{1}{\alpha^2} + \frac{v_\theta^2}{\alpha} = \frac{1}{\alpha^3}(-\alpha + \ell^2).$$

However, since the right-hand side of Eq. (A) vanishes at α, we have

$$2h\alpha^2 + 2\alpha - \ell^2 = 0,$$

so that

$$-\alpha + \ell^2 = (2h\alpha + 1)\alpha.$$

Since $\alpha < -1/2h$, it follows that $r'' = v'_r > 0$ when $r = \alpha$, so the r coordinate of solutions in $\Sigma_{h,\ell}$ reaches a minimum when the curve meets $r = \alpha$. Similarly, along $r = \beta$, the r coordinate reaches a maximum.

Hence solutions in $A_{\alpha,\beta}$ must behave as shown in Figure 13.3. As a remark, one can easily show that these curves are preserved by rotations about the origin, so all of these solutions behave symmetrically.

More, however, can be said. Each of these solutions actually lies on a closed orbit that traces out an ellipse in configuration space. To see this, we need to turn to analysis.

13.5 Kepler's First Law

For most nonlinear mechanical systems, the geometric analysis of the previous section is just about all we can hope for. In the Newtonian central force system,

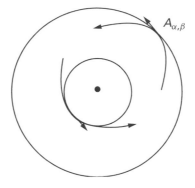

Figure 13.3   Solutions in $\Sigma_{h,\ell}$
that meet $r = \alpha$ or $r = \beta$.

however, we get lucky: As has been known for centuries, we can write down explicit solutions for this system.

Consider a particular solution curve of the differential equation. We have two constants of the motion for this system, namely, the angular momentum ℓ and total energy E. The case $\ell = 0$ yields collision-ejection solutions as we saw above. Hence we assume $\ell \neq 0$. We will show that in polar coordinates in configuration space, a solution with nonzero angular momentum lies on a curve given by $r(1 + \epsilon \cos\theta) = \kappa$ where ϵ and κ are constants. This equation defines a conic section, as can be seen by rewriting this equation in Cartesian coordinates. This fact is known as *Kepler's first law*.

To prove this, recall that $r^2\theta'$ is constant and nonzero. Hence the sign of θ' remains constant along each solution curve. Thus θ is always increasing or always decreasing in time. Therefore we may also regard r as a function of θ along the curve.

Let $W(t) = 1/r(t)$; then W is also a function of θ. Note that $W = -U$. The following proposition gives a convenient formula for kinetic energy.

Proposition. *The kinetic energy is given by*

$$K = \frac{\ell^2}{2}\left(\left(\frac{dW}{d\theta}\right)^2 + W^2\right).$$

Proof: In polar coordinates, we have

$$K = \frac{1}{2}\left((r')^2 + (r\theta')^2\right).$$

Since $r = 1/W$, we also have

$$r' = \frac{-1}{W^2}\frac{dW}{d\theta}\theta' = -\ell\frac{dW}{d\theta}.$$

Finally,

$$r\theta' = \frac{\ell}{r} = \ell W.$$

Substitution into the formula for K then completes the proof. ∎

Now we find a differential equation relating W and θ along the solution curve. Observe that $K = E - U = E + W$. From the proposition we get

$$\left(\frac{dW}{d\theta}\right)^2 + W^2 = \frac{2}{\ell^2}(E + W). \tag{B}$$

Differentiating both sides with respect to θ, dividing by $2\,dW/d\theta$, and using $dE/d\theta = 0$ (conservation of energy), we obtain

$$\frac{d^2 W}{d\theta^2} + W = \frac{1}{\ell^2}$$

where $1/\ell^2$ is a constant.

Note that this equation is just the equation for a harmonic oscillator with constant forcing $1/\ell^2$. From elementary calculus, solutions of this second-order equation can be written in the form

$$W(\theta) = \frac{1}{\ell^2} + A\cos\theta + B\sin\theta$$

or, equivalently,

$$W(\theta) = \frac{1}{\ell^2} + C\cos(\theta \mid \theta_0)$$

where the constants C and θ_0 are related to A and B.

If we substitute this expression into Eq. (B) and solve for C (at, say, $\theta + \theta_0 = \pi/2$), we find

$$C = \pm\frac{1}{\ell^2}\sqrt{1 + 2\ell^2 E}.$$

Inserting this into the solution above, we find

$$W(\theta) = \frac{1}{\ell^2}\left(1 \pm \sqrt{1 + 2E\ell^2}\,\cos(\theta + \theta_0)\right).$$

There is no need to consider both signs in front of the radical since

$$\cos(\theta + \theta_0 + \pi) = -\cos(\theta + \theta_0).$$

Moreover, by changing the variable θ to $\theta - \theta_0$ we can put any particular solution into the form

$$\frac{1}{\ell^2}\left(1 + \sqrt{1 + 2E\ell^2}\,\cos\theta\right).$$

This looks pretty complicated. However, recall from analytic geometry (or from Exercise 2 at the end of this chapter) that the equation of a conic in polar coordinates is

$$\frac{1}{r} = \frac{1}{\kappa}(1 + \epsilon\cos\theta).$$

Here κ is the *latus rectum* and $\epsilon \geq 0$ is the *eccentricity* of the conic. The origin is a focus and the three cases $\epsilon > 1$, $\epsilon = 1$, $\epsilon < 1$ correspond, respectively, to a hyperbola, parabola, and ellipse. The case $\epsilon = 0$ is a circle. In our case we have

$$\epsilon = \sqrt{1 + 2E\ell^2},$$

so the three different cases occur when $E > 0$, $E = 0$, or $E < 0$. We have proved:

Theorem. (Kepler's First Law) *The path of a particle moving under the influence of Newton's law of gravitation is a conic of eccentricity*

$$\sqrt{1 + 2E\ell^2}.$$

This path lies along a hyperbola, parabola, or ellipse according to whether $E > 0$, $E = 0$, or $E < 0$.   ∎

13.6 The Two-Body Problem

We now turn our attention briefly to what at first appears to be a more difficult problem, the *two-body problem*. In this system we assume that we have two masses that move in space according to their mutual gravitational attraction.

Let X_1, X_2 denote the positions of particles of mass m_1, m_2 in $\mathbb{R}^3$. So $X_1 = (x_1^1, x_2^1, x_3^1)$ and $X_2 = (x_1^2, x_2^2, x_3^2)$. From Newton's law of gravitation, we find the equations of motion

$$m_1 X_1'' = gm_1 m_2 \frac{X_2 - X_1}{|X_2 - X_1|^3}$$

$$m_2 X_2'' = gm_1 m_2 \frac{X_1 - X_2}{|X_1 - X_2|^3}.$$

Let's examine these equations from the perspective of a viewer living on the first mass. Let $X = X_2 - X_1$. We then have

$$X'' = X_2'' - X_1''$$
$$= gm_1 \frac{X_1 - X_2}{|X_1 - X_2|^3} - gm_2 \frac{X_2 - X_1}{|X_1 - X_2|^3}$$
$$= -g(m_1 + m_2) \frac{X}{|X|^3}.$$

But this is just the Newtonian central force problem, with a different choice of constants.

So, to solve the two-body problem, we first determine the solution of $X(t)$ of this central force problem. This then determines the right-hand side of the differential equations for both X_1 and X_2 as functions of t, and so we can simply integrate twice to find $X_1(t)$ and $X_2(t)$.

Another way to reduce the two-body problem to the Newtonian central force is as follows. The *center of mass* of the two-body system is the vector

$$X_c = \frac{m_1 X_1 + m_2 X_2}{m_1 + m_2}.$$

A computation shows that $X_c'' = 0$. Therefore we must have $X_c = At + B$ where A and B are fixed vectors in $\mathbb{R}^3$. This says that the center of mass of the system moves along a straight line with constant velocity.

We now change coordinates so that the origin of the system is located at X_c. That is, we set $Y_j = X_j - X_c$ for $j = 1, 2$. Therefore $m_1 Y_1(t) + m_2 Y_2(t) = 0$ for all t. Rewriting the differential equations in terms of the Y_j, we find

$$Y_1'' = -\frac{gm_2^3}{(m_1 + m_2)^3} \frac{Y_1}{|Y_1|^3}$$
$$Y_2'' = -\frac{gm_1^3}{(m_1 + m_2)^3} \frac{Y_2}{|Y_2|^3},$$

which yields a pair of central force problems. However, since we know that $m_1 Y_1(t) + m_2 Y_2(t) = 0$, we need only solve one of them.

13.7 Blowing Up the Singularity

The singularity at the origin in the Newtonian central force problem is the first time we have encountered such a situation. Usually our vector fields have been well defined on all of $\mathbb{R}^n$. In mechanics, such singularities can sometimes be removed by a combination of judicious changes of variables and time scalings. In the Newtonian central force system, this may be achieved using a change of variables introduced by McGehee [32].

We first introduce scaled variables

$$u_r = r^{1/2} v_r$$

$$u_\theta = r^{1/2} v_\theta.$$

In these variables the system becomes

$$r' = r^{-1/2} u_r$$

$$\theta' = r^{-3/2} u_\theta$$

$$u_r' = r^{-3/2} \left(\frac{1}{2} u_r^2 + u_\theta^2 - 1 \right)$$

$$u_\theta' = r^{-3/2} \left(-\frac{1}{2} u_r u_\theta \right).$$

We still have a singularity at the origin, but note that the last three equations are all multiplied by $r^{-3/2}$. We can remove these terms by simply multiplying the vector field by $r^{3/2}$. In doing so, solution curves of the system remain the same but are parameterized differently.

More precisely, we introduce a new time variable τ via the rule

$$\frac{dt}{d\tau} = r^{3/2}.$$

By the chain rule we have

$$\frac{dr}{d\tau} = \frac{dr}{dt} \frac{dt}{d\tau}$$

and similarly for the other variables. In this new timescale the system becomes

$$\dot{r} = r u_r$$

$$\dot{\theta} = u_\theta$$

$$\dot{u}_r = \frac{1}{2} u_r^2 + u_\theta^2 - 1$$

$$\dot{u}_\theta = -\frac{1}{2} u_r u_\theta$$

where the dot now indicates differentiation with respect to τ. Note that, when r is small, $dt/d\tau$ is close to zero, so "time" τ moves much more slowly than time t near the origin.

This system no longer has a singularity at the origin. We have "blown up" the singularity and replaced it with a new set given by $r=0$ with θ, u_r, and u_θ being arbitrary. On this set the system is now perfectly well defined. Indeed, the set $r=0$ is an invariant set for the flow since $\dot{r}=0$ when $r=0$. We have thus introduced a fictitious flow on $r=0$. While solutions on $r=0$ mean nothing in terms of the real system, by continuity of solutions, they can tell us a lot about how solutions behave near the singularity.

We need not concern ourselves with all of $r=0$ since the total energy relation in the new variables becomes

$$hr = \frac{1}{2}\left(u_r^2 + u_\theta^2\right) - 1.$$

On the set $r=0$, only the subset Λ defined by

$$u_r^2 + u_\theta^2 - 2, \; \theta \text{ arbitrary}$$

matters. The set Λ is called the *collision surface* for the system; how solutions behave on Λ dictates how solutions move near the singularity since any solution that approaches $r=0$ necessarily comes close to Λ in our new coordinates. Note that Λ is a two-dimensional torus: It is formed by a circle in the θ direction and a circle in the $u_r u_\theta$–plane.

On Λ the system reduces to

$$\dot{\theta} = u_\theta$$

$$\dot{u}_r = \frac{1}{2}u_\theta^2$$

$$\dot{u}_\theta = -\frac{1}{2}u_r u_\theta$$

where we have used the energy relation to simplify $\dot{u}_r$. This system is easy to analyze. We have $\dot{u}_r > 0$ provided $u_\theta \neq 0$. Hence the u_r coordinate must increase along any solution in Λ with $u_\theta \neq 0$.

On the other hand, when $u_\theta = 0$, the system has equilibrium points. There are two circles of equilibria, one given by $u_\theta = 0, u_r = \sqrt{2}$, and θ arbitrary, the other by $u_\theta = 0, u_r = -\sqrt{2}$, and θ arbitrary. Let $C^\pm$ denote these two circles with $u_r = \pm\sqrt{2}$ on $C^\pm$. All other solutions must travel from C^- to C^+ since v_θ increases along solutions.

To fully understand the flow on Λ, we introduce the angular variable ψ in each $u_r u_\theta$–plane via

$$u_r = \sqrt{2}\sin\psi$$
$$u_\theta = \sqrt{2}\cos\psi.$$

The torus is now parameterized by θ and ψ. In $\theta\psi$ coordinates, the system becomes

$$\dot\theta = \sqrt{2}\cos\psi$$
$$\dot\psi = \frac{1}{\sqrt{2}}\cos\psi.$$

The circles $C^\pm$ are now given by $\psi = \pm\pi/2$. Eliminating time from this equation, we find

$$\frac{d\psi}{d\theta} = \frac{1}{2}.$$

Thus all nonequilibrium solutions have constant slope $1/2$ when viewed in $\theta\psi$ coordinates. See Figure 13.4.

Now recall the collision-ejection solutions described in Section 13.4. Each of these solutions leaves the origin and then returns along a ray $\theta = \theta^*$ in configuration space. The solution departs with $v_r > 0$ (and so $u_r > 0$) and returns with $v_r < 0$ ($u_r < 0$). In our new four-dimensional coordinate system, it follows that this solution forms an unstable curve associated to the equilibrium point $(0, \theta^*, \sqrt{2}, 0)$ and a stable curve associated to $(0, \theta^*, -\sqrt{2}, 0)$. See Figure 13.5.

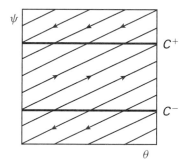

Figure 13.4   Solutions on Λ in $\theta\psi$ coordinates. Recall that θ and ψ are both defined mod 2π, so opposite sides of this square are identified to form a torus.

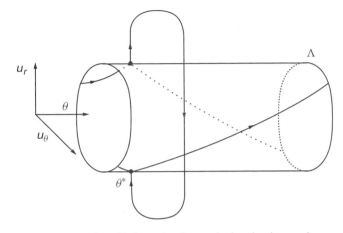

Figure 13.5   A collision-ejection solution in the region $r > 0$ leaving and returning to Λ and a connecting orbit on the collision surface.

What happens to nearby noncollision solutions? Well, they come close to the "lower" equilibrium point with $\theta - \theta^*, u_r = -\sqrt{2}$, then follow one of two branches of the unstable curve through this point up to the "upper" equilibrium point $\theta = \theta^*, u_r = +\sqrt{2}$, and then depart near the unstable curve leaving this equilibrium point. Interpreting this motion in configuration space, we see that each near-collision solution approaches the origin and then retreats after θ either increases or decreases by 2π units. Of course, we know this already, since these solutions whip around the origin in tight ellipses.

13.8 Exploration: Other Central Force Problems

In this exploration, we consider the (non-Newtonian) central force problem for which the potential energy is given by

$$U(X) = \frac{-1}{|X|^\nu}$$

where $\nu > 1$. The primary goal is to understand near-collision solutions.

1. Write this system in polar coordinates $(r, \theta, v_r, v_\theta)$ and state explicitly the formulas for total energy and angular momentum.
2. Using a computer, investigate the behavior of solutions of this system when $h < 0$ and $\ell \neq 0$.

3. Blow up the singularity at the origin via the change of variables

$$u_r = r^{\nu/2} v_r, \quad u_\theta = r^{\nu/2} v_\theta$$

and an appropriate change of timescale and write down the new system.

4. Compute the vector field on the collision surface Λ determined in $r = 0$ by the total energy relation.

5. Describe the bifurcation that occurs on Λ when $\nu = 2$.

6. Describe the structure of $\Sigma_{h,\ell}$ for all $\nu > 1$.

7. Describe the change of structure of $\Sigma_{h,\ell}$ that occurs when ν passes through the value 2.

8. Describe the behavior of solutions in $\Sigma_{h,\ell}$ when $\nu > 2$.

9. Suppose $1 < \nu < 2$. Describe the behavior of solutions as they pass close to the singularity at the origin.

10. Using the fact that solutions of this system are preserved by rotations about the origin, describe the behavior of solutions in $\Sigma_{h,\ell}$ when $h < 0$ and $\ell \neq 0$.

13.9 Exploration: Classical Limits of Quantum Mechanical Systems

In this exploration we investigate the anisotropic Kepler problem. This is a classical mechanical system with two degrees of freedom that depends on a parameter μ. When $\mu = 1$ the system reduces to the Newtonian central force system discussed in Section 13.4. When $\mu > 1$ some anisotropy is introduced into the system, so that we no longer have a central force field. We still have some collision-ejection orbits, as in the central force system, but the behavior of nearby orbits is quite different from those when $\mu = 1$.

The anisotropic Kepler problem was first introduced by Gutzwiller as a classical mechanical approximation to certain quantum mechanical systems. In particular, this system arises naturally when one looks for bound states of an electron near a donor impurity of a semiconductor. Here the potential is due to an ordinary Coulomb field, while the kinetic energy becomes anisotropic because of the electronic band structure in the solid. Equivalently, we can view this system as having an anisotropic potential energy function. Gutzwiller suggests that this situation is akin to an electron whose mass in one direction is larger than in other directions. For more background on the quantum mechanical applications of this work, refer to [22].

The anisotropic Kepler system is given by

$$x'' = \frac{-\mu x}{(\mu x^2 + y^2)^{3/2}}$$

$$y'' = \frac{-y}{(\mu x^2 + y^2)^{3/2}}$$

where μ is a parameter that we assume is greater than 1.

1. Show that this system is conservative with potential energy given by

$$U(x,y) = \frac{-1}{\sqrt{\mu x^2 + y^2}}$$

   and write down an explicit formula for total energy.
2. Describe the geometry of the energy surface Σ_h for energy $h < 0$.
3. Restricting to the case of negative energy, show that the only solutions that meet the zero velocity curve and are straight-line collision-ejection orbits for the system lie on the x- and y-axes in configuration space.
4. Show that angular momentum is no longer an integral for this system.
5. Rewrite this system in polar coordinates.
6. Using a change of variables and time rescaling as for the Newtonian central force problem (Section 13.7), blow up the singularity and write down a new system without any singularities at $r = 0$.
7. Describe the structure of the collision surface Λ (the intersection of Σ_h with $r = 0$ in the scaled coordinates). In particular, why would someone call this surface a "bumpy torus?"
8. Find all equilibrium points on Λ and determine the eigenvalues of the linearized system at these points. Determine which equilibria on Λ are sinks, sources, and saddles.
9. Explain the bifurcation that occurs on Λ when $\mu = 9/8$.
10. Find a function that is nondecreasing along all nonequilibrium solutions in Λ.
11. Determine the fate of the stable and unstable curves of the saddle points in the collision surface. *Hint:* Rewrite the equation on this surface to eliminate time and estimate the slope of solutions as they climb up Λ.
12. When $\mu > 9/8$, describe in qualitative terms what happens to solutions that approach collision close to the collision-ejection orbits on the x-axis. In particular, how do they retreat from the origin in the configuration space? How do solutions approach collision when traveling near the y-axis?

EXERCISES

1. Which of the following force fields on $\mathbb{R}^2$ are conservative?

   (a) $F(x, y) = (-x^2, -2y^2)$

(b) $F(x, y) = (x^2 - y^2, 2xy)$

(c) $F(x, y) = (x, 0)$

2. Prove that the equation

$$\frac{1}{r} = \frac{1}{h}(1 + \epsilon \cos \theta)$$

determines a hyperbola, parabola, and ellipse when $\epsilon > 1$, $\epsilon = 1$, and $\epsilon < 1$, respectively.

3. Consider the case of a particle moving directly away from the origin at time $t = 0$ in the Newtonian central force system. Find a specific formula for this solution and discuss the corresponding motion of the particle. For which initial conditions does the particle eventually reverse direction?

4. In the Newtonian central force system, describe the geometry of $\Sigma_{h,\ell}$ when $h > 0$ and $h = 0$.

5. Let $F(X)$ be a force field on $\mathbb{R}^3$. Let X_0, X_1 be points in $\mathbb{R}^3$ and let $Y(s)$ be a path in $\mathbb{R}^3$ with $s_0 \le s \le s_1$, parametrized by arc length s, from X_0 to X_1. The *work* done in moving a particle along this path is defined to be the integral

$$\int_{s_0}^{s_1} F(y(s)) \cdot y'(s) \, ds$$

where $Y'(s)$ is the (unit) tangent vector to the path. Prove that the force field is conservative if and only if the work is independent of the path. In fact, if $F = -\text{grad } V$, then the work done is $V(X_1) - V(X_0)$.

6. Describe solutions to the non-Newtonian central force system given by

$$X'' = -\frac{X}{|X|^4}.$$

7. Discuss solutions of the equation

$$X'' = \frac{X}{|X|^3}.$$

This equation corresponds to a repulsive rather than attractive force at the origin.

8. This and the next two problems deal with the two-body problem. Let the potential energy be

$$U = \frac{g m_1 m_2}{|X_2 - X_1|}$$

and

$$\mathrm{grad}_j(U) = \left(\frac{\partial U}{\partial x_1^j}, \frac{\partial U}{\partial x_2^j}, \frac{\partial U}{\partial x_3^j}\right).$$

Show that the equations for the two-body problem can be written

$$m_j X_j'' = -\mathrm{grad}_j(U).$$

9. Show that the total energy $K + U$ of the system is a constant of the motion, where

$$K = \frac{1}{2}\left(m_1|V_1|^2 + m_2|V_2|^2\right).$$

10. Define the angular momentum of the system by

$$\ell = m_1(X_1 \times V_1) + m_2(X_2 \times V_2)$$

and show that ℓ is also a first integral.

14

The Lorenz System

So far, in all of the differential equations we have studied, we have not encountered any "chaos." The reason is simple: The linear systems of the first few chapters always exhibit straightforward, predictable behavior. (OK, we may see solutions wrap densely around a torus as in the oscillators of Chapter 6, but this is not chaos.) Also, for the nonlinear planar systems of the last few chapters, the Poincaré-Bendixson theorem completely eliminates any possibility of chaotic behavior. So, to find chaotic behavior, we need to look at nonlinear, higher dimensional systems.

In this chapter we investigate the system that is, without doubt, the most famous of all chaotic differential equations, the Lorenz system from meteorology. First formulated in 1963 by E. N. Lorenz as a vastly oversimplified model of atmospheric convection, this system possesses what has come to be known as a "strange attractor." Before the Lorenz model started making headlines, the only types of stable attractors known in differential equations were equilibria and closed orbits. The Lorenz system truly opened up new horizons in all areas of science and engineering, because many of the phenomena present in the Lorenz system have later been found in all of the areas we have previously investigated (biology, circuit theory, mechanics, and elsewhere).

In the ensuing 40 years, much progress has been made in the study of chaotic systems. Be forewarned, however, that the analysis of the chaotic behavior of particular systems like the Lorenz system is usually extremely difficult. Most of the chaotic behavior that is readily understandable arises from geometric models for particular differential equations, rather than from the actual equations themselves. Indeed, this is the avenue we pursue here. We will present a geometric model for the Lorenz system that can be completely analyzed using tools from discrete dynamics. Although this model has been known for some 30 years, it is interesting to note the fact that this model was only shown to be equivalent to the Lorenz system in the year 1999.

14.1 Introduction to the Lorenz System

In 1963, E. N. Lorenz [29] attempted to set up a system of differential equations that would explain some of the unpredictable behavior of the weather. Most viable models for weather involve partial differential equations; Lorenz sought a much simpler and easier-to-analyze system.

The Lorenz model may be somewhat inaccurately thought of as follows. Imagine a planet whose "atmosphere" consists of a single fluid particle. As on earth, this particle is heated from below (and hence rises) and cooled from above (so then falls back down). Can a weather expert predict the "weather" on this planet? Sadly, the answer is no, which raises a lot of questions about the possibility of accurate weather prediction down here on earth, where we have quite a few more particles in our atmosphere.

A little more precisely, Lorenz looked at a two-dimensional fluid cell that was heated from below and cooled from above. The fluid motion can be described by a system of differential equations involving infinitely many variables. Lorenz made the tremendous simplifying assumption that all but three of these variables remained constant. The remaining independent variables then measured, roughly speaking, the rate of convective "overturning" (x), and the horizontal and vertical temperature variation (y and z, respectively). The resulting motion led to a three-dimensional system of differential equations that involved three parameters: the Prandtl number σ, the Rayleigh number r, and another parameter b that is related to the physical size of the system. When all of these simplifications were made, the system of differential equations involved only two nonlinear terms and was given by

$$x' = \sigma(y - x)$$
$$y' = rx - y - xz$$
$$z' = xy - bz.$$

In this system all three parameters are assumed to be positive and, moreover, $\sigma > b + 1$. We denote this system by $X' = \mathcal{L}(X)$. In Figure 14.1, we have displayed the solution curves through two different initial conditions $P_1 = (0, 2, 0)$ and $P_2 = (0, -2, 0)$ when the parameters are $\sigma = 10$, $b = 8/3$, and

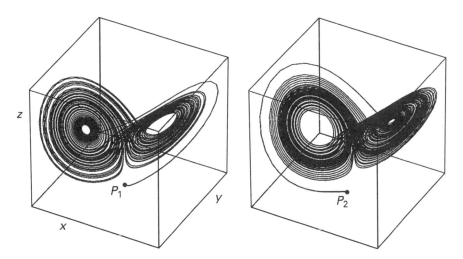

Figure 14.1 The Lorenz attractor. Two solutions with initial conditions $P_1 =$ (0, 2, 0) and $P_2 = (0, -2, 0)$.

$r = 28$. These are the original parameters that led to Lorenz's discovery. Note how both solutions start out very differently, but eventually have more or less the same fate: They both seem to wind around a pair of points, alternating at times which point they encircle. This is the first important fact about the Lorenz system: All nonequilibrium solutions tend eventually to the same complicated set, the so-called *Lorenz attractor*.

There is another important ingredient lurking in the background here. In Figure 14.1, we started with two relatively far apart initial conditions. Had we started with two very close initial conditions, we would not have observed the "transient behavior" apparent in Figure 14.1. Rather, more or less the same picture would have resulted for each solution. This, however, is misleading. When we plot the actual coordinates of the solutions, we see that these two solutions actually move quite far apart during their journey around the Lorenz attractor. This is illustrated in Figure 14.2, where we have graphed the x coordinates of two solutions that start out nearby, one at $(0, 2, 0)$, the other (in gray) at $(0, 2.01, 0)$. These graphs are nearly identical for a certain time period, but then they differ considerably as one solution travels around one of the lobes of the attractor while the other solution travels around the other. No matter how close two solutions start, they always move apart in this manner when they are close to the attractor. This is *sensitive dependence on initial conditions*, one of the main features of a chaotic system.

We will describe in detail the concept of an attractor and chaos in this chapter. But first, we need to investigate some of the more familiar features of the system.

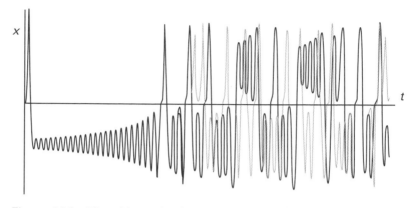

Figure 14.2   The $x(t)$ graphs for two nearby initial conditions $P_1 = (0, 2, 0)$ and $P_2 = (0, 2.01, 0)$.

14.2 Elementary Properties of the Lorenz System

As usual, to analyze this system, we begin by finding the equilibria. Some easy algebra yields three equilibrium points, the origin, and

$$Q_\pm = (\pm\sqrt{b(r-1)}, \pm\sqrt{b(r-1)}, r-1).$$

The latter two equilibria only exist when $r > 1$, so already we see that we have a bifurcation when $r = 1$.

Linearizing, we find the system

$$Y' = \begin{pmatrix} -\sigma & \sigma & 0 \\ r-z & -1 & -x \\ y & x & -b \end{pmatrix} Y.$$

At the origin, the eigenvalues of this matrix are $-b$ and

$$\lambda_\pm = \frac{1}{2}\left(-(\sigma+1)\pm\sqrt{(\sigma+1)^2 - 4\sigma(1-r)}\right).$$

Note that both $\lambda_\pm$ are negative when $0 \le r < 1$. Hence the origin is a sink in this case.

The Lorenz vector field $\mathcal{L}(X)$ possesses a symmetry. If we let $S(x, y, z) = (-x, -y, z)$, then we have $S(\mathcal{L}(X)) = \mathcal{L}(S(X))$. That is, reflection through the z-axis preserves the vector field. In particular, if $(x(t), y(t), z(t))$ is a solution of the Lorenz equations, then so is $(-x(t), -y(t), z(t))$.

When $x = y = 0$, we have $x' = y' = 0$, so the z-axis is invariant. On this axis, we have simply $z' = -bz$, so all solutions tend to the origin on this axis. In fact, the solution through any point in $\mathbb{R}^3$ tends to the origin when $r < 1$, for we have:

Proposition. *Suppose $r < 1$. Then all solutions of the Lorenz system tend to the equilibrium point at the origin.*

Proof: We construct a strict Liapunov function on all of $\mathbb{R}^3$. Let

$$L(x, y, z) = x^2 + \sigma y^2 + \sigma z^2.$$

Then we have

$$\dot{L} = -2\sigma \left(x^2 + y^2 - (1 + r)xy \right) - 2\sigma bz^2.$$

We therefore have $\dot{L} < 0$ away from the origin provided that

$$g(x, y) = x^2 + y^2 - (1 + r)xy > 0$$

for $(x, y) \neq (0, 0)$. This is clearly true along the y-axis. Along any other straight line $y = mx$ in the xy-plane we have

$$g(x, mx) = x^2(m^2 - (1 + r)m + 1).$$

But the quadratic term $m^2 \quad (1 + r)m + 1$ is positive for all m if $r < 1$, as is easily checked. Hence $g(x, y) > 0$ for $(x, y) \neq (0, 0)$. ∎

When r increases through 1, two things happen. First, the eigenvalue λ_+ at the origin becomes positive, so the origin is now a saddle with a two-dimensional stable surface and an unstable curve. Second, the two equilibria $Q_\pm$ are born at the origin when $r = 1$ and move away as r increases.

Proposition. *The equilibrium points $Q_\pm$ are sinks provided*

$$1 < r < r^* = \sigma \left(\frac{\sigma + b + 3}{\sigma - b - 1} \right).$$

Proof: From the linearization, we calculate that the eigenvalues at $Q_\pm$ satisfy the cubic polynomial

$$f_r(\lambda) = \lambda^3 + (1 + b + \sigma)\lambda^2 + b(\sigma + r)\lambda + 2b\sigma(r - 1) = 0.$$

When $r = 1$ the polynomial f_1 has distinct roots at 0, $-b$, and $-\sigma - 1$. These roots are distinct since $\sigma > b + 1$ so that

$$-\sigma - 1 < -\sigma + 1 < -b < 0.$$

Hence for r close to but greater than 1, f_r has three real roots close to these values. Note that $f_r(\lambda) > 0$ for $\lambda \geq 0$ and $r > 1$. Looking at the graph of f_r, it follows that, at least for r close to 1, the three roots of f_r must be real and negative.

We now let r increase and ask what is the lowest value of r for which f_r has an eigenvalue with zero real part. Note that this eigenvalue must in fact be of the form $\pm i\omega$ with $\omega \neq 0$, since f_r is a real polynomial that has no roots equal to 0 when $r > 1$. Solving $f_r(i\omega) = 0$ by equating both real and imaginary parts to zero then yields the result (recall that we have assumed $\sigma > b + 1$). ∎

We remark that a Hopf bifurcation is known to occur at r^*, but proving this is beyond the scope of this book.

When $r > 1$ it is no longer true that all solutions tend to the origin. However, we can say that solutions that start far from the origin do at least move closer in. To be precise, let

$$V(x, y, z) = rx^2 + \sigma y^2 + \sigma(z - 2r)^2.$$

Note that $V(x, y, z) = \nu > 0$ defines an ellipsoid in $\mathbb{R}^3$ centered at $(0, 0, 2r)$. We will show:

Proposition. *There exists ν^* such that any solution that starts outside the ellipsoid $V = \nu^*$ eventually enters this ellipsoid and then remains trapped therein for all future time.*

Proof: We compute

$$\dot{V} = -2\sigma\left(rx^2 + y^2 + b(z^2 - 2rz)\right)$$
$$= -2\sigma\left(rx^2 + y^2 + b(z - r)^2 - br^2\right).$$

The equation

$$rx^2 + y^2 + b(z - r)^2 = \mu$$

also defines an ellipsoid when $\mu > 0$. When $\mu > br^2$ we have $\dot{V} < 0$. Thus we may choose ν^* large enough so that the ellipsoid $V = \nu^*$ strictly contains the ellipsoid

$$rx^2 + y^2 + b(z - r)^2 = br^2$$

in its interior. Then $\dot{V} < 0$ for all $v \geq v^*$. ∎

As a consequence, all solutions starting far from the origin are attracted to a set that sits inside the ellipsoid $V = v^*$. Let Λ denote the set of all points whose solutions remain for all time (forward and backward) in this ellipsoid. Then the ω-limit set of any solution of the Lorenz system must lie in Λ. Theoretically, Λ could be a large set, perhaps bounding an open region in $\mathbb{R}^3$. However, for the Lorenz system, this is not the case.

To see this, recall from calculus that the *divergence* of a vector field $F(X)$ on $\mathbb{R}^3$ is given by

$$\text{div } F = \sum_{i=1}^{3} \frac{\partial F_i}{\partial x_i}(X).$$

The divergence of F measures how fast volumes change under the flow ϕ_t of F. Suppose D is a region in $\mathbb{R}^3$ with a smooth boundary, and let $D(t) = \phi_t(D)$, the image of D under the time t map of the flow. Let $V(t)$ be the volume of $D(t)$. Then Liouville's theorem asserts that

$$\frac{dV}{dt} = \int_{D(t)} \text{div } F \, dx \, dy \, dz.$$

For the Lorenz system, we compute immediately that the divergence is the constant $-(\sigma + 1 + b)$ so that volume decreases at a constant rate

$$\frac{dV}{dt} = -(\sigma + 1 + b)V.$$

Solving this simple differential equation yields

$$V(t) = e^{-(\sigma+1+b)t} V(0)$$

so that any volume must shrink exponentially fast to 0. In particular, we have:

Proposition. *The volume of Λ is zero.* ∎

The natural question is what more can we say about the structure of the "attractor" Λ? In dimension two, such a set would consist of a collection of limit cycles, equilibrium points, and solutions connecting them. In higher dimensions, these attractors may be much "stranger," as we show in the next section.

14.3 The Lorenz Attractor

The behavior of the Lorenz system as the parameter r increases is the subject of much contemporary research; we are decades (if not centuries) away from rigorously understanding all of the fascinating dynamical phenomena that occur as the parameters change. Sparrow has written an entire book devoted to this subject [44].

In this section we will deal with one specific set of parameters where the Lorenz system has an attractor. Roughly speaking, an attractor for the flow is an invariant set that "attracts" all nearby solutions. To be more precise:

Definition
Let $X' = F(X)$ be a system of differential equations in $\mathbb{R}^n$ with flow ϕ_t. A set Λ is called an *attractor* if

1. Λ is compact and invariant;
2. There is an open set U containing Λ such that for each $X \in U$, $\phi_t(X) \in U$ for all $t \geq 0$ and $\cap_{t \geq 0} \phi_t(U) = \Lambda$;
3. (Transitivity) Given any points $Y_1, Y_2 \in \Lambda$ and any open neighborhoods U_j about Y_j in U, there is a solution curve that begins in U_1 and later passes through U_2.

The transitivity condition in this definition may seem a little strange. Basically, we include it to guarantee that we are looking at a single attractor rather than a collection of dynamically different attractors. For example, the transitivity condition rules out situations such as that given by the planar system

$$x' = x - x^3$$
$$y' = -y.$$

The phase portrait of this system is shown in Figure 14.3. Note that any solution of this system enters the set marked U and then tends to one of the three equilibrium points: either to one of the sinks at $(\pm 1, 0)$ or to the saddle $(0, 0)$. The forward intersection of the flow ϕ_t applied to U is the interval $-1 \leq x \leq 1$. This interval meets conditions 1 and 2 in the definition, but

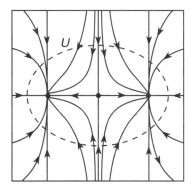

Figure 14.3 The interval on the x-axis between the two sinks is not an attractor for this system, despite the fact that all solutions enter U.

condition 3 is violated, because none of the solution curves passes close to points in both the left and right half of this interval. We choose not to consider this set an attractor since most solutions tend to one of the two sinks. We really have two distinct attractors in this case.

As a remark, there is no universally accepted definition of an attractor in mathematics; some people choose to say that a set Λ that meets only conditions 1 and 2 is an attractor, while if Λ also meets condition 3, it would be called a transitive attractor. For planar systems, condition 3 is usually easily verified; in higher dimensions, however, this can be much more difficult, as we shall see.

For the rest of this chapter, we restrict attention to the very special case of the Lorenz system where the parameters are given by $\sigma = 10$, $b = 8/3$, and $r = 28$. Historically, these are the values Lorenz used when he first encountered chaotic phenomena in this system. Thus, the specific Lorenz system we consider is

$$X' = \mathcal{L}(X) = \begin{pmatrix} 10(y - x) \\ 28x - y - xz \\ xy - (8/3)z \end{pmatrix}.$$

As in the previous section, we have three equilibria: the origin and $Q_\pm = (\pm 6\sqrt{2}, \pm 6\sqrt{2}, 27)$. At the origin we find eigenvalues $\lambda_1 = -8/3$ and

$$\lambda_\pm = -\frac{11}{2} \pm \frac{\sqrt{1201}}{2}.$$

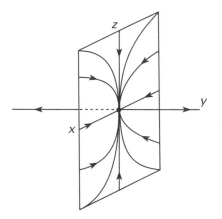

Figure 14.4   Linearization at the
origin for the Lorenz system.

For later use, note that these eigenvalues satisfy

$$\lambda_- < -\lambda_+ < \lambda_1 < 0 < \lambda_+.$$

The linearized system at the origin is then

$$Y' = \begin{pmatrix} \lambda_- & 0 & 0 \\ 0 & \lambda_+ & 0 \\ 0 & 0 & \lambda_1 \end{pmatrix} Y.$$

The phase portrait of the linearized system is shown in Figure 14.4. Note that all solutions in the stable plane of this system tend to the origin tangentially to the z-axis.

At $Q_\pm$ a computation shows that there is a single negative real eigenvalue and a pair of complex conjugate eigenvalues with positive real parts. Note that the symmetry in the system forces the rotations about Q_+ and Q_- to have opposite orientations.

In Figure 14.5, we have displayed a numerical computation of a portion of the left- and right-hand branches of the unstable curve at the origin. Note that the right-hand portion of this curve comes close to Q_- and then spirals away. The left portion behaves symmetrically under reflection through the z-axis. In Figure 14.6, we have displayed a significantly larger portion of these unstable curves. Note that they appear to circulate around the two equilibria, sometimes spiraling around Q_+, sometimes about Q_-. In particular, these curves continually reintersect the portion of the plane $z = 27$ containing $Q_\pm$ in which the vector field points downward. This suggests that we may construct

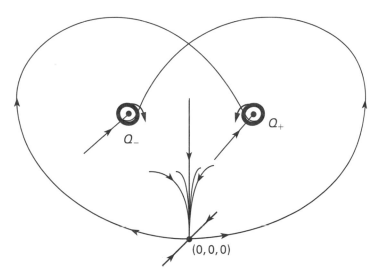

Figure 14.5   The unstable curve at the origin.

a Poincaré map on a portion of this plane. As we have seen before, computing a Poincaré map is often impossible, and this case is no different. So we will content ourselves with building a simplified model that exhibits much of the behavior we find in the Lorenz system. As we shall see in the following section, this model provides a computable means to assess the chaotic behavior of the system.

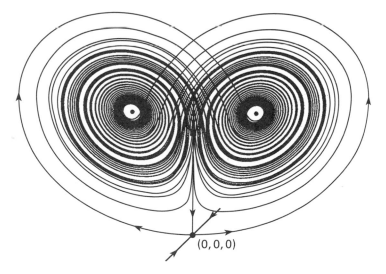

Figure 14.6   More of the unstable curve at the origin.

14.4 A Model for the Lorenz Attractor

In this section we describe a geometric model for the Lorenz attractor originally proposed by Guckenheimer and Williams [20]. Tucker [46] showed that this model does indeed correspond to the Lorenz system for certain parameters. Rather than specify the vector field exactly, we give instead a qualitative description of its flow, much as we did in Chapter 11. The specific numbers we use are not that important; only their relative sizes matter.

We will assume that our model is symmetric under the reflection $(x, y, z) \rightarrow (-x, -y, z)$, as is the Lorenz system. We first place an equilibrium point at the origin in $\mathbb{R}^3$ and assume that, in the cube S given by $|x|, |y|, |z| \leq 5$, the system is linear. Rather than use the eigenvalues λ_1 and $\lambda_{\pm}$ from the actual Lorenz system, we simplify the computations a bit by assuming that the eigenvalues are $-1, 2$, and -3, and that the system is given in the cube by

$$x' = -3x$$
$$y' = 2y$$
$$z' = -z.$$

Note that the phase portrait of this system agrees with that in Figure 14.4 and that the relative magnitudes of the eigenvalues are the same as in the Lorenz case.

We need to know how solutions make the transit near $(0, 0, 0)$. Consider a rectangle $\mathcal{R}_1$ in the plane $z = 1$ given by $|x| \leq 1$, $0 < y \leq \epsilon < 1$. As time moves forward, all solutions that start in $\mathcal{R}_1$ eventually reach the rectangle $\mathcal{R}_2$ in the plane $y = 1$ defined by $|x| \leq 1$, $0 < z \leq 1$. Hence we have a function $h\colon \mathcal{R}_1 \rightarrow \mathcal{R}_2$ defined by following solution curves as they pass from $\mathcal{R}_1$ to $\mathcal{R}_2$. We leave it as an exercise to check that this function assumes the form

$$h\begin{pmatrix} x \\ y \end{pmatrix} = \begin{pmatrix} x_1 \\ z_1 \end{pmatrix} = \begin{pmatrix} xy^{3/2} \\ y^{1/2} \end{pmatrix}.$$

It follows that h takes lines $y = c$ in $\mathcal{R}_1$ to lines $z = c^{1/2}$ in $\mathcal{R}_2$. Also, since $x_1 = xz_1^3$, we have that h maps lines $x = c$ to curves of the form $x_1 = cz_1^3$. Each of these image curves meet the xy–plane perpendicularly, as depicted in Figure 14.7.

Mimicking the Lorenz system, we place two additional equilibria in the plane $z = 27$, one at $Q_- = (-10, -20, 27)$ and the other at $Q_+ = (10, 20, 27)$. We assume that the lines given by $y = \pm 20$, $z = 27$ form portions of the stable lines at $Q_{\pm}$, and that the other two eigenvalues at these points are complex with positive real part.

Let Σ denote the square $|x|, |y| \leq 20$, $z = 27$. We assume that the vector field points downward in the interior of this square. Hence solutions spiral

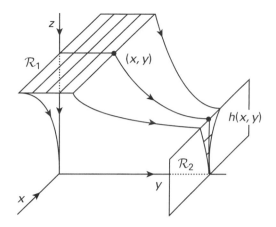

Figure 14.7 Solutions making the transit near (0, 0, 0).

away from $Q_\pm$ in the same manner as in the Lorenz system. We also assume that the stable surface of $(0, 0, 0)$ first meets Σ in the line of intersection of the xz–planc and Σ.

Let $\zeta^\pm$ denote the two branches of the unstable curve at the origin. We assume that these curves make a passage around Σ and then enter this square as shown in Figure 14.8. We denote the first point of intersection of $\zeta^\pm$ with Σ by $\rho^\pm = (\pm x^*, \mp y^*)$.

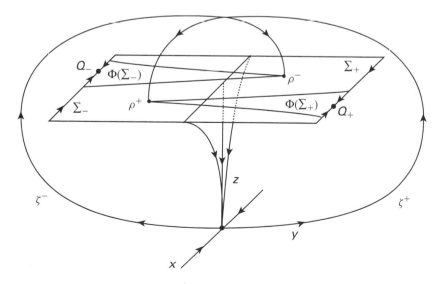

Figure 14.8 The solutions $\zeta^\pm$ and their intersection with Σ in the model for the Lorenz attractor.

Now consider a straight line $y = v$ in Σ. If $v = 0$, all solutions beginning at points on this line tend to the origin as time moves forward. Hence these solutions never return to Σ. We assume that all other solutions originating in Σ do return to Σ as time moves forward. How these solutions return leads to our major assumptions about this model:

1. *Return condition:* Let $\Sigma_+ = \Sigma \cap \{y > 0\}$ and $\Sigma_- = \Sigma \cap \{y < 0\}$. We assume that the solutions through any point in $\Sigma_\pm$ return to Σ in forward time. Hence we have a Poincaré map $\Phi : \Sigma_+ \cup \Sigma_- \to \Sigma$. We assume that the images $\Phi(\Sigma_\pm)$ are as depicted in Figure 14.8. By symmetry, we have $\Phi(x, y) = -\Phi(-x, -y)$.
2. *Contracting direction:* For each $v \neq 0$ we assume that Φ maps the line $y = v$ in Σ into the line $y = g(v)$ for some function g. Moreover we assume that Φ contracts this line in the x direction.
3. *Expanding direction:* We assume that Φ stretches Σ_+ and Σ_- in the y direction by a factor strictly greater than $\sqrt{2}$, so that $g'(y) > \sqrt{2}$.
4. *Hyperbolicity condition:* Besides the expansion and contraction, we assume that $D\Phi$ maps vectors tangent to $\Sigma_\pm$ whose slopes are ± 1 to vectors whose slopes have magnitude larger than $\mu > 1$.

Analytically, these assumptions imply that the map Φ assumes the form

$$\Phi(x, y) = (f(x, y), g(y))$$

where $g'(y) > \sqrt{2}$ and $0 < \partial f / \partial x < c < 1$. The hyperbolicity condition implies that

$$g'(y) > \mu \left| \frac{\partial f}{\partial x} \pm \frac{\partial f}{\partial y} \right|.$$

Geometrically, this condition implies that the sectors in the tangent planes given by $|y| \geq |x|$ are mapped by $D\Phi$ strictly inside a sector with steeper slopes. Note that this condition holds if $|\partial f / \partial y|$ and c are sufficiently small throughout $\Sigma_\pm$.

Technically, $\Phi(x, 0)$ is not defined, but we do have

$$\lim_{y \to 0^\pm} \Phi(x, y) = \rho^\pm$$

where we recall that $\rho^\pm$ is the first point of intersection of $\zeta^\pm$ with Σ. We call $\rho^\pm$ the *tip* of $\Phi(\Sigma_\pm)$. In fact, our assumptions on the eigenvalues guarantee that $g'(y) \to \infty$ as $y \to 0$ (see Exercise 3 at the end of this chapter).

To find the attractor, we may restrict attention to the rectangle $R \subset \Sigma$ given by $|y| \leq y^*$, where we recall that $\mp y^*$ is the y coordinate of the tips $\rho^\pm$. Let $R_\pm = R \cap \Sigma_\pm$. It is easy to check that any solution starting in the interior of $\Sigma_\pm$

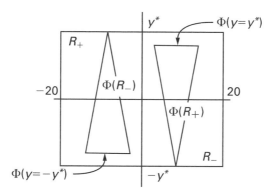

Figure 14.9   The Poincaré map Φ on R.

but outside R must eventually meet R, so it suffices to consider the behavior of Φ on R. A planar picture of the action of Φ on R is displayed in Figure 14.9. Note that $\Phi(R) \subset R$.

Let Φ^n denote the nth iterate of Φ, and let

$$A = \bigcap_{n=0}^{\infty} \overline{\Phi^n(R)}.$$

Here $\overline{U}$ denotes the closure of the set U. The set A will be the intersection of the attractor for the flow with R. That is, let

$$\mathcal{A} = \left(\bigcup_{t \in \mathbb{R}} \phi_t(A) \right) \cup \{(0,0,0)\}.$$

We add the origin here so that $\mathcal{A}$ will be a closed set. We will prove:

Theorem. $\mathcal{A}$ *is an attractor for the model Lorenz system.*

Proof: The proof that $\mathcal{A}$ is an attractor for the flow follows immediately from the fact that A is an attractor for the mapping Φ (where an attractor for a mapping is defined completely analogously to that for a flow). Clearly A is closed. Technically, A itself is not invariant under Φ since Φ is not defined along $y = 0$. However, for the flow, the solutions through all such points do lie in $\mathcal{A}$ and so $\mathcal{A}$ is invariant. If we let $\mathcal{O}$ be the open set given by the interior of Σ, then for any $(x, y) \in \mathcal{O}$, there is an n such that $\Phi^n(x, y) \in R$. Hence

$$\bigcap_{n=0}^{\infty} \Phi^n(\mathcal{O}) \subset A$$

By definition, $A = \cap_{n \geq 0} \Phi^n(R)$, and so $A = \cap_{n \geq 0} \Phi^n(\mathcal{O})$ as well. Therefore conditions 1 and 2 in the definition of an attractor hold for Φ.

It remains to show the transitivity property. We need to show that if P_1 and P_2 are points in A, and W_j are open neighborhoods of P_j in $\mathcal{O}$, then there exists an $n \geq 0$ such that $\Phi^n(W_1) \cap W_2 \neq \emptyset$.

Given a set $U \subset R$, let $\Pi_y(U)$ denote the projection of U onto the y-axis. Also let $\ell_y(U)$ denote the length of $\Pi_y(U)$, which we call the y length of U. In the following, U will be a finite collection of connected sets, so $\ell_y(U)$ is well defined.

We need a lemma. ■

Lemma. *For any open set $W \subset R$, there exists $n > 0$ such that $\Pi_y(\Phi^n(W))$ is the interval $[-y^*, y^*]$. Equivalently, $\Phi^n(W)$ meets each line $y = c$ in R.*

Proof: First suppose that W contains a connected subset W' that extends from one of the tips to $y = 0$. Hence $\ell_y(W') = y^*$. Then $\Phi(W')$ is connected and we have $\ell_y(\Phi(W')) > \sqrt{2}y^*$. Moreover, $\Phi(W')$ also extends from one of the tips, but now crosses $y = 0$ since its y length exceeds y^*.

Now apply Φ again. Note that $\Phi^2(W')$ contains two pieces, one of which extends to ρ^+, the other to ρ^-. Moreover, $\ell_y(\Phi^2(W')) > 2y^*$. Thus it follows that $\Pi_y(\Phi^2(W')) = [-y^*, y^*]$ and so we are done in this special case.

For the general case, suppose first that W is connected and does not cross $y = 0$. Then we have $\ell_y(\Phi(W)) > \sqrt{2}\ell_y(W)$ as above, so the y length of $\Phi(W)$ grows by a factor of more than $\sqrt{2}$.

If W does cross $y = 0$, then we may find a pair of connected sets $W^\pm$ with $W^\pm \subset \{R^\pm \cap W\}$ and $\ell_y(W^+ \cup W^-) = \ell_y(W)$. The images $\Phi(W^\pm)$ extend to the tips $\rho^\pm$. If either of these sets also meets $y = 0$, then we are done by the above. If neither $\Phi(W^+)$ nor $\Phi(W^-)$ crosses $y = 0$, then we may apply Φ again. Both $\Phi(W^+)$ and $\Phi(W^-)$ are connected sets, and we have $\ell_y(\Phi^2(W^\pm)) > 2\ell_y(W^\pm)$. Hence for one of W^+ or W^- we have $\ell_y(\Phi^2(W^\pm)) > \ell_y(W)$ and again the y length of W grows under iteration.

Thus if we continue to iterate Φ or Φ^2 and choose the appropriate largest subset of $\Phi^j(W)$ at each stage as above, then we see that the y lengths of these images grow without bound. This completes the proof of the lemma. ■

We now complete the proof of the theorem. We must find a point in W_1 whose image under an iterate of Φ lies in W_2. Toward that end, note that

$$\left| \Phi^k(x_1, y) - \Phi^k(x_2, y) \right| \leq c^k |x_1 - x_2|$$

since $\Phi^j(x_1, y)$ and $\Phi^j(x_2, y)$ lie on the same straight line parallel to the x-axis for each j and Φ contracts distances in the x direction by a factor of $c < 1$.

We may assume that W_2 is a disk of diameter ϵ. Recalling that the width of R in the x direction is 40, we choose m such that $40c^m < \epsilon$. Consider $\Phi^{-m}(P_2)$. Note that $\Phi^{-m}(P_2)$ is defined since $P_2 \in \cap_{n \geq 0} \Phi^n(R)$. Say $\Phi^{-m}(P_2) = (\xi, \eta)$.

From the lemma, we know that there exists n such that

$$\Pi_y(\Phi^n(W_1)) = [-y^*, y^*].$$

Hence we may choose a point $(\xi_1, \eta) \in \Phi^n(W_1)$. Say $(\xi_1, \eta) = \Phi^n(\tilde{x}, \tilde{y})$ where $(\tilde{x}, \tilde{y}) \in W_1$, so that $\Phi^n(\tilde{x}, \tilde{y})$ and $\Phi^{-m}(P_2)$ have the same y coordinate. Then we have

$$|\Phi^{m+n}(\tilde{x}, \tilde{y}) - P_2| = |\Phi^m(\xi_1, \eta) - P_2|$$
$$= |\Phi^m(\xi_1, \eta) - \Phi^m(\xi, \eta)|$$
$$\leq 40c^m < \epsilon.$$

We have found a point $(\tilde{x}, \tilde{y}) \in W_1$ whose solution passes through W_2. This concludes the proof. ■

Note that, in the above proof, the solution curve that starts near P_1 and comes close to P_2 need not lie in the attractor. However, it is possible to find such a solution that does lie in $\mathcal{A}$ (see Exercise 4).

14.5 The Chaotic Attractor

In the previous section we reduced the study of the behavior of solutions of the Lorenz system to the analysis of the dynamics of the Poincaré map Φ. In the process, we dropped from a three-dimensional system of differential equations to a two-dimensional mapping. But we can do better. According to our assumptions, two points that share the same y coordinate in Σ are mapped to two new points whose y coordinates are given by $g(y)$ and hence are again the same. Moreover, the distance between these points is contracted. It follows that, under iteration of Φ, we need not worry about all points on a line $y = $ constant; we need only keep track of how the y coordinate changes under iteration of g. Then, as we shall see, the Poincaré map Φ is completely determined by the dynamics of the one-dimensional function g defined on the interval $[-y^*, y^*]$. Indeed, iteration of this function completely determines the behavior of all solutions in the attractor. In this section, we begin to analyze

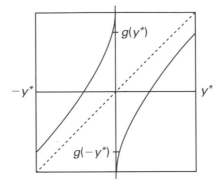

Figure 14.10 The graph of the one-dimensional function g on $I = [-y^*, y^*]$.

the dynamics of this *one-dimensional discrete dynamical system*. In Chapter 15, we plunge more deeply into this topic.

Let I be the interval $[-y^*, y^*]$. Recall that g is defined on I except at $y = 0$ and satisfies $g(-y) = -g(y)$. From the results of the previous section, we have $g'(y) > \sqrt{2}$, and $0 < g(y^*) < y^*$ and $-y^* < g(-y^*) < 0$. Also,

$$\lim_{y \to 0^{\pm}} g(y) = \mp y^*.$$

Hence the graph of g resembles that shown in Figure 14.10. Note that all points in the interval $[g(-y^*), g(y^*)]$ have two preimages, while points in the intervals $(-y^*, g(-y^*))$ and $(g(y^*), y^*)$ have only one. The endpoints of I, namely, $\pm y^*$, have no preimages in I since $g(0)$ is undefined.

Let $y_0 \in I$. We will investigate the structure of the set $A \cap \{y = y_0\}$. We define the *(forward) orbit of y_0* to be the set $(y_0, y_1, y_2, \ldots)$ where $y_n = g(y_{n-1}) = g^n(y_0)$. For each y_0, the forward orbit of y_0 is uniquely determined, though it terminates if $g^n(y_0) = 0$.

A *backward orbit of y_0* is a sequence of the form $(y_0, y_{-1}, y_{-2}, \ldots)$ where $g(y_{-k}) = y_{-k+1}$. Unlike forward orbits of g, there are infinitely many distinct backward orbits for a given y_0 except in the case where $y_0 = \pm y^*$ (since these two points have no preimages in I). To see this, suppose first that y_0 does not lie on the forward orbit of $\pm y^*$. Then each point y_{-k} must have either one or two distinct preimages, since $y_{-k} \neq \pm y^*$. If y_{-k} has only one preimage y_{-k-1}, then y_{-k} lies in either $(-y^*, g(-y^*))$ or $(g(y^*), y^*)$. But then the graph of g shows that y_{-k-1} must have two preimages. So no two consecutive points in a given backward orbit can have only one preimage, and this shows that y_0 has infinitely many distinct backward orbits.

If we happen to have $y_{-k} = \pm y^*$ for some $k > 0$, then this backward orbit stops since $\pm y^*$ has no preimage in I. However, y_{-k+1} must have two preimages, one of which is the endpoint and the other is a point in I that does not equal the other endpoint. Hence we can continue taking preimages of this second backward orbit as before, thereby generating infinitely many distinct backward orbits as before.

We claim that each of these infinite backward orbits of y_0 corresponds to a unique point in $A \cap \{y = y_0\}$. To see this, consider the line J_{-k} given by $y = y_{-k}$ in R. Then $\Phi^k(J_{-k})$ is a closed subinterval of J_0 for each k. Note that $\Phi(J_{-k-1}) \subset J_{-k}$, since $\Phi(y = y_{-k-1})$ is a proper subinterval of $y = y_{-k}$. Hence the nested intersection of the sets $\Phi^k(J_{-k})$ is nonempty, and any point in this intersection has backward orbit $(y_0, y_{-1}, y_{-2}, \ldots)$ by construction. Furthermore, the intersection point is unique, since each application of Φ contracts the intervals $y = y_{-k}$ by a factor of $c < 1$.

In terms of our model, we therefore see that the attractor $\mathcal{A}$ is a complicated set. We have proved the following:

Proposition.   *The attractor $\mathcal{A}$ for the model Lorenz system meets each of the lines $y = y_0 \neq y^*$ in R infinitely many distinct points. In forward time all of the solution curves through each point on this line either:*

1. *Meet the line $y = 0$, in which case the solution curves all tend to the equilibrium point at $(0, 0, 0)$, or*
2. *Continually reintersect R, and the distances between these intersection points on the line $y = y_k$ tend to 0 as time increases.* ■

Now we turn to the dynamics of Φ in R. We first discuss the behavior of the one-dimensional function g, and then use this information to understand what happens for Φ. Given any point $y_0 \in I$, note that nearby forward orbits of g move away from the orbit of y_0 since $g' > \sqrt{2}$. More precisely, we have:

Proposition.   *Let $0 < v < y^*$. Let $y_0 \in I = [-y^*, y^*]$. Given any $\epsilon > 0$, we may find $u_0, v_0 \in I$ with $|u_0 - y_0| < \epsilon$ and $|v_0 - y_0| < \epsilon$ and $n > 0$ such that $|g^n(u_0) - g^n(v_0)| \geq 2v$.*

Proof: Let J be the interval of length 2ϵ centered at y_0. Each iteration of g expands the length of J by a factor of at least $\sqrt{2}$, so there is an iteration for which $g^n(J)$ contains 0 in its interior. Then $g^{n+1}(J)$ contains points arbitrarily close to both $\pm y^*$, and hence there are points in $g^{n+1}(J)$ whose distance from each other is at least $2v$. This completes the proof. ■

Let's interpret the meaning of this proposition in terms of the attractor A. Given any point in the attractor, we may always find points arbitrarily nearby whose forward orbits move apart just about as far as they possibly can. This is the hallmark of a chaotic system: We call this behavior *sensitive dependence on initial conditions*. A tiny change in the initial position of the orbit may result in drastic changes in the eventual behavior of the orbit. Note that we must have a similar sensitivity for the flow in $\mathcal{A}$; certain nearby solution curves in $\mathcal{A}$ must also move far apart. This is the behavior we witnessed in Figure 14.2.

This should be contrasted with the behavior of points in A that lie on the same line $y = $ constant with $-y^* < y < y^*$. As we saw previously, there are infinitely many such points in A. Under iteration of Φ, the successive images of all of these points move closer together rather than separating.

Recall now that a subset of I is dense if its closure is all of I. Equivalently, a subset of I is dense if there are points in the subset arbitrarily close to any point whatsoever in I. Also, a *periodic point* for g is a point y_0 for which $g^n(y_0) = y_0$ for some $n > 0$. Periodic points for g correspond to periodic solutions of the flow.

Proposition.   *The periodic points of g are dense in I.*

Proof: As in the proof that A is an attractor in the last section, given any subinterval J of $I - \{0\}$, we may find n so that g^n maps some subinterval $J' \subset J$ in one-to-one fashion over either $(-y^*, 0]$ or $[0, y^*)$. Thus either $g^n(J')$ contains J', or the next iteration, $g^{n+1}(J')$, contains J'. In either case, the graphs of g^n or g^{n+1} cross the diagonal line $y = x$ over J'. This yields a periodic point for g in J.   ∎

Now let us interpret this result in terms of the attractor A. We claim that periodic points for Φ are also dense in A. To see this, let $P \in A$ and U be an open neighborhood of P. We assume that U does not cross the line $y = 0$ (otherwise just choose a smaller neighborhood nearby that is disjoint from $y = 0$). For small enough $\epsilon > 0$, we construct a rectangle $W \subset U$ centered at P and having width 2ϵ (in the x direction) and height ϵ (in the y direction).

Let $W_1 \subset W$ be a smaller square centered at P with sidelength $\epsilon/2$. By the transitivity result of the previous section, we may find a point $Q_1 \in W_1$ such that $\Phi^n(Q_1) = Q_2 \in W_1$. By choosing a subset of W_1 if necessary, we may assume that $n > 4$ and furthermore that n is so large that $c^n < \epsilon/8$. It follows that the image of $\Phi^n(W)$ (not $\Phi^n(W_1)$) crosses through the interior of W nearly vertically and extends beyond its top and bottom boundaries, as depicted in Figure 14.11. This fact uses the hyperbolicity condition.

Now consider the lines $y = c$ in W. These lines are mapped to other such lines in R by Φ^n. Since the vertical direction is expanded, some of the lines must be mapped above W and some below. It follows that one such line $y = c_0$ must be mapped inside itself by Φ^n, and therefore there must be a fixed point for Φ^n on this line. Since this line is contained in W, we have produced a periodic point for Φ in W. This proves density of periodic points in A.

In terms of the flow, a solution beginning at a periodic point of Φ is a closed orbit. Hence the set of points lying on closed orbits is a dense subset of $\mathcal{A}$. The

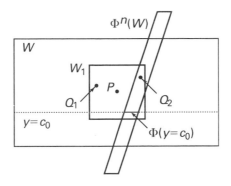

Figure 14.11 Φ maps W across itself.

structure of these closed orbits is quite interesting from a topological point of view, because many of these closed curves are actually "knotted." See [9] and Exercise 10.

Finally, we say that a function g is *transitive* on I if, for any pair of points y_1 and y_2 in I and neighborhoods U_i of y_i, we can find $\tilde{y} \in U_1$ and n such that $g^n(\tilde{y}) \in U_2$. Just as in the proof of density of periodic points, we may use the fact that g is expanding in the y direction to prove:

Proposition. *The function g is transitive on I.*

We leave the details to the reader. In terms of Φ, we almost proved the corresponding result when we showed that A was an attractor. The only detail we did not provide was the fact that we could find a point in A whose orbit made the transit arbitrarily close to any given pair of points in A. For this detail, we refer to Exercise 4.

Thus we can summarize the dynamics of Φ on the attractor A of the Lorenz model as follows.

Theorem. (Dynamics of the Lorenz Model) *The Poincaré map Φ restricted to the attractor A for the Lorenz model has the following properties:*

1. *Φ has sensitive dependence on initial conditions;*
2. *Periodic points of Φ are dense in A;*
3. *Φ is transitive on A.* ■

We say that a mapping with the above properties is *chaotic.* We caution the reader that, just as in the definition of an attractor, there are many definitions of chaos around. Some involve exponential separation of orbits, others involve *positive Liapunov exponents,* and others do not require density of periodic points. It is an interesting fact that, for continuous functions of the real line, density of periodic points and transitivity are enough to guarantee sensitive dependence. See [8]. We will delve more deeply into chaotic behavior of discrete systems in the next chapter.

14.6 Exploration: The Rössler Attractor

In this exploration, we investigate a three-dimensional system similar in many respects to the Lorenz system. The Rössler system [37] is given by

$$x' = -y - z$$
$$y' = x + ay$$
$$z' = b + z(x - c)$$

where a, b, and c are real parameters. For simplicity, we will restrict attention to the case where $a = 1/4$, $b = 1$, and c ranges from 0 to 7.

As with the Lorenz system, it is difficult to prove specific results about this system, so much of this exploration will center on numerical experimentation and the construction of a model.

1. First find all equilibrium points for this system.
2. Describe the bifurcation that occurs at $c = 1$.
3. Investigate numerically the behavior of this system as c increases. What bifurcations do you observe?
4. In Figure 14.12 we have plotted a single solution for $c = 5.5$. Compute other solutions for this parameter value, and display the results from other

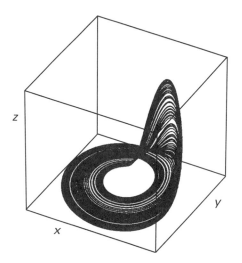

Figure 14.12 The Rössler attractor.

viewpoints in $\mathbb{R}^3$. What conjectures do you make about the behavior of this system?

5. Using techniques described in this chapter, devise a geometric model that mimics the behavior of the Rössler system for this parameter value.

6. Construct a model mapping on a two-dimensional region whose dynamics might explain the behavior observed in this system.

7. As in the Lorenz system, describe a possible way to reduce this function to a mapping on an interval.

8. Give an explicit formula for this one-dimensional model mapping. What can you say about the chaotic behavior of your model?

9. What other bifurcations do you observe in the Rössler system as c rises above 5.5?

EXERCISES

1. Consider the system

$$x' = -3x$$
$$y' = 2y$$
$$z' = -z.$$

Recall from Section 14.4 that there is a function $h\colon \mathcal{R}_1 \to \mathcal{R}_2$ where $\mathcal{R}_1$ is given by $|x| \le 1$, $0 < y \le \epsilon < 1$ and $z = 1$, and $\mathcal{R}_2$ is given by $|x| \le 1$, $0 < z \le 1$, and $y = 1$. Show that h is given by

$$h\begin{pmatrix} x \\ y \end{pmatrix} = \begin{pmatrix} x_1 \\ z_1 \end{pmatrix} = \begin{pmatrix} xy^{3/2} \\ y^{1/2} \end{pmatrix}.$$

2. Suppose that the roles of x and z are reversed in the previous exercise. That is, suppose $x' = -x$ and $z' = -3z$. Describe the image of $h(x, y)$ in $\mathcal{R}_2$ in this case.

3. For the Poincaré map $\Phi(x, y) = (f(x, y), g(y))$ for the model attractor, use the results of Exercise 1 to show that $g'(y) \to \infty$ as $y \to 0$.

4. Show that it is possible to verify the transitivity condition for the Lorenz model with a solution that actually lies in the attractor.

5. Prove that arbitrarily close to any point in the model Lorenz attractor, there is a solution that eventually tends to the equilibrium point at $(0, 0, 0)$.

6. Prove that there is a periodic solution γ of the model Lorenz system that meets the rectangle R in precisely two distinct points.

7. Prove that arbitrarily close to any point in the model Lorenz attractor, there is a solution that eventually tends to the periodic solution γ from the previous exercise.

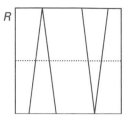

Figure 14.13 Φ maps
R completely across
itself.

8. Consider a map Φ on a rectangle R as shown in Figure 14.13. where Φ has properties similar to the model Lorenz Φ. How many periodic points of period n does Φ have?

9. Consider the system

$$x' = 10(y - x)$$
$$y' = 28x - y + xz$$
$$z' = xy - (8/3)z.$$

Show that this system is *not* chaotic in the region where x, y, and z are all positive. (Note the $+xz$ term in the equation for y'.) *Hint:* Show that most solutions tend to ∞.

10. A simple closed curve in $\mathbb{R}^3$ is *knotted* if it cannot be continuously deformed into the "unknot," the unit circle in the xy–plane, without having self-intersections along the way. Using the model Lorenz attractor, sketch a curve that follows the dynamics of Φ (so should approximate a real solution) and is knotted. (You might want to use some string for this!)

11. Use a computer to investigate the behavior of the Lorenz system as r increases from 1 to 28 (with $\sigma = 10$ and $b = 8/3$). Describe in qualitative terms any bifurcations you observe.

15

Discrete Dynamical Systems

Our goal in this chapter is to begin the study of discrete dynamical systems. As we have seen at several stages in this book, it is sometimes possible to reduce the study of the flow of a differential equation to that of an iterated function, namely, a Poincaré map. This reduction has several advantages. First and foremost, the Poincaré map lives on a lower dimensional space, which therefore makes visualization easier. Secondly, we do not have to integrate to find "solutions" of discrete systems. Rather, given the function, we simply iterate the function over and over to determine the behavior of the orbit, which then dictates the behavior of the corresponding solution.

Given these two simplifications, it then becomes much easier to comprehend the complicated chaotic behavior that often arises for systems of differential equations. While the study of discrete dynamical systems is a topic that could easily fill this entire book, we will restrict attention here primarily to the portion of this theory that helps us understand chaotic behavior in one dimension. In the following chapter we will extend these ideas to higher dimensions.

15.1 Introduction to Discrete Dynamical Systems

Throughout this chapter we will work with real functions $f : \mathbb{R} \to \mathbb{R}$. As usual, we assume throughout that f is C^∞, although there will be several special examples where this is not the case.

Let f^n denote the nth iterate of f. That is, f^n is the n-fold composition of f with itself. Given $x_0 \in \mathbb{R}$, the *orbit* of x_0 is the sequence

$$x_0, \ x_1 = f(x_0), \ x_2 = f^2(x_0), \ldots, \ x_n = f^n(x_0), \ldots.$$

The point x_0 is called the *seed* of the orbit.

Example. Let $f(x) = x^2 + 1$. Then the orbit of the seed 0 is the sequence

$$x_0 = 0$$
$$x_1 = 1$$
$$x_2 = 2$$
$$x_3 = 5$$
$$x_4 = 26$$

$$\vdots$$

$$x_n = \text{big}$$
$$x_{n+1} = \text{bigger}$$

$$\vdots$$

and so forth, so we see that this orbit tends to ∞ as $n \to \infty$.   ■

In analogy with equilibrium solutions of systems of differential equations, *fixed points* play a central role in discrete dynamical systems. A point x_0 is called a fixed point if $f(x_0) = x_0$. Obviously, the orbit of a fixed point is the constant sequence $x_0, x_0, x_0, \ldots$.

The analog of closed orbits for differential equations is given by *periodic points of period n*. These are seeds x_0 for which $f^n(x_0) = x_0$ for some $n > 0$. As a consequence, like a closed orbit, a periodic orbit repeats itself:

$$x_0, x_1, \ldots, x_{n-1}, x_0, x_1, \ldots, x_{n-1}, x_0 \ldots.$$

Periodic orbits of period n are also called *n-cycles*. We say that the periodic point x_0 has *minimal period n* if n is the least positive integer for which $f^n(x_0) = x_0$.

Example. The function $f(x) = x^3$ has fixed points at $x = 0, \pm 1$. The function $g(x) = -x^3$ has a fixed point at 0 and a periodic point of period 2 at $x = \pm 1$, since $g(1) = -1$ and $g(-1) = 1$, so $g^2(\pm 1) = \pm 1$. The function

$$h(x) = (2 - x)(3x + 1)/2$$

has a 3-cycle given by $x_0 = 0, x_1 = 1, x_2 = 2, x_3 = x_0 = 0 \ldots$. ∎

A useful way to visualize orbits of one-dimensional discrete dynamical systems is via *graphical iteration*. In this picture, we superimpose the curve $y = f(x)$ and the diagonal line $y = x$ on the same graph. We display the orbit of x_0 as follows: Begin at the point (x_0, x_0) on the diagonal and draw a vertical line to the graph of f, reaching the graph at $(x_0, f(x_0)) = (x_0, x_1)$. Then draw a horizontal line back to the diagonal, ending at (x_1, x_1). This procedure moves us from a point on the diagonal directly over the seed x_0 to a point directly over the next point on the orbit, x_1. Then we continue from (x_1, x_1): First go vertically to the graph to the point (x_1, x_2), then horizontally back to the diagonal at (x_2, x_2). On the x-axis this moves us from x_1 to the next point on the orbit, x_2. Continuing, we produce a sequence of pairs of lines, each of which terminates on the diagonal at a point of the form (x_n, x_n).

In Figure 15.1a, graphical iteration shows that the orbit of x_0 tends to the fixed point z_0 under iteration of f. In Figure 15.1b, the orbit of x_0 under g lies on a 3-cycle: $x_0, x_1, x_2, x_0, x_1, \ldots$.

As in the case of equilibrium points of differential equations, there are different types of fixed points for a discrete dynamical system. Suppose that x_0 is a fixed point for f. We say that x_0 is a *sink* or an *attracting fixed point* for f if there is a neighborhood $\mathcal{U}$ of x_0 in $\mathbb{R}$ having the property that, if $y_0 \in \mathcal{U}$, then

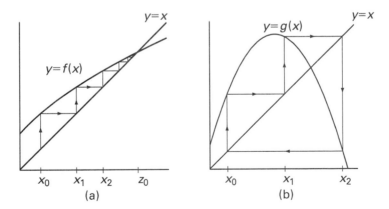

Figure 15.1 (a) The orbit of x_0 tends to the fixed point at z_0 under iteration of f, while (b) the orbit of x_0 lies on a 3-cycle under iteration of g.

$f^n(y_0) \in \mathcal{U}$ for all n and, moreover, $f^n(y_0) \to x_0$ as $n \to \infty$. Similarly, x_0 is a *source* or a *repelling fixed point* if all orbits (except x_0) leave $\mathcal{U}$ under iteration of f. A fixed point is called *neutral* or *indifferent* if it is neither attracting nor repelling.

For differential equations, we saw that it was the derivative of the vector field at an equilibrium point that determined the type of the equilibrium point. This is also true for fixed points, although the numbers change a bit.

Proposition.   *Suppose f has a fixed point at x_0. Then*

1. *x_0 is a sink if $|f'(x_0)| < 1$;*
2. *x_0 is a source if $|f'(x_0)| > 1$;*
3. *we get no information about the type of x_0 if $f'(x_0) = \pm 1$.*

Proof: We first prove case (1). Suppose $|f'(x_0)| = \nu < 1$. Choose K with $\nu < K < 1$. Since f' is continuous, we may find $\delta > 0$ so that $|f'(x)| < K$ for all x in the interval $I = [x_0 - \delta, x_0 + \delta]$. We now invoke the mean value theorem. Given any $x \in I$, we have

$$\frac{f(x) - x_0}{x - x_0} = \frac{f(x) - f(x_0)}{x - x_0} = f'(c)$$

for some c between x and x_0. Hence we have

$$|f(x) - x_0| < K|x - x_0|.$$

It follows that $f(x)$ is closer to x_0 than x and so $f(x) \in I$. Applying this result again, we have

$$|f^2(x) - x_0| < K|f(x) - x_0| < K^2|x - x_0|,$$

and, continuing, we find

$$|f^n(x) - x_0| < K^n|x - x_0|,$$

so that $f^n(x) \to x_0$ in I as required, since $0 < K < 1$.

The proof of case (2) follows similarly. In case (3), we note that each of the functions

1. $f(x) = x + x^3$;
2. $g(x) = x - x^3$;
3. $h(x) = x + x^2$

has a fixed point at 0 with $f'(0) = 1$. But graphical iteration (Figure 15.2) shows that f has a source at 0; g has a sink at 0; and 0 is attracting from one side and repelling from the other for the function h. ∎

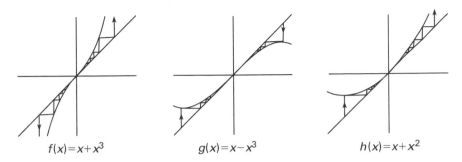

$f(x)=x+x^3$ $g(x)=x-x^3$ $h(x)=x+x^2$

Figure 15.2   In each case, the derivative at 0 is 1, but f has a source at 0; g has a sink; and h has neither.

Note that, at a fixed point x_0 for which $f'(x_0) < 0$, the orbits of nearby points jump from one side of the fixed point to the other at each iteration. See Figure 15.3. This is the reason why the output of graphical iteration is often called a *web diagram*.

Since a periodic point x_0 of period n for f is a fixed point of f^n, we may classify these points as sinks or sources depending on whether $|(f^n)'(x_0)| < 1$ or $|(f^n)'(x_0)| > 1$. One may check that $(f^n)'(x_0) = (f^n)'(x_j)$ for any other point x_j on the periodic orbit, so this definition makes sense (see Exercise 6 at the end of this chapter).

Example. The function $f(x) = x^2 - 1$ has a 2-cycle given by 0 and -1. One checks easily that $(f^2)'(0) = 0 = (f^2)'(-1)$, so this cycle is a sink. In Figure 15.4, we show a graphical iteration of f with the graph of f^2 superimposed. Note that 0 and -1 are attracting fixed points for f^2.   ∎

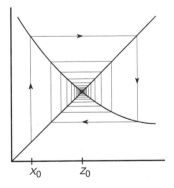

x_0 z_0

Figure 15.3   Since $-1 < f'(z_0) < 0$, the orbit of x_0 "spirals" toward the attracting fixed point at z_0.

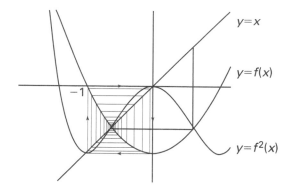

Figure 15.4   The graphs of $f(x) = x^2 - 1$ and f^2 showing that 0 and -1 lie on an attracting 2-cycle for f.

15.2 Bifurcations

Discrete dynamical systems undergo bifurcations when parameters are varied just as differential equations do. We deal in this section with several types of bifurcations that occur for one-dimensional systems.

Example. Let $f_c(x) = x^2 + c$ where c is a parameter. The fixed points for this family are given by solving the equation $x^2 + c = x$, which yields

$$p_\pm = \frac{1}{2} \pm \frac{\sqrt{1 - 4c}}{2}.$$

Hence there are no fixed points if $c > 1/4$; a single fixed point at $x = 1/2$ when $c = 1/4$; and a pair of fixed points at $p_\pm$ when $c < 1/4$. Graphical iteration shows that all orbits of f_c tend to ∞ if $c > 1/4$. When $c = 1/4$, the fixed point at $x = 1/2$ is neutral, as is easily seen by graphical iteration. See Figure 15.5. When $c < 1/4$, we have $f_c'(p_+) = 1 + \sqrt{1 - 4c} > 1$, so p_+ is always repelling. A straightforward computation also shows that $-1 < f_c'(p_-) < 1$ provided $-3/4 < c < 1/4$. For these c-values, p_- is attracting. When $-3/4 < c < 1/4$, all orbits in the interval $(-p_+, p_+)$ tend to p_- (though, technically, the orbit of $-p_-$ is *eventually fixed*, since it maps directly onto p_-, as do the orbits of certain other points in this interval when $c < 0$). Thus as c decreases through the bifurcation value $c = 1/4$, we see the birth of a single neutral fixed point, which then immediately splits into two fixed points, one attracting and one repelling. This is an example of a *saddle-node* or *tangent bifurcation*. Graphically, this bifurcation is essentially the same as its namesake for first-order differential equations as described in Chapter 8. See Figure 15.5. ∎

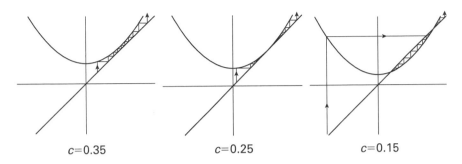

$c{=}0.35$ $c{=}0.25$ $c{=}0.15$

Figure 15.5   The saddle-node bifurcation for $f_c(x) = x^2 + c$ at $c = 1/4$.

Note that, in this example, at the bifurcation point, the derivative at the fixed point equals 1. This is no accident, for we have:

Theorem. (**The Bifurcation Criterion**) *Let f_λ be a family of functions depending smoothly on the parameter λ. Suppose that $f_{\lambda_0}(x_0) = x_0$ and $f'_{\lambda_0}(x_0) \neq 1$. Then there are intervals I about x_0 and J about λ_0 and a smooth function $p: J \to I$ such that $p(\lambda_0) = x_0$ and $f_\lambda(p(\lambda)) = p(\lambda)$. Moreover, f_λ has no other fixed points in I.*

Proof: Consider the function defined by $G(x, \lambda) = f_\lambda(x) - x$. By hypothesis, $G(x_0, \lambda_0) = 0$ and

$$\frac{\partial G}{\partial x}(x_0, \lambda_0) = f'_{\lambda_0}(x_0) - 1 \neq 0.$$

By the implicit function theorem, there are intervals I about x_0 and J about λ_0, and a smooth function $p: J \to I$ such that $p(\lambda_0) = x_0$ and $G(p(\lambda), \lambda) \equiv 0$ for all $\lambda \in J$. Moreover, $G(x, \lambda) \neq 0$ unless $x = p(\lambda)$. This concludes the proof. ■

As a consequence of this result, f_λ may undergo a bifurcation involving a change in the number of fixed points only if f_λ has a fixed point with derivative equal to 1. The typical bifurcation that occurs at such parameter values is the saddle-node bifurcation (see Exercises 18 and 19). However, many other types of bifurcations of fixed points may occur.

Example. Let $f_\lambda(x) = \lambda x(1 - x)$. Note that $f_\lambda(0) = 0$ for all λ. We have $f'_\lambda(0) = \lambda$, so we have a possible bifurcation at $\lambda = 1$. There is a second fixed point for f_λ at $x_\lambda = (\lambda - 1)/\lambda$. When $\lambda < 1$, x_λ is negative, and when $\lambda > 1$, x_λ is positive. When $\lambda = 1$, x_λ coalesces with the fixed point at 0 so there is a single fixed point for f_1. A computation shows that 0 is repelling and x_λ is attracting if $\lambda > 1$ (and $\lambda < 3$), while the reverse is true if $\lambda < 1$. For this reason, this type of bifurcation is known as an *exchange bifurcation*. ■

Example. Consider the family of functions $f_\mu(x) = \mu x + x^3$. When $\mu = 1$ we have $f_1(0) = 0$ and $f_1'(0) = 1$ so we have the possibility for a bifurcation. The fixed points are 0 and $\pm\sqrt{1 - \mu}$, so we have three fixed points when $\mu < 1$ but only one fixed point when $\mu \geq 1$, so a bifurcation does indeed occur as μ passes through 1. ■

The only other possible bifurcation value for a one-dimensional discrete system occurs when the derivative at the fixed (or periodic) point is equal to -1, since at these values the fixed point may change from a sink to a source or from a source to a sink. At all other values of the derivative, the fixed point simply remains a sink or source and there are no other periodic orbits nearby. Certain portions of a periodic orbit may come close to a source, but the entire orbit cannot lie close by (see Exercise 7). In the case of derivative -1 at the fixed point, the typical bifurcation is a *period doubling* bifurcation.

Example. As a simple example of this type of bifurcation, consider the family $f_\lambda(x) = \lambda x$ near $\lambda_0 = -1$. There is a fixed point at 0 for all λ. When $-1 < \lambda < 1$, 0 is an attracting fixed point and all orbits tend to 0. When $|\lambda| > 1$, 0 is repelling and all nonzero orbits tend to $\pm\infty$. When $\lambda = -1$, 0 is a neutral fixed point and all nonzero points lie on 2-cycles. As λ passes through -1, the type of the fixed point changes from attracting to repelling; meanwhile, a family of 2-cycles appears. ■

Generally, when a period doubling bifurcation occurs, the 2-cycles do not all exist for a single parameter value. A more typical example of this bifurcation is provided next.

Example. Again consider $f_c(x) = x^2 + c$, this time with c near $c = -3/4$. There is a fixed point at

$$ p_- = \frac{1}{2} - \frac{\sqrt{1 - 4c}}{2}. $$

We have seen that $f_{-3/4}'(p_-) = -1$ and that p_- is attracting when c is slightly larger than $-3/4$ and repelling when c is less than $-3/4$. Graphical iteration shows that more happens as c descends through $-3/4$: We see the birth of an (attracting) 2-cycle as well. This is the period doubling bifurcation. See Figure 15.6. Indeed, one can easily solve for the period two points and check that they are attracting (for $-5/4 < c < -3/4$; see Exercise 8). ■

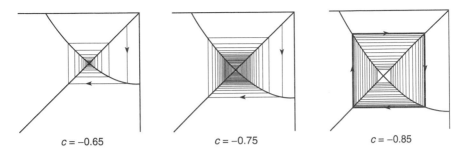

$$c = -0.65 \qquad c = -0.75 \qquad c = -0.85$$

Figure 15.6   The period doubling bifurcation for $f_c(x) = x^2 + c$ at $c = -3/4$. The fixed point is attracting for $c \geq -0.75$ and repelling for $c < -0.75$.

15.3 The Discrete Logistic Model

In Chapter 1 we introduced one of the simplest nonlinear first-order differential equations, the logistic model for population growth

$$x' = ax(1 - x).$$

In this model we took into account the fact that there is a carrying capacity for a typical population, and we saw that the resulting solutions behaved quite simply: All nonzero solutions tended to the "ideal" population. Now something about this model may have bothered you way back then: Populations generally are not continuous functions of time! A more natural type of model would measure populations at specific times, say, every year or every generation. Here we introduce just such a model, the *discrete logistic model* for population growth.

   Suppose we consider a population whose members are counted each year (or at other specified times). Let x_n denote the population at the end of year n. If we assume that no overcrowding can occur, then one such population model is the *exponential growth* model where we assume that

$$x_{n+1} = kx_n$$

for some constant $k > 0$. That is, the next year's population is directly proportional to this year's. Thus we have

$$x_1 = kx_0$$

$$x_2 = kx_1 = k^2 x_0$$

$$x_3 = kx_2 = k^3 x_0$$

$$\vdots$$

Clearly, $x_n = k^n x_0$ so we conclude that the population explodes if $k > 1$, becomes extinct if $0 \leq k < 1$, or remains constant if $k = 1$.

This is an example of a first-order *difference equation*. This is an equation that determines x_n based on the value of x_{n-1}. A second-order difference equation would give x_n based on x_{n-1} and x_{n-2}. From our point of view, the successive populations are given by simply iterating the function $f_k(x) = kx$ with the seed x_0.

A more realistic assumption about population growth is that there is a maximal population M such that, if the population exceeds this amount, then all resources are used up and the entire population dies out in the next year.

One such model that reflects these assumptions is the *discrete logistic population model*. Here we assume that the populations obey the rule

$$x_{n+1} = kx_n \left(1 - \frac{x_n}{M}\right)$$

where k and M are positive parameters. Note that, if $x_n \geq M$, then $x_{n+1} \leq 0$, so the population does indeed die out in the ensuing year.

Rather than deal with actual population numbers, we will instead let x_n denote the fraction of the maximal population, so that $0 \leq x_n \leq 1$. The logistic difference equation then becomes

$$x_{n+1} = \lambda x_n (1 - x_n)$$

where $\lambda > 0$ is a parameter. We may therefore predict the fate of the initial population x_0 by simply iterating the quadratic function $f_\lambda(x) = \lambda x(1 - x)$ (also called the *logistic map*). Sounds easy, right? Well, suffice it to say that this simple quadratic iteration was only completely understood in the late 1990s, thanks to the work of hundreds of mathematicians. We will see why the discrete logistic model is so much more complicated than its cousin, the logistic differential equation, in a moment, but first let's do some simple cases.

We consider only the logistic map on the unit interval I. We have $f_\lambda(0) = 0$, so 0 is a fixed point. The fixed point is attracting in I for $0 < \lambda \leq 1$, and repelling thereafter. The point 1 is eventually fixed, since $f_\lambda(1) = 0$. There is a second fixed point $x_\lambda = (\lambda - 1)/\lambda$ in I for $\lambda > 1$. The fixed point x_λ is attracting for $1 < \lambda \leq 3$ and repelling for $\lambda > 3$. At $\lambda = 3$ a period doubling bifurcation occurs (see Exercise 4). For λ-values between 3 and approximately 3.4, the only periodic points present are the two fixed points and the 2-cycle.

When $\lambda = 4$, the situation is much more complicated. Note that $f_\lambda'(1/2) = 0$ and that $1/2$ is the only critical point for f_λ for each λ. When $\lambda = 4$, we have $f_4(1/2) = 1$, so $f_4^2(1/2) = 0$. Therefore f_4 maps each of the half-intervals $[0, 1/2]$ and $[1/2, 1]$ onto the entire interval I. Consequently, there exist points $y_0 \in [0, 1/2]$ and $y_1 \in [1/2, 1]$ such that $f_4(y_j) = 1/2$ and hence $f_4^2(y_j) = 1$.

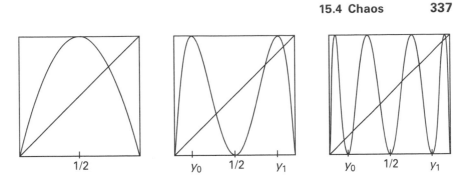

Figure 15.7 The graphs of the logistic function $f_\lambda(x) = \lambda\, x(1 - x)$ as well as f_λ^2 and f_λ^3 over the interval I.

Therefore we have

$$f_4^2[0, y_0] = f_4^2[y_0, 1/2] = I$$

and

$$f_4^2[1/2, y_1] = f_4^2[y_1, 1] = I.$$

Since the function f_4^2 is a quartic, it follows that the graph of f_4^2 is as depicted in Figure 15.7. Continuing in this fashion, we find 2^3 subintervals of I that are mapped onto I by f_4^3, 2^4 subintervals mapped onto I by f_4^4, and so forth. We therefore see that f_4 has two fixed points in I; f_4^2 has four fixed points in I; f_4^3 has 2^3 fixed points in I; and, inductively, f_4^n has 2^n fixed points in I. The fixed points for f_4 occur at 0 and 3/4. The four fixed points for f_4^2 include these two fixed points plus a pair of periodic points of period 2. Of the eight fixed points for f_4^3, two must be the fixed points and the other six must lie on a pair of 3-cycles. Among the 16 fixed points for f_4^4 are two fixed points, two periodic points of period 2, and twelve periodic points of period 4. Clearly, a lot has changed as λ varies from 3.4 to 4.

On the other hand, if we choose a random seed in the interval I and plot the orbit of this seed under iteration of f_4 using graphical iteration, we rarely see any of these cycles. In Figure 15.8 we have plotted the orbit of 0.123 under iteration of f_4 using 200 and 500 iterations. Presumably, there is something "chaotic" going on.

15.4 Chaos

In this section we introduce several quintessential examples of chaotic one-dimensional discrete dynamical systems. Recall that a subset $U \subset W$ is said

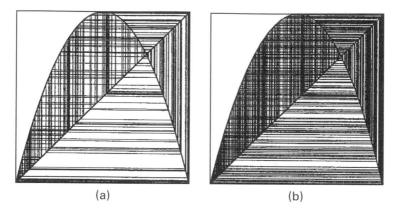

(a)   (b)

Figure 15.8   The orbit of the seed 0.123 under f_4 using (a) 200
iterations and (b) 500 iterations.

to be *dense* in W if there are points in U arbitrarily close to any point in the
larger set W. As in the Lorenz model, we say that a map f, which takes an
interval $I = [\alpha, \beta]$ to itself, is *chaotic* if

1. Periodic points of f are dense in I;
2. f is transitive on I; that is, given any two subintervals U_1 and U_2 in I,
   there is a point $x_0 \in U_1$ and an $n > 0$ such that $f^n(x_0) \in U_2$;
3. f has sensitive dependence in I; that is, there is a *sensitivity constant* β
   such that, for any $x_0 \in I$ and any open interval U about x_0, there is some
   seed $y_0 \in U$ and $n > 0$ such that

$$|f^n(x_0) - f^n(y_0)| > \beta.$$

It is known that the transitivity condition is equivalent to the existence of
an orbit that is dense in I. Clearly, a dense orbit implies transitivity, for such
an orbit repeatedly visits any open subinterval in I. The other direction relies
on the Baire category theorem from analysis, so we will not prove this here.

Curiously, for maps of an interval, condition 3 in the definition of chaos is
redundant [8]. This is somewhat surprising, since the first two conditions in
the definition are topological in nature, while the third is a metric property (it
depends on the notion of distance).

We now discuss several classical examples of chaotic one-dimensional maps.

Example. (The Doubling Map) Define the discontinuous function
$D\colon [0, 1) \to [0, 1)$ by $D(x) = 2x \bmod 1$. That is,

$$D(x) = \begin{cases} 2x & \text{if } 0 \le x < 1/2 \\ 2x - 1 & \text{if } 1/2 \le x < 1 \end{cases}$$

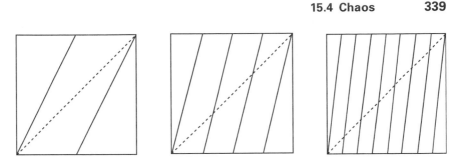

Figure 15.9 The graph of the doubling map D and its higher iterates D^2 and D^3 on $[0, 1)$.

An easy computation shows that $D^n(x) = 2^n x$ mod 1, so that the graph of D^n consists of 2^n straight lines with slope 2^n, each extending over the entire interval $[0, 1)$. See Figure 15.9.

To see that the doubling function is chaotic on $[0, 1)$, note that D^n maps any interval of the form $[k/2^n, (k+1)/2^n)$ for $k = 0, 1, \ldots 2^n - 2$ onto the interval $[0, 1)$. Hence the graph of D^n crosses the diagonal $y = x$ at some point in this interval, and so there is a periodic point in any such interval. Since the lengths of these intervals are $1/2^n$, it follows that periodic points are dense in $[0, 1)$. Transitivity also follows, since, given any open interval J, we may always find an interval of the form $[k/2^n, (k+1)/2^n)$ inside J for sufficiently large n. Hence D^n maps J onto all of $[0, 1)$. This also proves sensitivity, where we choose the sensitivity constant $1/2$.

We remark that it is possible to write down all of the periodic points for D explicitly (see Exercise 5a). It is also interesting to note that, if you use a computer to iterate the doubling function, then it appears that all orbits are eventually fixed at 0, which, of course, is false! See Exercise 5c for an explanation of this phenomenon. ∎

Example. (The Tent Map) Now consider a continuous cousin of the doubling map given by

$$T(x) = \begin{cases} 2x & \text{if } 0 \leq x < 1/2 \\ -2x + 2 & \text{if } 1/2 \leq x \leq 1. \end{cases}$$

T is called the *tent map*. See Figure 15.10. The fact that T is chaotic on $[0, 1]$ follows exactly as in the case of the doubling function, using the graphs of T^n (see Exercise 15).

Looking at the graphs of the tent function T and the logistic function $f_4(x) = 4x(1 - x)$ that we discussed in Section 15.3, it appears that they should share many of the same properties under iteration. Indeed, this is the case. To understand this, we need to reintroduce the notion of conjugacy, this time for discrete systems.

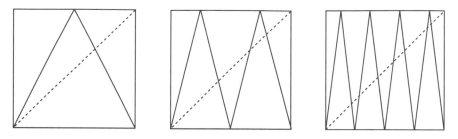

Figure 15.10 The graph of the tent map T and its higher iterates T^2 and T^3 on [0, 1].

Suppose I and J are intervals and $f: I \to I$ and $g: J \to J$. We say that f and g are *conjugate* if there is a homeomorphism $h: I \to J$ such that h satisfies the *conjugacy equation* $h \circ f = g \circ h$. Just as in the case of flows, a conjugacy takes orbits of f to orbits of g. This follows since we have $h(f^n(x)) = g^n(h(x))$ for all $x \in I$, so h takes the nth point on the orbit of x under f to the nth point on the orbit of $h(x)$ under g. Similarly, h^{-1} takes orbits of g to orbits of f. ∎

Example. Consider the logistic function $f_4: [0, 1] \to [0, 1]$ and the quadratic function $g: [-2, 2] \to [-2, 2]$ given by $g(x) = x^2 - 2$. Let $h(x) = -4x + 2$ and note that h takes $[0, 1]$ to $[-2, 2]$. Moreover, we have $h(4x(1 - x)) = (h(x))^2 - 2$, so h satisfies the conjugacy equation and f_4 and g are conjugate. ∎

From the point of view of chaotic systems, conjugacies are important since they map one chaotic system to another.

Proposition. *Suppose $f: I \to I$ and $g: J \to J$ are conjugate via h, where both I and J are closed intervals in $\mathbb{R}$ of finite length. If f is chaotic on I, then g is chaotic on J.*

Proof: Let U be an open subinterval of J and consider $h^{-1}(U) \subset I$. Since periodic points of f are dense in I, there is a periodic point $x \in h^{-1}(U)$ for f. Say x has period n. Then

$$g^n(h(x)) = h(f^n(x)) = h(x)$$

by the conjugacy equation. This gives a periodic point $h(x)$ for g in U and shows that periodic points of g are dense in J.

If U and V are open subintervals of J, then $h^{-1}(U)$ and $h^{-1}(V)$ are open intervals in I. By transitivity of f, there exists $x_1 \in h^{-1}(U)$ such that

$f^m(x_1) \in h^{-1}(V)$ for some m. But then $h(x_1) \in U$ and we have $g^m(h(x_1)) = h(f^m(x_1)) \in V$, so g is transitive also.

For sensitivity, suppose that f has sensitivity constant β. Let $I = [\alpha_0, \alpha_1]$. We may assume that $\beta < \alpha_1 - \alpha_0$. For any $x \in [\alpha_0, \alpha_1 - \beta]$, consider the function $|h(x + \beta) - h(x)|$. This is a continuous function on $[\alpha_0, \alpha_1 - \beta]$ that is positive. Hence it has a minimum value β'. It follows that h takes intervals of length β in I to intervals of length at least β' in J. Then it is easy to check that β' is a sensitivity constant for g. This completes the proof. ∎

It is not always possible to find conjugacies between functions with equivalent dynamics. However, we can relax the requirement that the conjugacy be one to one and still salvage the preceding proposition. A continuous function h that is at most n to one and that satisfies the conjugacy equation $f \circ h = h \circ g$ is called a *semiconjugacy* between g and f. It is easy to check that a semiconjugacy also preserves chaotic behavior on intervals of finite length (see Exercise 12). A semiconjugacy need not preserve the minimal periods of cycles, but it does map cycles to cycles.

Example. The tent function T and the logistic function f_4 are semiconjugate on the unit interval. To see this, let

$$h(x) = \frac{1}{2}(1 - \cos 2\pi x).$$

Then h maps the interval $[0, 1]$ in two-to-one fashion over itself, except at $1/2$, which is the only point mapped to 1. Then we compute

$$h(T(x)) = \frac{1}{2}(1 - \cos 4\pi x)$$

$$= \frac{1}{2} - \frac{1}{2}(2\cos^2 2\pi x - 1)$$

$$= 1 - \cos^2 2\pi x$$

$$= 4\left(\frac{1}{2} - \frac{1}{2}\cos 2\pi x\right)\left(\frac{1}{2} + \frac{1}{2}\cos 2\pi x\right)$$

$$= f_4(h(x)).$$

Thus h is a semiconjugacy between T and f_4. As a remark, recall that we may find arbitrarily small subintervals mapped onto all of $[0, 1]$ by T. Hence f_4 maps the images of these intervals under h onto all of $[0, 1]$. Since h is continuous, the images of these intervals may be chosen arbitrarily small. Hence we may choose 1/2 as a sensitivity constant for f_4 as well. We have proven the following proposition: ■

Proposition. *The logistic function $f_4(x) = 4x(1 - x)$ is chaotic on the unit interval.* ■

15.5 Symbolic Dynamics

We turn now to one of the most useful tools for analyzing chaotic systems, *symbolic dynamics*. We give just one example of how to use symbolic dynamics here; several more are included in the next chapter.

Consider the logistic map $f_\lambda(x) = \lambda x(1 - x)$ where $\lambda > 4$. Graphical iteration seems to imply that almost all orbits tend to $-\infty$. See Figure 15.11. Of course, this is not true, because we have fixed points and other periodic points for this function. In fact, there is an unexpectedly "large" set called a Cantor set that is filled with chaotic behavior for this function, as we shall see.

Unlike the case $\lambda \leq 4$, the interval $I = [0, 1]$ is no longer invariant when $\lambda > 4$: Certain orbits escape from I and then tend to $-\infty$. Our goal is to understand the behavior of the nonescaping orbits. Let Λ denote the set of points in I whose orbits never leave I. As shown in Figure 15.12a, there is an open interval A_0 on which $f_\lambda > 1$. Hence $f_\lambda^2(x) < 0$ for any $x \in A_0$ and, as a consequence, the orbits of all points in A_0 tend to $-\infty$. Note that any orbit that leaves I must first enter A_0 before departing toward $-\infty$. Also, the

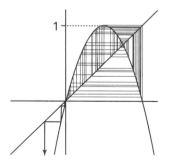

Figure 15.11   Typical orbits for the logistic function f_λ with $\lambda > 4$ seem to tend to $-\infty$ after wandering around the unit interval for a while.

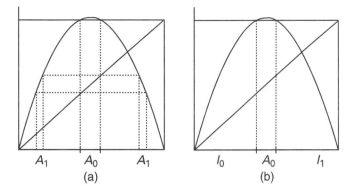

Figure 15.12   (a) The exit set in I consists of a collection of disjoint open intervals. (b) The intervals I_0 and I_1 lie to the left and right of A_0.

orbits of the endpoints of A_0 are eventually fixed at 0, so these endpoints are contained in Λ. Now let A_1 denote the preimage of A_0 in I: A_1 consists of two open intervals in I, one on each side of A_0. All points in A_1 are mapped into A_0 by f_λ, and hence their orbits also tend to $-\infty$. Again, the endpoints of A_1 are eventual fixed points. Continuing, we see that each of the two open intervals in A_1 has as a preimage a pair of disjoint intervals, so there are four open intervals that consist of points whose first iteration lies in A_1, the second in A_0, and so, again, all of these points have orbits that tend to $-\infty$. Call these four intervals A_2. In general, let A_n denote the set of points in I whose nth iterate lies in A_0. A_n consists of set 2^n disjoint open intervals in I. Any point whose orbit leaves I must lie in one of the A_n. Hence we see that

$$\Lambda = I - \bigcup_{n=0}^{\infty} A_n.$$

To understand the dynamics of f_λ on I, we introduce *symbolic dynamics*. Toward that end, let I_0 and I_1 denote the left and right closed interval respectively in $I - A_0$. See Figure 15.12b. Given $x_0 \in \Lambda$, the entire orbit of x_0 lies in $I_0 \cup I_1$. Hence we may associate an infinite sequence $S(x_0) = (s_0 s_1 s_2 \ldots)$ consisting of 0's and 1's to the point x_0 via the rule

$$s_j = k \text{ if and only if } f_\lambda^j(x_0) \in I_k.$$

That is, we simply watch how $f_\lambda^j(x_0)$ bounces around I_0 and I_1, assigning a 0 or 1 at the jth stage depending on which interval $f_\lambda^j(x_0)$ lies in. The sequence $S(x_0)$ is called the *itinerary* of x_0.

Example. The fixed point 0 has itinerary $S(0) = (000\ldots)$. The fixed point x_λ in I_1 has itinerary $S(x_\lambda) = (111\ldots)$. The point $x_0 = 1$ is eventually fixed and has itinerary $S(1) = (1000\ldots)$. A 2-cycle that hops back and forth between I_0 and I_1 has itinerary $(\overline{01}\ldots)$ or $(\overline{10}\ldots)$ where $\overline{01}$ denotes the infinitely repeating sequence consisting of repeated blocks 01.

Let Σ denote the set of all possible sequences of 0's and 1's. A "point" in the space Σ is therefore an infinite sequence of the form $s = (s_0 s_1 s_2 \ldots)$. To visualize Σ, we need to tell how far apart different points in Σ are. To do this, let $s = (s_0 s_1 s_2 \ldots)$ and $t = (t_0 t_1 t_2 \ldots)$ be points in Σ. A *distance function* or *metric* on Σ is a function $d = d(s, t)$ that satisfies

1. $d(s, t) \geq 0$ and $d(s, t) = 0$ if and only if $s = t$;
2. $d(s, t) = d(t, s)$;
3. the triangle inequality: $d(s, u) \leq d(s, t) + d(t, u)$.

Since Σ is not naturally a subset of a Euclidean space, we do not have a Euclidean distance to use on Σ. Hence we must concoct one of our own. Here is the distance function we choose:

$$d(s, t) = \sum_{i=0}^{\infty} \frac{|s_i - t_i|}{2^i}.$$

Note that this infinite series converges: The numerators in this series are always either 0 or 1, so this series converges by comparison to the geometric series:

$$d(s, t) \leq \sum_{i=0}^{\infty} \frac{1}{2^i} = \frac{1}{1 - 1/2} = 2.$$

It is straightforward to check that this choice of d satisfies the three requirements to be a distance function (see Exercise 13). While this distance function may look a little complicated at first, it is often easy to compute.   ∎

Example.

$$(1)\ \ d\left((\overline{0}), (\overline{1})\right) = \sum_{i=0}^{\infty} \frac{|0 - 1|}{2^i} = \sum_{i=0}^{\infty} \frac{1}{2^i} = 2$$

$$(2)\ \ d\left((\overline{01}), (\overline{10})\right) = \sum_{i=0}^{\infty} \frac{1}{2^i} = 2$$

$$(3)\ \ d\left((\overline{01}), (\overline{1})\right) = \sum_{i=0}^{\infty} \frac{1}{4^i} = \frac{1}{1 - 1/4} = \frac{4}{3}.$$

∎

The importance of having a distance function on Σ is that we now know when points are close together or far apart. In particular, we have

Proposition. *Suppose* $s = (s_0 s_1 s_2 \ldots)$ *and* $t = (t_0 t_1 t_2 \ldots) \in \Sigma$.

1. *If* $s_j = t_j$ *for* $j = 0, \ldots, n$, *then* $d(s, t) \leq 1/2^n$;
2. *Conversely, if* $d(s, t) < 1/2^n$, *then* $s_j = t_j$ *for* $j = 0, \ldots, n$.

Proof: In case (1), we have

$$d(s, t) = \sum_{i=0}^{n} \frac{|s_i - s_i|}{2^i} + \sum_{i=n+1}^{\infty} \frac{|s_i - t_i|}{2^i}$$

$$\leq 0 + \frac{1}{2^{n+1}} \sum_{i=0}^{\infty} \frac{1}{2^i}$$

$$= \frac{1}{2^n}.$$

If, on the other hand, $d(s, t) < 1/2^n$, then we must have $s_j = t_j$ for any $j \leq n$, because otherwise $d(s, t) \geq |s_j - t_j|/2^j = 1/2^j \geq 1/2^n$. ∎

Now that we have a notion of closeness in Σ, we are ready to prove the main theorem of this chapter:

Theorem. *The itinerary function* $S \colon \Lambda \to \Sigma$ *is a homeomorphism provided* $\lambda > 4$.

Proof: Actually, we will only prove this for the case in which λ is sufficiently large that $|f_\lambda'(x)| > K > 1$ for some K and for all $x \in I_0 \cup I_1$. The reader may check that $\lambda > 2 + \sqrt{5}$ suffices for this. For the more complicated proof in the case where $4 < \lambda \leq 2 + \sqrt{5}$, see [25].

We first show that S is one to one. Let $x, y \in \Lambda$ and suppose $S(x) = S(y)$. Then, for each n, $f_\lambda^n(x)$ and $f_\lambda^n(y)$ lie on the same side of $1/2$. This implies that f_λ is monotone on the interval between $f_\lambda^n(x)$ and $f_\lambda^n(y)$. Consequently, all points in this interval remain in $I_0 \cup I_1$ when we apply f_λ. Now $|f_\lambda'| > K > 1$ at all points in this interval, so, as in Section 15.1, each iteration of f_λ expands this interval by a factor of K. Hence the distance between $f_\lambda^n(x)$ and $f_\lambda^n(y)$ grows without bound, so these two points must eventually lie on opposite sides of A_0. This contradicts the fact that they have the same itinerary.

To see that S is onto, we first introduce the following notation. Let $J \subset I$ be a closed interval. Let

$$f_\lambda^{-n}(J) = \{x \in I \mid f_\lambda^n(x) \in J\},$$

so that $f_\lambda^{-n}(J)$ is the preimage of J under f_λ^n. A glance at the graph of f_λ when $\lambda > 4$ shows that, if $J \subset I$ is a closed interval, then $f_\lambda^{-1}(J)$ consists of two closed subintervals, one in I_0 and one in I_1.

Now let $s = (s_0 s_1 s_2 \ldots)$. We must produce $x \in \Lambda$ with $S(x) = s$. To that end we define

$$I_{s_0 s_1 \ldots s_n} = \{x \in I \mid x \in I_{s_0}, f_\lambda(x) \in I_{s_1}, \ldots, f_\lambda^n(x) \in I_{s_n}\}$$
$$= I_{s_0} \cap f_\lambda^{-1}(I_{s_1}) \cap \ldots \cap f_\lambda^{-n}(I_{s_n}).$$

We claim that the $I_{s_0 \ldots s_n}$ form a nested sequence of nonempty closed intervals. Note that

$$I_{s_0 s_1 \ldots s_n} = I_{s_0} \cap f_\lambda^{-1}(I_{s_1 \ldots s_n}).$$

By induction, we may assume that $I_{s_1 \ldots s_n}$ is a nonempty subinterval, so that, by the observation above, $f_\lambda^{-1}(I_{s_1 \ldots s_n})$ consists of two closed intervals, one in I_0 and one in I_1. Hence $I_{s_0} \cap f_\lambda^{-1}(I_{s_1 \ldots s_n})$ is a single closed interval. These intervals are nested because

$$I_{s_0 \ldots s_n} = I_{s_0 \ldots s_{n-1}} \cap f_\lambda^{-n}(I_{s_n}) \subset I_{s_0 \ldots s_{n-1}}.$$

Therefore we conclude that

$$\bigcap_{n \geq 0}^{\infty} I_{s_0 s_1 \ldots s_n}$$

is nonempty. Note that if $x \in \cap_{n \geq 0} I_{s_0 s_1 \ldots s_n}$, then $x \in I_{s_0}$, $f_\lambda(x) \in I_{s_1}$, etc. Hence $S(x) = (s_0 s_1 \ldots)$. This proves that S is onto.

Observe that $\cap_{n \geq 0} I_{s_0 s_1 \ldots s_n}$ consists of a unique point. This follows immediately from the fact that S is one to one. In particular, we have that the diameter of $I_{s_0 s_1 \ldots s_n}$ tends to 0 as $n \to \infty$.

To prove continuity of S, we choose $x \in \Lambda$ and suppose that $S(x) = (s_0 s_1 s_2 \ldots)$. Let $\epsilon > 0$. Pick n so that $1/2^n < \epsilon$. Consider the closed subintervals $I_{t_0 t_1 \ldots t_n}$ defined above for all possible combinations $t_0 t_1 \ldots t_n$. These subintervals are all disjoint, and Λ is contained in their union. There are 2^{n+1} such subintervals, and $I_{s_0 s_1 \ldots s_n}$ is one of them. Hence we may choose δ such that $|x - y| < \delta$ and $y \in \Lambda$ implies that $y \in I_{s_0 s_1 \ldots s_n}$. Therefore, $S(y)$ agrees with $S(x)$ in the first $n + 1$ terms. So, by the previous proposition, we have

$$d(S(x), S(y)) \leq \frac{1}{2^n} < \epsilon.$$

This proves the continuity of S. It is easy to check that S^{-1} is also continuous. Thus, S is a homeomorphism. ∎

15.6 The Shift Map

We now construct a map $\sigma: \Sigma \to \Sigma$ with the following properties:

1. σ is chaotic;
2. σ is conjugate to f_λ on Λ;
3. σ is completely understandable from a dynamical systems point of view.

The meaning of this last item will become clear as we proceed.

We define the *shift map* $\sigma: \Sigma \to \Sigma$ by

$$\sigma(s_0 s_1 s_2 \ldots) = (s_1 s_2 s_3 \ldots).$$

That is, the shift map simply drops the first digit in each sequence in Σ. Note that σ is a two-to-one map onto Σ. This follows since, if $(s_0 s_1 s_2 \ldots) \in \Sigma$, then we have

$$\sigma(0 s_0 s_1 s_2 \ldots) = \sigma(1 s_0 s_1 s_2 \ldots) = (s_0 s_1 s_2 \ldots).$$

Proposition. *The shift map $\sigma: \Sigma \to \Sigma$ is continuous.*

Proof: Let $s = (s_0 s_1 s_2 \ldots) \in \Sigma$, and let $\epsilon > 0$. Choose n so that $1/2^n < \epsilon$. Let $\delta = 1/2^{n+1}$. Suppose that $d(s, t) < \delta$, where $t = (t_0 t_1 t_2 \ldots)$. Then we have $s_i = t_i$ for $i = 0, \ldots, n + 1$.

Now $\sigma(t) = (s_1 s_2 \ldots s_n s_{n+1} s_{n+2} \ldots)$ so that $d(\sigma(s), \sigma(t)) \leq 1/2^n < \epsilon$. This proves that σ is continuous. ∎

Note that we can easily write down all of the periodic points of any period for the shift map. Indeed, the fixed points are $(\overline{0})$ and $(\overline{1})$. The 2 cycles are $(\overline{01})$ and $(\overline{10})$. In general, the periodic points of period n are given by repeating sequences that consist of repeated blocks of length n: $(\overline{s_0 \ldots s_{n-1}})$. Note how much nicer σ is compared to f_λ: Just try to write down explicitly all of the periodic points of period n for f_λ someday! They are there and we know roughly where they are, because we have:

Theorem. *The itinerary function $S: \Lambda \to \Sigma$ provides a conjugacy between f_λ and the shift map σ.*

Proof: In the previous section we showed that S is a homeomorphism. So it suffices to show that $S \circ f_\lambda = \sigma \circ S$. To that end, let $x_0 \in \Lambda$ and suppose that $S(x_0) = (s_0 s_1 s_2 \ldots)$. Then we have $x_0 \in I_{s_0}$, $f_\lambda(x_0) \in I_{s_1}$, $f_\lambda^2(x_0) \in I_{s_2}$, and so forth. But then the fact that $f_\lambda(x_0) \in I_{s_1}$, $f_\lambda^2(x_0) \in I_{s_2}$, etc., says that $S(f_\lambda(x_0)) = (s_1 s_2 s_3 \ldots)$, so $S(f_\lambda(x_0)) = \sigma(S(x_0))$, which is what we wanted to prove.   ∎

Now, not only can we write down all periodic points for σ, but we can in fact write down explicitly a point in Σ whose orbit is dense. Here is such a point:

$$s^* = (\ \underbrace{0\ 1}_{\text{1 blocks}}\ |\ \underbrace{00\ 01\ 10\ 11}_{\text{2 blocks}}\ |\ \underbrace{000\ 001\cdots}_{\text{3 blocks}}\ |\ \underbrace{\cdots}_{\text{4 blocks}}\ ).$$

The sequence s^* is constructed by successively listing all possible blocks of 0's and 1's of length 1, length 2, length 3, and so forth. Clearly, some iterate of σ applied to s^* yields a sequence that agrees with any given sequence in an arbitrarily large number of initial places. That is, given $t = (t_0 t_1 t_2 \ldots) \in \Sigma$, we may find k so that the sequence $\sigma^k(s^*)$ begins

$$(t_0 \ldots t_n s_{n+1} s_{n+2} \ldots)$$

so that

$$d(\sigma^k(s^*), t) \leq 1/2^n.$$

Hence the orbit of s^* comes arbitrarily close to every point in Σ. This proves that the orbit of s^* under σ is dense in Σ and so σ is transitive. Note that we may construct a multitude of other points with dense orbits in Σ by just rearranging the blocks in the sequence s^*. Again, think about how difficult it would be to identify a seed whose orbit under a quadratic function like f_4 is dense in $[0, 1]$. This is what we meant when we said earlier that the dynamics of σ are "completely understandable."

The shift map also has sensitive dependence. Indeed, we may choose the sensitivity constant to be 2, which is the largest possible distance between two points in Σ. The reason for this is, if $s = (s_0 s_1 s_2 \ldots) \in \Sigma$ and $\hat{s}_j$ denotes "not s_j" (that is, if $s_j = 0$, then $\hat{s}_j = 1$, or if $s_j = 1$ then $\hat{s}_j = 0$), then the point $s' = (s_0 s_1 \ldots s_n \hat{s}_{n+1} \hat{s}_{n+2} \ldots)$ satisfies:

1. $d(s, s') = 1/2^n$, but
2. $d(\sigma^{n+1}(s), \sigma^{n+1}(s')) = 2$.

As a consequence, we have proved the following:

Theorem. *The shift map σ is chaotic on Σ, and so by the conjugacy in the previous theorem, the logistic map f_λ is chaotic on Λ when $\lambda > 4$.*   ∎

Thus symbolic dynamics provides us with a computable model for the dynamics of f_λ on the set Λ, despite the fact that f_λ is chaotic on Λ.

15.7 The Cantor Middle-Thirds Set

We mentioned earlier that Λ was an example of a Cantor set. Here we describe the simplest example of such a set, the *Cantor middle-thirds set C*. As we shall see, this set has some unexpectedly interesting properties.

To define C, we begin with the closed unit interval $I = [0, 1]$. The rule is, each time we see a closed interval, we remove its open middle third. Hence, at the first stage, we remove $(1/3, 2/3)$, leaving us with two closed intervals, $[0, 1/3]$ and $[2/3, 1]$. We now repeat this step by removing the middle thirds of these two intervals. We are left with four closed intervals $[0, 1/9]$, $[2/9, 1/3]$, $[2/3, 7/9]$, and $[8/9, 1]$. Removing the open middle thirds of these intervals leaves us with 2^3 closed intervals, each of length $1/3^3$. Continuing in this fashion, at the nth stage we are left with 2^n closed intervals each of length $1/3^n$. The Cantor middle thirds set C is what is left when we take this process to the limit as $n \to \infty$. Note how similar this construction is to that of Λ in Section 15.5. In fact, it can be proved that Λ is homeomorphic to C (see Exercises 16 and 17).

What points in I are left in C after removing all of these open intervals? Certainly 0 and 1 remain in C, as do the endpoints $1/3$ and $2/3$ of the first removed interval. Indeed, each endpoint of a removed open interval lies in C because such a point never lies in an open middle-third subinterval. At first glance, it appears that these are the only points in the Cantor set, but in fact, that is far from the truth. Indeed, most points in C are **not** endpoints!

To see this, we attach an address to each point in C. The address will be an infinite string of L's or R's determined as follows. At each stage of the construction, our point lies in one of two small closed intervals, one to the left of the removed open interval or one to its right. So at the nth stage we may assign an L or R to the point depending on its location left or right of the interval removed at that stage. For example, we associate $LLL\ldots$ to 0 and $RRR\ldots$ to 1. The endpoints $1/3$ and $2/3$ have addresses $LRRR\ldots$ and $RLLL\ldots$, respectively. At the next stage, $1/9$ has address $LLRRR\ldots$ since $1/9$ lies in $[0, 1/3]$ and $[0, 1/9]$ at the first two stages, but then always lies in the right-hand interval. Similarly, $2/9$ has address $LRLLL\ldots$, while $7/9$ and $8/9$ have addresses $RLRRR\ldots$ and $RRLLL\ldots$.

Notice what happens at each endpoint of C. As the above examples indicate, the address of an endpoint always ends in an infinite string of all L's or all R's. But there are plenty of other possible addresses for points in C. For example, there is a point with address $LRLRLR\ldots$ This point lies in

$$[0, 1/3] \cap [2/9, 1/3] \cap [2/9, 7/27] \cap [20/81, 7/27]\ldots$$

Note that this point lies in the nested intersection of closed intervals of length $1/3^n$ for each n, and it is the unique such point that does so. This shows that most points in C are not endpoints, for the typical address will not end in all L's or all R's.

We can actually say quite a bit more: The Cantor middle-thirds set contains uncountably many points. Recall that an infinite set is *countable* if it can be put in one-to-one correspondence with the natural numbers; otherwise, the set is *uncountable*.

Proposition.   *The Cantor middle-thirds set is uncountable.*

Proof: Suppose that C is countable. This means that we can pair each point in C with a natural number in some fashion, say as

$$\begin{array}{rcl}
1 & : & LLLLL\ldots \\
2 & : & RRRR\ldots \\
3 & : & LRLR\ldots \\
4 & : & RLRL\ldots \\
5 & : & LRRLRR\ldots
\end{array}$$

and so forth. But now consider the address whose first entry is the opposite of the first entry of sequence 1, whose second entry is the opposite of the second entry of sequence 2; and so forth. This is a new sequence of L's and R's (which, in the example above, begins with $RLRRL\ldots$). Thus we have created a sequence of L's and R's that disagrees in the nth spot with the nth sequence on our list. Hence this sequence is not on our list and so we have failed in our construction of a one-to-one correspondence with the natural numbers. This contradiction establishes the result. ∎

We can actually determine the points in the Cantor middle-thirds set in a more familiar way. To do this we change the address of a point in C from a sequence of L's and R's to a sequence of 0's and 2's; that is, we replace each L with a 0 and each R with a 2. To determine the numerical value of a point $x \in C$ we approach x from below by starting at 0 and moving $s_n/3^n$ units to the right for each n, where $s_n = 0$ or 2 depending on the nth digit in the address for $n = 1, 2, 3\ldots$.

For example, 1 has address $RRR\ldots$ or $222\ldots$, so 1 is given by

$$\frac{2}{3} + \frac{2}{3^2} + \frac{2}{3^3} + \cdots = \frac{2}{3}\sum_{n=0}^{\infty}\frac{1}{3^n} = \frac{2}{3}\left(\frac{1}{1-1/3}\right) = 1.$$

Similarly, 1/3 has address $LRRR\ldots$ or $0222\ldots$, which yields

$$\frac{0}{3} + \frac{2}{3^2} + \frac{2}{3^3} + \cdots = \frac{2}{9}\sum_{n=0}^{\infty}\frac{1}{3^n} = \frac{2}{9}\cdot\frac{3}{2} = \frac{1}{3}.$$

Finally, the point with address $LRLRLR\ldots$ or $020202\ldots$ is

$$\frac{0}{3} + \frac{2}{3^2} + \frac{0}{3^3} + \frac{2}{3^4} + \cdots = \frac{2}{9}\sum_{n=0}^{\infty}\frac{1}{9^n} = \frac{2}{9}\left(\frac{1}{1-1/9}\right) = \frac{1}{4}.$$

Note that this is one of the non-endpoints in C referred to earlier.

The astute reader will recognize that the address of a point x in C with 0's and 2's gives the *ternary expansion* of x. A point $x \in I$ has ternary expansion $a_1 a_2 a_3 \ldots$ if

$$x = \sum_{i=1}^{\infty}\frac{a_i}{3^i}$$

where each a_i is either 0, 1, or 2. Thus we see that points in the Cantor middle-thirds set have ternary expansions that may be written with no 1's among the digits.

We should be a little careful here. The ternary expansion of 1/3 is $1000\ldots$. But 1/3 also has ternary expansion $0222\ldots$ as we saw above. So 1/3 may be written in ternary form in a way that contains no 1's. In fact, every endpoint in C has a similar pair of ternary representations, one of which contains no 1's.

We have shown that C contains uncountably many points, but we can say even more:

Proposition. *The Cantor middle-thirds set contains as many points as the interval* $[0, 1]$.

Proof: C consists of all points whose ternary expansion $a_0 a_1 a_2 \ldots$ contains only 0's or 2's. Take this expansion and change each 2 to a 1 and then think of this string as a binary expansion. We get every possible binary expansion in this manner. We have therefore made a correspondence (at most two to one) between the points in C and the points in $[0, 1]$, since every such point has a binary expansion. ∎

Finally, we note that

Proposition.   *The Cantor middle-thirds set has length 0.*

Proof: We compute the "length" of C by adding up the lengths of the intervals removed at each stage to determine the length of the complement of C. These removed intervals have successive lengths $1/3, 2/9, 4/27\ldots$ and so the length of $I - C$ is

$$\frac{1}{3} + \frac{2}{9} + \frac{4}{27} + \cdots = \frac{1}{3}\sum_{n=0}^{\infty}\left(\frac{2}{3}\right)^{n} = 1. \qquad \blacksquare$$

   This fact may come as no surprise since C consists of a "scatter" of points. But now consider the Cantor middle-fifths set, obtained by removing the open middle-fifth of each closed interval in similar fashion to the construction of C. The length of this set is nonzero, yet it is homeomorphic to C. These Cantor sets have, as we said earlier, unexpectedly interesting properties! And remember, the set Λ on which f_4 is chaotic is just this kind of object.

15.8 Exploration: Cubic Chaos

In this exploration, you will investigate the behavior of the discrete dynamical system given by the family of cubic functions $f_\lambda(x) = \lambda x - x^3$. You should attempt to prove rigorously everything outlined below.

1. Describe the dynamics of this family of functions for all $\lambda < -1$.
2. Describe the bifurcation that occurs at $\lambda = -1$. *Hint:* Note that f_λ is an odd function. In particular, what happens when the graph of f_λ crosses the line $y = -x$?
3. Describe the dynamics of f_λ when $-1 < \lambda < 1$.
4. Describe the bifurcation that occurs at $\lambda = 1$.
5. Find a λ-value, λ^*, for which f_{λ^*} has a pair of invariant intervals $[0, \pm x^*]$ on each of which the behavior of f_λ mimics that of the logistic function $4x(1 - x)$.
6. Describe the change in dynamics that occurs when λ increases through λ^*.
7. Describe the dynamics of f_λ when λ is very large. Describe the set of points Λ_λ whose orbits do not escape to $\pm\infty$ in this case.
8. Use symbolic dynamics to set up a sequence space and a corresponding shift map for λ large. Prove that f_λ is chaotic on Λ_λ.
9. Find the parameter value $\lambda' > \lambda^*$ above, which the results of the previous two investigations hold true.
10. Describe the bifurcation that occurs as λ increases through λ'.

15.9 Exploration: The Orbit Diagram

Unlike the previous exploration, this exploration is primarily experimental. It is designed to acquaint you with the rich dynamics of the logistic family as the parameter increases from 0 to 4. Using a computer and whatever software seems appropriate, construct the *orbit diagram* for the logistic family $f_\lambda(x) = \lambda x(1 - x)$ as follows: Choose N equally spaced λ-values $\lambda_1, \lambda_2, \ldots, \lambda_N$ in the interval $0 \le \lambda_j \le 4$. For example, let $N = 800$ and set $\lambda_j = 0.005j$. For each λ_j, compute the orbit of 0.5 under f_{λ_j} and plot this orbit as follows.

Let the horizontal axis be the λ-axis and let the vertical axis be the x-axis. Over each λ_j, plot the points $(\lambda_j, f_{\lambda_j}^k(0.5))$ for, say, $50 \le k \le 250$. That is, compute the first 250 points on the orbit of 0.5 under f_{λ_j}, but display only the last 200 points on the vertical line over $\lambda = \lambda_j$. Effectively, you are displaying the "fate" of the orbit of 0.5 in this way.

You will need to magnify certain portions of this diagram; one such magnification is displayed in Figure 15.13, where we have displayed only that portion of the orbit diagram for λ in the interval $3 \le \lambda \le 4$.

1. The region bounded by $0 \le \lambda < 3.57\ldots$ is called the *period 1 window*. Describe what you see as λ increases in this window. What type of bifurcations occur?
2. Near the bifurcations in the previous question, you sometimes see a smear of points. What causes this?
3. Observe the *period 3 window* bounded approximately by $3.828\ldots < \lambda < 3.857\ldots$. Investigate the bifurcation that gives rise to this window as λ increases.

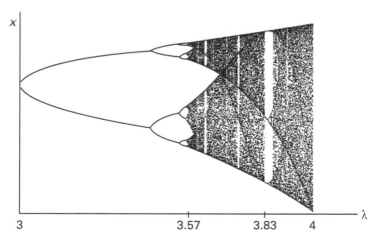

Figure 15.13   The orbit diagram for the logistic family with $3 \le \lambda \le 4$.

4. There are many other period n windows (named for the least period of the cycle in that window). Discuss any pattern you can find in how these windows are arranged as λ increases. In particular, if you magnify portions between the period 1 and period 3 windows, how are the larger windows in each successive enlargement arranged?

5. You observe "darker" curves in this orbit diagram. What are these? Why does this happen?

EXERCISES

1. Find all periodic points for each of the following maps and classify them as attracting, repelling, or neither.

(a) $Q(x) = x - x^2$ (b) $Q(x) = 2(x - x^2)$

(c) $C(x) = x^3 - \frac{1}{9}x$ (d) $C(x) = x^3 - x$

(e) $S(x) = \frac{1}{2}\sin(x)$ (f) $S(x) = \sin(x)$

(g) $E(x) = e^{x-1}$ (h) $E(x) = e^x$

(i) $A(x) = \arctan x$ (j) $A(x) = -\frac{\pi}{4}\arctan x$

2. Discuss the bifurcations that occur in the following families of maps at the indicated parameter value

(a) $S_\lambda(x) = \lambda \sin x, \quad \lambda = 1$

(b) $C_\mu(x) = x^3 + \mu x, \quad \mu = -1$ (*Hint:* Exploit the fact that C_μ is an odd function.)

(c) $G_\nu(x) = x + \sin x + \nu, \quad \nu = 1$

(d) $E_\lambda(x) = \lambda e^x, \quad \lambda = 1/e$

(e) $E_\lambda(x) = \lambda e^x, \quad \lambda = -e$

(f) $A_\lambda(x) = \lambda \arctan x, \quad \lambda = 1$

(g) $A_\lambda(x) = \lambda \arctan x, \quad \lambda = -1$

3. Consider the linear maps $f_k(x) = kx$. Show that there are four open sets of parameters for which the behavior of orbits of f_k is similar. Describe what happens in the exceptional cases.

4. For the function $f_\lambda(x) = \lambda x(1 - x)$ defined on $\mathbb{R}$:

(a) Describe the bifurcations that occur at $\lambda = -1$ and $\lambda = 3$.

(b) Find all period 2 points.

(c) Describe the bifurcation that occurs at $\lambda = -1.75$.

5. For the doubling map D on $[0, 1)$:

(a) List all periodic points explicitly.

(b) List all points whose orbits end up landing on 0 and are thereby eventually fixed.

(c) Let $x \in [0, 1)$ and suppose that x is given in binary form as $a_0 a_1 a_2 \ldots$ where each a_j is either 0 or 1. First give a formula for the binary representation of $D(x)$. Then explain why this causes orbits of D generated by a computer to end up eventually fixed at 0.

6. Show that, if x_0 lies on a cycle of period n, then

$$(f^n)'(x_0) = \prod_{i=0}^{n-1} f'(x_i).$$

Conclude that

$$(f^n)'(x_0) = (f^n)'(x_j)$$

for $j = 1, \ldots, n-1$.

7. Prove that if f_{λ_0} has a fixed point at x_0 with $|f'_{\lambda_0}(x_0)| > 1$, then there is an interval I about x_0 and an interval J about λ_0 such that, if $\lambda \in J$, then f_λ has a unique fixed source in I and no other orbits that lie entirely in I.

8. Verify that the family $f_c(x) = x^2 + c$ undergoes a period doubling bifurcation at $c = -3/4$ by

(a) Computing explicitly the period two orbit.

(b) Showing that this orbit is attracting for $-5/4 < c < -3/4$.

9. Show that the family $f_c(x) = x^2 + c$ undergoes a second period doubling bifurcation at $c = -5/4$ by using the graphs of f_c^2 and f_c^4.

10. Find an example of a bifurcation in which more than three fixed points are born.

11. Prove that $f_3(x) = 3x(1 - x)$ on I is conjugate to $f(x) = x^2 - 3/4$ on a certain interval in $\mathbb{R}$. Determine this interval.

12. Suppose $f, g \colon [0, 1] \to [0, 1]$ and that there is a semiconjugacy from f to g. Suppose that f is chaotic. Prove that g is also chaotic on $[0, 1]$.

13. Prove that the function $d(s, t)$ on Σ satisfies the three properties required for d to be a distance function or metric.

14. Identify the points in the Cantor middle-thirds set C whose addresses are

(a) *LLRLLRLLR*...

(b) *LRRLLRRLLRRL*...

15. Consider the tent map

$$T(x) = \begin{cases} 2x & \text{if } 0 \le x < 1/2 \\ -2x + 2 & \text{if } 1/2 \le x \le 1. \end{cases}$$

Prove that T is chaotic on $[0, 1]$.

16. Consider a different "tent function" defined on all of $\mathbb{R}$ by

$$T(x) = \begin{cases} 3x & \text{if } x \leq 1/2 \\ -3x + 3 & \text{if } 1/2 \leq x. \end{cases}$$

Identify the set of points Λ whose orbits do not go to $-\infty$. What can you say about the dynamics of this set?

17. Use the results of the previous exercise to show that the set Λ in Section 15.5 is homeomorphic to the Cantor middle-thirds set.

18. Prove the following saddle-node bifurcation theorem: Suppose that f_λ depends smoothly on the parameter λ and satisfies:

(a) $f_{\lambda_0}(x_0) = x_0$

(b) $f'_{\lambda_0}(x_0) = 1$

(c) $f''_{\lambda_0}(x_0) \neq 0$

(d) $\left.\dfrac{\partial f_\lambda}{\partial \lambda}\right|_{\lambda=\lambda_0}(x_0) \neq 0$

Then there is an interval I about x_0 and a smooth function $\mu : I \to \mathbb{R}$ satisfying $\mu(x_0) = \lambda_0$ and such that

$$f_{\mu(x)}(x) = x.$$

Moreover, $\mu'(x_0) = 0$ and $\mu''(x_0) \neq 0$. *Hint:* Apply the implicit function theorem to $G(x, \lambda) = f_\lambda(x) - x$ at (x_0, λ_0).

19. Discuss why the saddle-node bifurcation is the "typical" bifurcation involving only fixed points.

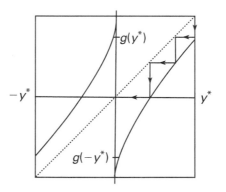

Figure 15.14 The graph of the one-dimensional function g on $[-y^*, y^*]$.

20. Recall that comprehending the behavior of the Lorenz system in Chapter 14 could be reduced to understanding the dynamics of a certain one-dimensional function g on an interval $[-y^*, y^*]$ whose graph is shown in Figure 15.14. Recall also $|g'(y)| > 1$ for all $y \neq 0$ and that g is undefined at 0. Suppose now that $g^3(y^*) = 0$ as displayed in this graph. By symmetry, we also have $g^3(-y^*) = 0$. Let $I_0 = [-y^*, 0)$ and $I_1 = (0, y^*]$ and define the usual itinerary map on $[-y^*, y^*]$.

(a) Describe the set of possible itineraries under g.

(b) What are the possible periodic points for g?

(c) Prove that g is chaotic on $[-y^*, y^*]$.

16

Homoclinic Phenomena

In this chapter we investigate several other three-dimensional systems of differential equations that display chaotic behavior. These systems include the Shil'nikov system and the double scroll attractor. As with the Lorenz system, our principal means of studying these systems involves reducing them to lower dimensional discrete dynamical systems, and then invoking symbolic dynamics. In these cases the discrete system is a planar map called the *horseshoe map*. This was one of the first chaotic systems to be analyzed completely.

16.1 The Shil'nikov System

In this section we investigate the behavior of a nonlinear system of differential equations that possesses a homoclinic solution to an equilibrium point that is a spiral saddle. While we deal primarily with a model system here, the work of Shil'nikov and others [4, 40, 41], shows that the phenomena described in this chapter hold in many actual systems of differential equations. Indeed, in the exploration at the end of this chapter, we investigate the system of differential equations governing the Chua circuit, which, for certain parameter values, has a pair of such homoclinic solutions.

For this example, we do not specify the full system of differential equations. Rather, we first set up a linear system of differential equations in a certain cylindrical neighborhood of the origin. This system has a two-dimensional stable surface in which solutions spiral toward the origin and a one-dimensional unstable curve. We then make the simple but crucial dynamical assumption that one of the two branches of the unstable curve is a homoclinic solution and thus eventually enters the stable surface. We do not write down a specific differential equation having this behavior. Although it is possible to do so, having the equations is not particularly useful for understanding the global dynamics of the system. In fact, the phenomena we study here depend only on the qualitative properties of the linear system described previously a key inequality involving the eigenvalues of this linear system, and the homoclinic assumption.

The first portion of the system is defined in the cylindrical region S of $\mathbb{R}^3$ given by $x^2 + y^2 \leq 1$ and $|z| \leq 1$. In this region consider the linear system

$$X' = \begin{pmatrix} -1 & 1 & 0 \\ -1 & -1 & 0 \\ 0 & 0 & 2 \end{pmatrix} X.$$

The associated eigenvalues are $-1 \pm i$ and 2. Using the results of Chapter 6, the flow ϕ_t of this system is easily derived:

$$x(t) = x_0 e^{-t} \cos t + y_0 e^{-t} \sin t$$

$$y(t) = -x_0 e^{-t} \sin t + y_0 e^{-t} \cos t$$

$$z(t) = z_0 e^{2t}.$$

Using polar coordinates in the xy–plane, solutions in S are given more succinctly by

$$r(t) = r_0 e^{-t}$$

$$\theta(t) = \theta_0 - t$$

$$z(t) = z_0 e^{2t}.$$

This system has a two-dimensional stable plane (the xy–plane) and a pair of unstable curves $\zeta^{\pm}$ lying on the positive and negative z-axis, respectively.

We remark that there is nothing special about our choice of eigenvalues for this system. Everything below works fine for eigenvalues $\alpha \pm i\beta$ and λ where $\alpha < 0$, $\beta \neq 0$, and $\lambda > 0$ subject only to the important condition that $\lambda > -\alpha$.

The boundary of S consists of three pieces: the upper and lower disks $D^{\pm}$ given by $z = \pm 1$, $r \leq 1$, and the cylindrical boundary C given by $r = 1$, $|z| \leq 1$. The stable plane meets C along the circle $z = 0$ and divides C into two pieces, the upper and lower halves given by C^+ and C^-, on which $z > 0$ and $z < 0$, respectively. We may parametrize $D^{\pm}$ by r and θ and C by θ and z. We will concentrate in this section on C^+.

Any solution of this system that starts in C^+ must eventually exit from S through D^+. Hence we can define a map $\psi_1 : C^+ \to D^+$ given by following solution curves that start in C^+ until they first meet D^+. Given $(\theta_0, z_0) \in C^+$, let $\tau = \tau(\theta_0, z_0)$ denote the time it takes for the solution through (θ_0, z_0) to make the transit to D^+. We compute immediately using $z(t) = z_0 e^{2t}$ that $\tau = -\log(\sqrt{z_0})$. Therefore

$$\psi_1 \begin{pmatrix} 1 \\ \theta_0 \\ z_0 \end{pmatrix} = \begin{pmatrix} r_1 \\ \theta_1 \\ 1 \end{pmatrix} = \begin{pmatrix} \sqrt{z_0} \\ \theta_0 + \log(\sqrt{z_0}) \\ 1 \end{pmatrix}.$$

For simplicity, we will regard ψ_1 as a map from the (θ_0, z_0) cylinder to the (r_1, θ_1) plane. Note that a vertical line given by $\theta_0 = \theta^*$ in C^+ is mapped by ψ_1 to the spiral

$$z_0 \to \left(\sqrt{z_0}, \theta^* + \log\left(\sqrt{z_0}\right) \right),$$

which spirals down to the point $r = 0$ in $D^{\pm}$, since $\log \sqrt{z_0} \to -\infty$ as $z_0 \to 0$.

To define the second piece of the system, we assume that the branch ζ^+ of the unstable curve leaving the origin through D^+ is a homoclinic solution. That is, ζ^+ eventually returns to the stable plane. See Figure 16.1. We assume that ζ^+ first meets the cylinder C at the point $r = 1$, $\theta = 0$, $z = 0$. More precisely, we assume that there is a time t_1 such that $\phi_{t_1}(0, \theta, 1) = (1, 0, 0)$ in r, θ, z coordinates.

Therefore we may define a second map ψ_2 by following solutions beginning near $r = 0$ in D^+ until they reach C. We will assume that ψ_2 is, in fact, defined on all of D^+. In Cartesian coordinates on D^+, we assume that ψ_2 takes $(x, y) \in D^+$ to $(\theta_1, z_1) \in C$ via the rule

$$\psi_2 \begin{pmatrix} x \\ y \end{pmatrix} = \begin{pmatrix} \theta_1 \\ z_1 \end{pmatrix} = \begin{pmatrix} y/2 \\ x/2 \end{pmatrix}.$$

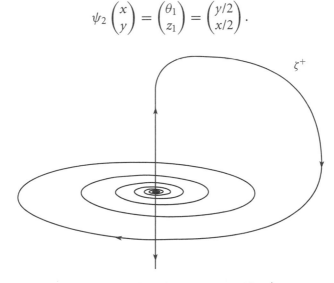

Figure 16.1   The homoclinic orbit ζ^+.

In polar coordinates, ψ_2 is given by

$$\theta_1 = (r \sin \theta)/2$$
$$z_1 = (r \cos \theta)/2.$$

Of course, this is a major assumption, since writing down such a map for a particular nonlinear system would be virtually impossible.

Now the composition $\Phi = \psi_2 \circ \psi_1$ defines a Poincaré map on C^+. The map ψ_1 is defined on C^+ and takes values in D^+, and then ψ_2 takes values in C. We have $\Phi : C^+ \to C$ where

$$\Phi \begin{pmatrix} \theta_0 \\ z_0 \end{pmatrix} = \begin{pmatrix} \theta_1 \\ z_1 \end{pmatrix} = \begin{pmatrix} \frac{1}{2} \sqrt{z_0} \sin \left(\theta_0 + \log(\sqrt{z_0}) \right) \\ \frac{1}{2} \sqrt{z_0} \cos \left(\theta_0 + \log(\sqrt{z_0}) \right) \end{pmatrix}.$$

See Figure 16.2.

As in the Lorenz system, we have now reduced the study of the flow of this three-dimensional system to the study of a planar discrete dynamical system. As we shall see in the next section, this type of mapping has incredibly rich dynamics that may be (partially) analyzed using symbolic dynamics. For a little taste of what is to come, we content ourselves here with just finding the fixed points of Φ. To do this we need to solve

$$\theta_0 = \frac{1}{2} \sqrt{z_0} \sin \left(\theta_0 + \log(\sqrt{z_0}) \right)$$

$$z_0 = \frac{1}{2} \sqrt{z_0} \cos \left(\theta_0 + \log(\sqrt{z_0}) \right).$$

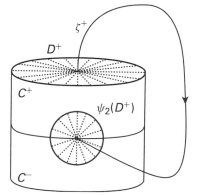

Figure 16.2   The map ψ_2: $D^+ \to C$.

These equations look pretty formidable. However, if we square both equations and add them, we find

$$\theta_0^2 + z_0^2 = \frac{z_0}{4}$$

so that

$$\theta_0 = \pm \frac{1}{2}\sqrt{z_0 - 4z_0^2},$$

which is well defined provided that $0 \leq z_0 \leq 1/4$. Substituting this expression into the second equation above, we find that we need to solve

$$\cos\left(\pm\frac{1}{2}\sqrt{z_0 - 4z_0^2} + \log\left(\sqrt{z_0}\right)\right) = 2\sqrt{z_0}.$$

Now the term $\sqrt{z_0 - 4z_0^2}$ tends to zero as $z_0 \rightarrow 0$, but $\log(\sqrt{z_0}) \rightarrow -\infty$. Therefore the graph of the left-hand side of this equation oscillates infinitely many times between ± 1 as $z_0 \rightarrow 0$. Hence there must be infinitely many places where this graph meets that of $2\sqrt{z_0}$, and so there are infinitely many solutions of this equation. This, in turn, yields infinitely many fixed points for Φ. Each of these fixed points then corresponds to a periodic solution of the system that starts in C^+, winds a number of times around the z-axis near the origin, and then travels around close to the homoclinic orbit until closing up when it returns to C^+. See Figure 16.3.

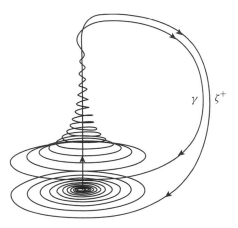

Figure 16.3 A periodic solution γ near the homoclinic solution ζ^+.

We now describe the geometry of this map; in the next section we use these ideas to investigate the dynamics of a simplified version of this map. First note that the circles $z_0 = \alpha$ in C^+ are mapped by ψ_1 to circles $r = \sqrt{\alpha}$ centered at $r = 0$ in D^+ since

$$\psi_1 \begin{pmatrix} \theta_0 \\ \alpha \end{pmatrix} = \begin{pmatrix} r_1 \\ \theta_1 \end{pmatrix} = \begin{pmatrix} \sqrt{\alpha} \\ \theta_0 + \log(\sqrt{\alpha}) \end{pmatrix}.$$

Then ψ_2 maps these circles to circles of radius $\sqrt{\alpha}/2$ centered at $\theta_1 = z_1 = 0$ in C. (To be precise, these are circles in the θz–plane; in the cylinder, these circles are "bent.") In particular, we see that "one-half" of the domain C^+ is mapped into the lower part of the cylinder C^- and therefore no longer comes into play.

Let H denote the half-disk $\Phi(C^+) \cap \{z \geq 0\}$. Half-disk H has center at $\theta_1 = z_1 = 0$ and radius $1/2$. The preimage of H in C^+ consists of all points (θ_0, z_0) whose images satisfy $z_1 \geq 0$, so that we must have

$$z_1 = \frac{1}{2}\sqrt{z_0} \cos\left(\theta_0 + \log(\sqrt{z_0})\right) \geq 0.$$

It follows that the preimage of H is given by

$$\Phi^{-1}(H) = \{(\theta_0, z_0) \mid -\pi/2 \leq \theta_0 + \log(\sqrt{z_0}) \leq \pi/2\}$$

where $0 < z_0 \leq 1$. This is a region bounded by the two curves $\theta_0 + \log(\sqrt{z_0}) = \pm\pi/2$, each of which spirals downward in C^+ toward the circle $z = 0$. See Figure 16.4. This follows since, as $z_0 \to 0$, we must have $\theta_0 \to \infty$. More generally, consider the curves ℓ_α given by

$$\theta_0 + \log(\sqrt{z_0}) = \alpha$$

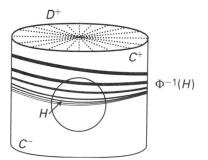

Figure 16.4   The half-disk H and its preimage in C^+.

for $-\pi/2 \le \alpha \le \pi/2$. These curves fill the preimage $\Phi^{-1}(H)$ and each spirals around C just as the boundary curves do. Now we have

$$\Phi(\ell_\alpha) = \frac{\sqrt{z_0}}{2} \begin{pmatrix} \sin\alpha \\ \cos\alpha \end{pmatrix},$$

so Φ maps each ℓ_α to a ray that emanates from $\theta = z = 0$ in C^+ and is parameterized by $\sqrt{z_0}$. In particular, Φ maps each of the boundary curves $\ell_{\pm\pi/2}$ to $z = 0$ in C.

Since the curves $\ell_{\pm\pi/2}$ spiral down toward the circle $z = 0$ in C, it follows that $\Phi^{-1}(H)$ meets H in infinitely many strips, which are nearly horizontal close to $z = 0$. See Figure 16.4. We denote these strips by H_k for k sufficiently large. More precisely, let H_k denote the component of $\Phi^{-1}(H) \cap H$ for which we have

$$2k\pi - \frac{1}{2} \le \theta_0 \le 2k\pi + \frac{1}{2}.$$

The top boundary of H_k is given by a portion of the spiral $\ell_{\pi/2}$ and the bottom boundary by a piece of $\ell_{-\pi/2}$. Using the fact that

$$-\frac{\pi}{2} \le \theta_0 + \log\left(\sqrt{z_0}\right) \le \frac{\pi}{2},$$

we find that, if $(\theta_0, z_0) \in H_k$, then

$$-(4k+1)\pi - 1 \le -\pi - 2\theta_0 \le 2\log\sqrt{z_0} \le \pi - 2\theta_0 \le -(4k-1)\pi + 1$$

from which we conclude that

$$\exp(-(4k+1)\pi - 1) \le z_0 \le \exp(-(4k-1)\pi + 1).$$

Now consider the image of H_k under Φ. The upper and lower boundaries of H_k are mapped to $z = 0$. The curves $\ell_\alpha \cap H_k$ are mapped to arcs in rays emanating from $\theta = z = 0$. These rays are given as above by

$$\frac{\sqrt{z_0}}{2} \begin{pmatrix} \sin\alpha \\ \cos\alpha \end{pmatrix}.$$

In particular, the curve ℓ_0 is mapped to the vertical line $\theta_1 = 0$, $z_1 = \sqrt{z_0}/2$. Using the above estimate of the size of z_0 in H_k, one checks easily that the image of ℓ_0 lies completely above H_k when $k \ge 2$. Therefore the image of $\Phi(H_k)$ is a "horseshoe-shaped" region that crosses H_k twice as shown in Figure 16.5. In particular, if k is large, the curves $\ell_\alpha \cap H_k$ meet the horseshoe $\Phi(H_k)$ in nearly horizontal subarcs.

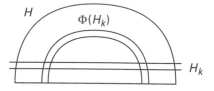

Figure 16.5   The image of H_k is a horseshoe that crosses H_k twice in C^+.

Such a map is called a *horseshoe map*; in the next section we discuss the prototype of such a function.

16.2 The Horseshoe Map

Symbolic dynamics, which played such a crucial role in our understanding of the one-dimensional logistic map, can also be used to study higher dimensional phenomena. In this section, we describe an important example in $\mathbb{R}^2$, the horseshoe map [43]. We shall see that this map has much in common with the Poincaré map described in the previous section.

To define the horseshoe map, we consider a region D consisting of three components: a central square S with sides of length 1, together with two semicircles D_1 and D_2 at the top and bottom. D is shaped like a "stadium."

The horseshoe map F takes D inside itself according to the following prescription. First, F linearly contracts S in the horizontal direction by a factor $\delta < 1/2$ and expands it in the vertical direction by a factor of $1/\delta$ so that S is long and thin, and then F curls S back inside D in a horseshoe-shaped figure as displayed in Figure 16.6. We stipulate that F maps S linearly onto the two vertical "legs" of the horseshoe.

We assume that the semicircular regions D_1 and D_2 are mapped inside D_1 as depicted. We also assume that there is a fixed point in D_1 that attracts all other orbits in D_1. Note that $F(D) \subset D$ and that F is one-to-one. However, since F is not onto, the inverse of F is not globally defined. The remainder of this section is devoted to the study of the dynamics of F in D.

Note first that the preimage of S consists of two horizontal rectangles, H_0 and H_1, which are mapped linearly onto the two vertical components V_0 and V_1 of $F(S) \cap S$. The width of V_0 and V_1 is therefore δ, as is the height of H_0 and H_1. See Figure 16.7. By linearity of $F : H_0 \to V_0$ and $F : H_1 \to V_1$, we know that F takes horizontal and vertical lines in H_j to horizontal and vertical lines in V_j for $j = 1, 2$. As a consequence, if both h and $F(h)$ are horizontal line segments in S, then the length of $F(h)$ is δ times the length of h. Similarly,

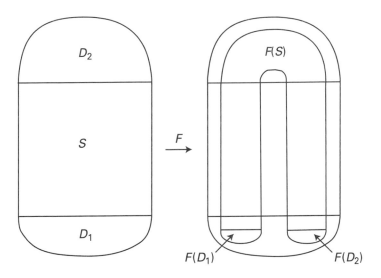

Figure 16.6   The first iterate of the horseshoe map.

if v is a vertical line segment in S whose image also lies in S, then the length of $F(v)$ is $1/\delta$ times the length of v.

We now describe the *forward orbit* of each point $X \in D$. Recall that the forward orbit of X is given by $\{F^n(X) \mid n \geq 0\}$. By assumption, F has a unique fixed point X_0 in D_1 and $\lim_{n \to \infty} F^n(X) = X_0$ for all $X \in D_1$. Also, since $F(D_2) \subset D_1$, all forward orbits in D_2 behave likewise. Similarly, if $X \in S$ but $F^k(X) \notin S$ for some $k > 0$, then we must have that $F^k(X) \in D_1 \cup D_2$ so that $F^n(X) \to X_0$ as $n \to \infty$ as well. Consequently, we understand the forward orbits of any $X \in D$ whose orbit enters D_1, so it suffices to consider the set of

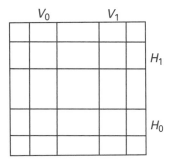

Figure 16.7   The rectangles H_0 and H_1 and their images V_0 and V_1.

points whose forward orbits never enter D_1 and so lie completely in S. Let

$$\Lambda_+ = \{X \in S \mid F^n(X) \in S \text{ for } n = 0, 1, 2, \ldots\}.$$

We claim that Λ_+ has properties similar to the corresponding set for the one-dimensional logistic map described in Chapter 15.

If $X \in \Lambda_+$, then $F(X) \in S$, so we must have that either $X \in H_0$ or $X \in H_1$, for all other points in S are mapped into D_1 or D_2. Since $F^2(X) \in S$ as well, we must also have $F(X) \in H_0 \cup H_1$, so that $X \in F^{-1}(H_0 \cup H_1)$. Here $F^{-1}(W)$ denotes the preimage of a set W lying in D. In general, since $F^n(X) \in S$, we have $X \in F^{-n}(H_0 \cup H_1)$. Thus we may write

$$\Lambda_+ = \bigcap_{n=0}^{\infty} F^{-n}(H_0 \cup H_1).$$

Now if H is any horizontal strip connecting the left and right boundaries of S with height h, then $F^{-1}(H)$ consists of a pair of narrower horizontal strips of height δh, one in each of H_0 and H_1. The images under F of these narrower strips are given by $H \cap V_0$ and $H \cap V_1$. In particular, if $H = H_i$, $F^{-1}(H_i)$ is a pair of horizontal strips, each of height δ^2, with one in H_0 and the other in H_1. Similarly, $F^{-1}(F^{-1}(H_i)) = F^{-2}(H_i)$ consists of four horizontal strips, each of height δ^3, and $F^{-n}(H_i)$ consists of 2^n horizontal strips of width δ^{n+1}. Hence the same procedure we used in Section 15.5 shows that Λ_+ is a Cantor set of line segments, each extending horizontally across S.

The main difference between the horseshoe and the logistic map is that, in the horseshoe case, there is a single backward orbit rather than infinitely many such orbits. The *backward orbit* of $X \in S$ is $\{F^{-n}(X) \mid n = 1, 2, \ldots\}$, provided $F^{-n}(X)$ is defined and in D. If $F^{-n}(X)$ is not defined, then the backward orbit of X terminates. Let Λ_- denote the set of points whose backward orbit is defined for all n and lies entirely in S. If $X \in \Lambda_-$, then we have $F^{-n}(X) \in S$ for all $n \geq 1$, which implies that $X \in F^n(S)$ for all $n \geq 1$. As above, this forces $X \in F^n(H_0 \cup H_1)$ for all $n \geq 1$. Therefore we may also write

$$\Lambda_- = \bigcap_{n=1}^{\infty} F^n(H_0 \cup H_1).$$

On the other hand, if $X \in S$ and $F^{-1}(X) \in S$, then we must have $X \in F(S) \cap S$, so that $X \in V_0$ or $X \in V_1$. Similarly, if $F^{-2}(X) \in S$ as well, then $X \in F^2(S) \cap S$, which consists of four narrower vertical strips, two in V_0 and two in V_1. In Figure 16.8 we show the image of D under F^2. Arguing entirely analogously as above, it is easy to check that Λ_- consists of a Cantor set of vertical lines.

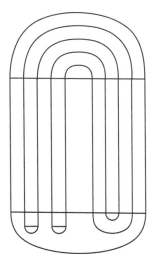

Figure 16.8   The second iterate of the horseshoe map.

Let

$$\Lambda = \Lambda_+ \cap \Lambda_-$$

be the intersection of these two sets. Any point in Λ has its entire orbit (both the backward and forward orbit) in S.

To introduce symbolic dynamics into this picture, we will assign a doubly infinite sequence of 0's and 1's to each point in Λ. If $X \in \Lambda$, then, from the above, we have

$$X \in \bigcap_{n=-\infty}^{\infty} F^n(H_0 \cup H_1).$$

Thus we associate to X the *itinerary*

$$S(X) = (\ldots s_{-2} s_{-1} \cdot s_0 s_1 s_2 \ldots)$$

where $s_j = 0$ or 1 and $s_j = k$ if and only if $F^j(X) \in H_k$. This then provides us with the symbolic dynamics on Λ. Let Σ_2 denote the set of all doubly infinite sequences of 0's and 1's:

$$\Sigma_2 = \{(\mathbf{s}) = (\ldots s_{-2} s_{-1} \cdot s_0 s_1 s_2 \ldots) \mid s_j = 0 \text{ or } 1\}.$$

We impose a distance function on Σ_2 by defining

$$d((\mathbf{s}),(\mathbf{t})) = \sum_{i=-\infty}^{\infty} \frac{|s_i - t_i|}{2^{|i|}}$$

as in Section 15.5. Thus two sequences in Σ_2 are "close" if they agree in all k spots where $|k| \le n$ for some (large) n. Define the (two-sided) *shift map* σ by

$$\sigma(\ldots s_{-2}s_{-1} \cdot s_0 s_1 s_2 \ldots) = (\ldots s_{-2}s_{-1}s_0 \cdot s_1 s_2 \ldots).$$

That is, σ simply shifts each sequence in Σ_2 one unit to the left (equivalently, σ shifts the decimal point one unit to the right). Unlike our previous (one-sided) shift map, this map has an inverse. Clearly, shifting one unit to the right gives this inverse. It is easy to check that σ is a homeomorphism on Σ_2 (see Exercise 2 at the end of this chapter).

The shift map is now the model for the restriction of F to Λ. Indeed, the itinerary map S gives a conjugacy between F on Λ and σ on Σ_2. For if $X \in \Lambda$ and $S(X) = (\ldots s_{-2}s_{-1} \cdot s_0 s_1 s_2 \ldots)$, then we have $X \in H_{s_0}$, $F(X) \in H_{s_1}$, $F^{-1}(X) \in H_{s_{-1}}$, and so forth. But then we have $F(X) \in H_{s_1}$, $F(F(X)) \in H_{s_2}$, $X = F^{-1}(F(X)) \in H_{s_0}$, and so forth. This tells us that the itinerary of $F(X)$ is $(\ldots s_{-1}s_0 \cdot s_1 s_2 \ldots)$, so that

$$S(F(x)) = (\ldots s_{-1}s_0 \cdot s_1 s_2 \ldots) = \sigma(S(X)),$$

which is the conjugacy equation. We leave the proof of the fact that S is a homeomorphism to the reader (see Exercise 3).

All of the properties that held for the old one-sided shift from the previous chapter hold for the two-sided shift σ as well. For example, there are precisely 2^n periodic points of period n for σ and there is a dense orbit for σ. Moreover, F is chaotic on Λ (see Exercises 4 and 5). But new phenomena are present as well. We say that two points X_1 and X_2 are *forward asymptotic* if $F^n(X_1), F^n(X_2) \in D$ for all $n \ge 0$ and

$$\lim_{n \to \infty} |F^n(X_1) - F^n(X_2)| = 0.$$

Points X_1 and X_2 are *backward asymptotic* if their backward orbits are defined for all n and the above limit is zero as $n \to -\infty$. Intuitively, two points in D are forward asymptotic if their orbits approach each other as $n \to \infty$. Note that any point that leaves S under forward iteration of F is forward asymptotic to the fixed point $X_0 \in D_1$. Also, if X_1 and X_2 lie on the same horizontal line in Λ_+, then X_1 and X_2 are forward asymptotic. If X_1 and X_2 lie on the same vertical line in Λ_-, then they are backward asymptotic.

We define the *stable set* of X to be

$$W^s(X) = \left\{Z \mid |F^n(Z) - F^n(X)| \to 0 \text{ as } n \to \infty\right\}.$$

The *unstable set* of X is given by

$$W^u(X) = \left\{Z \mid |F^{-n}(X) - F^{-n}(Z)| \to 0 \text{ as } n \to \infty\right\}.$$

Equivalently, a point Z lies in $W^s(X)$ if X and Z are forward asymptotic. As above, any point in S whose orbit leaves S under forward iteration of the horseshoe map lies in the stable set of the fixed point in D_1.

The stable and unstable sets of points in Λ are more complicated. For example, consider the fixed point X^*, which lies in H_0 and therefore has the itinerary $(\ldots 00 \cdot 000 \ldots)$. Any point that lies on the horizontal segment ℓ_s through X^* lies in $W^s(X^*)$. But there are many other points in this stable set. Suppose the point Y eventually maps into ℓ_s. Then there is an integer n such that $|F^n(Y) - X^*| < 1$. Hence

$$|F^{n+k}(Y) - X^*| < \delta^k$$

and it follows that $Y \in W^s(X^*)$. Thus the union of horizontal intervals given by $F^{-k}(\ell_s)$ for $k = 1, 2, 3, \ldots$ all lie in $W^s(X^*)$. The reader may easily check that there are 2^k such intervals.

Since $F(D) \subset D$, the unstable set of the fixed point X^* assumes a somewhat different form. The vertical line segment ℓ_u through X^* in D clearly lies in $W^u(X^*)$. As above, all of the forward images of ℓ_u also lie in D. One may easily check that $F^k(\ell_u)$ is a "snake-like" curve in D that cuts vertically across S exactly 2^k times. See Figure 16.9. The union of these forward images is then a very complicated curve that passes through S infinitely often. The closure of this curve in fact contains all points in Λ as well as all of their unstable curves (see Exercise 12).

The stable and unstable sets in Λ are easy to describe on the shift level. Let

$$\mathbf{s}^* = (\ldots s_{-2}^* s_{-1}^* \cdot s_0^* s_1^* s_2^* \ldots) \in \Sigma_2.$$

Clearly, if $\mathbf{t}$ is a sequence whose entries agree with those of $\mathbf{s}^*$ to the right of some entry, then $\mathbf{t} \in W^s(\mathbf{s}^*)$. The converse of this is also true, as is shown in Exercise 6.

A natural question that arises is the relationship between the set Λ for the one-dimensional logistic map and the corresponding Λ for the horseshoe map. Intuitively, it may appear that the Λ for the horseshoe has many more points. However, both Λ's are actually homeomorphic! This is best seen on the shift level.

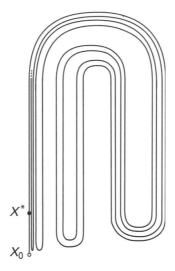

Figure 16.9   The unstable
set for X^* in D.

Let Σ_2^1 denote the set of one-sided sequences of 0's and 1's and Σ_2 the set of two-sided such sequences. Define a map

$$\Phi\colon \Sigma_2^1 \to \Sigma_2$$

by

$$\Phi(s_0 s_1 s_2 \ldots) = (\ldots s_5 s_3 s_1 \cdot s_0 s_2 s_4 \ldots).$$

It is easy to check that Φ is a homeomorphism between Σ_2^1 and Σ_2 (see Exercise 11).

Finally, to return to the subject of Section 16.1, note that the return map investigated in that section consists of infinitely many pieces that resemble the horseshoe map of this section. Of course, the horseshoe map here was effectively linear in the region where the map was chaotic, so the results in this section do not go over immediately to prove that the return maps near the homoclinic orbit have similar properties. This can be done; however, the techniques for doing so (involving a generalized notion of hyperbolicity) are beyond the scope of this book. See [13] or [36] for details.

16.3 The Double Scroll Attractor

In this section we continue the study of behavior near homoclinic solutions in a three-dimensional system. We return to the system described in Section 16.1,

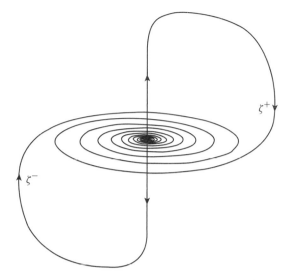

Figure 16.10   The homoclinic orbits $\zeta^{\pm}$.

only now we assume that the vector field is skew-symmetric about the origin. In particular, this means that both branches of the unstable curve at the orgin, $\zeta^{\pm}$, now yield homoclinic solutions as depicted in Figure 16.10. We assume that ζ^{+} meets the cylinder C given by $r = 1$, $|z| \leq 1$ at the point $\theta = 0$, $z = 0$, so that ζ^{-} meets the cylinder at the diametrically opposite point, $\theta = \pi$, $z = 0$.

As in the Section 16.1, we have a Poincaré map Φ defined on the cylinder C. This time, however, we cannot disregard solutions that reach C in the region $z < 0$; now these solutions follow the second homoclinic solution ζ^{-} around until they reintersect C. Thus Φ is defined on all of $C - \{z = 0\}$.

As before, the Poincaré map Φ^{+} defined in the top half of the cylinder, C^{+}, is given by

$$\Phi^{+} \begin{pmatrix} \theta_0 \\ z_0 \end{pmatrix} = \begin{pmatrix} \theta_1 \\ z_1 \end{pmatrix} = \begin{pmatrix} \frac{1}{2}\sqrt{z_0} \sin\left(\theta_0 + \log(\sqrt{z_0})\right) \\ \frac{1}{2}\sqrt{z_0} \cos\left(\theta_0 + \log(\sqrt{z_0})\right) \end{pmatrix}.$$

Invoking the symmetry, a computation shows that Φ^{-} on C^{-} is given by

$$\Phi^{-} \begin{pmatrix} \theta_0 \\ z_0 \end{pmatrix} = \begin{pmatrix} \theta_1 \\ z_1 \end{pmatrix} = \begin{pmatrix} \pi - \frac{1}{2}\sqrt{-z_0} \sin\left(\theta_0 + \log(\sqrt{-z_0})\right) \\ \frac{1}{2}\sqrt{-z_0} \cos\left(\theta_0 + \log(\sqrt{-z_0})\right) \end{pmatrix}.$$

where $z_0 < 0$ and θ_0 is arbitrary. Hence $\Phi(C^{+})$ is the disk of radius $1/2$ centered at $\theta = 0$, $z = 0$, while $\Phi(C^{-})$ is a similar disk centered at $\theta = \pi$, $z = 0$.

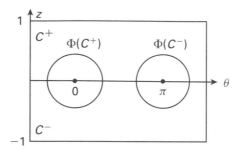

Figure 16.11 $\Phi(C^{\pm}) \cap C$, where we have displayed the cylinder C as a strip.

The centers of these disks do not lie in the image, because these are the points where $\zeta^{\pm}$ enters C. See Figure 16.11.

Now let $X \in C$. Either the solution through X lies on the stable surface of the origin, or else $\Phi(X)$ is defined, so that the solution through X returns to C at some later time. As a consequence, each point $X \in C$ has the property that

1. Either the solution through X crosses C infinitely many times as $t \to \infty$, so that $\Phi^n(X)$ is defined for all $n \geq 0$, or
2. The solution through X eventually meets $z = 0$ and hence lies on the stable surface through the origin.

In backward time, the situation is different: only those points that lie in $\Phi(C^{\pm})$ have solutions that return to C; strictly speaking, we have not defined the backward solution of points in $C - \Phi(C^{\pm})$, but we think of these solutions as being defined in $\mathbb{R}^3$ and eventually meeting C, after which time these solutions continually revisit C.

As in the case of the Lorenz attractor, we let

$$A = \bigcap_{n=0}^{\infty} \overline{\Phi^n(C)}$$

where $\overline{\Phi^n(C)}$ denotes the closure of the set $\Phi^n(C)$. Then we set

$$A = \left(\bigcup_{t \in \mathbb{R}} \phi_t(A) \right) \bigcup \{(0, 0, 0)\}.$$

Note that $\overline{\Phi^n(C)} - \Phi^n(C)$ is just the two intersection points of the homoclinic solutions $\zeta^{\pm}$ with C. Therefore we only need to add the origin to $\mathcal{A}$ to ensure that $\mathcal{A}$ is a closed set.

The proof of the following result is similar in spirit to the corresponding result for the Lorenz attractor in Section 14.4.

Proposition. *The set $\mathcal{A}$ has the following properties:*

1. *$\mathcal{A}$ is compact and invariant;*
2. *There is an open set U containing $\mathcal{A}$ such that for each $X \in U$, $\phi_t(X) \in U$ for all $t \geq 0$ and $\cap_{t \geq 0} \phi_t(U) = \mathcal{A}$.* ∎

Thus $\mathcal{A}$ has all of the properties of an attractor except the transitivity property. Nonetheless, $\mathcal{A}$ is traditionally called a *double scroll attractor*.

We cannot compute the divergence of the double scroll vector field as we did in the Lorenz case for the simple reason that we have not written down the formula for this system. However, we do have an expression for the Poincaré map Φ. A straightforward computation shows that $\det D\Phi = 1/8$. That is, the Poincaré map Φ shrinks areas by a factor of 1/8 at each iteration. Hence $A = \cap_{n \geq 0} \overline{\Phi^n(C)}$ has area 0 in C and we have:

Proposition. *The volume of the double scroll attractor $\mathcal{A}$ is zero.* ∎

16.4 Homoclinic Bifurcations

In higher dimensions, bifurcations associated with homoclinic orbits may lead to horribly (or wonderfully, depending on your point of view) complicated behavior. In this section we give a brief indication of some of the ramifications of this type of bifurcation. We deal here with a specific perturbation of the double scroll vector field that breaks both of the homoclinic connections.

The full picture of this bifurcation involves understanding the "unfolding" of infinitely many horseshoe maps. By this we mean the following. Consider a family of maps F_λ defined on a rectangle R with parameter $\lambda \in [0, 1]$. The image of $F_\lambda(R)$ is a horseshoe as displayed in Figure 16.12. When $\lambda = 0$, $F_\lambda(R)$ lies below R. As λ increases, $F_\lambda(R)$ rises monotonically. When $\lambda = 1$, $F_\lambda(R)$ crosses R twice and we assume that F_1 is the horseshoe map described in Section 16.2.

Clearly, F_0 has no periodic points whatsoever in R, but by the time λ has reached 1, infinitely many periodic points have been born, and other chaotic behavior has appeared. The family F_λ has undergone infinitely many bifurcations en route to the horseshoe map. How these bifurcations occur is the subject of much contemporary research in mathematics.

The situation here is significantly more complex than the bifurcations that occur for the one-dimensional logistic family $f_\lambda(x) = \lambda x(1 - x)$ with $0 \leq \lambda \leq 4$. The bifurcation structure of the logistic family has recently been completely determined; the planar case is far from being resolved.

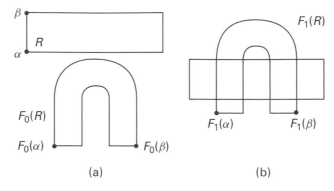

Figure 16.12   The images $F_\lambda(R)$ for (a) $\lambda = 0$ and
(b) $\lambda = 1$.

We now introduce a parameter ϵ into the double scroll system. When $\epsilon = 0$ the system will be the double scroll system considered in the previous section. When $\epsilon \neq 0$ we change the system by simply translating $\zeta^+ \cap C$ (and the corresponding transit map) in the z direction by ϵ. More precisely, we assume that the system remains unchanged in the cylindrical region $r \leq 1$, $|z| \leq 1$, but we change the transit map defined on the upper disk D^+ by adding $(0, \epsilon)$ to the image. That is, the new Poincaré map is given on C^+ by

$$\Phi_\epsilon^+(\theta, z) = \begin{pmatrix} \frac{1}{2}\sqrt{z}\sin\left(\theta + \log(\sqrt{z})\right) \\ \frac{1}{2}\sqrt{z}\cos\left(\theta + \log(\sqrt{z})\right) + \epsilon \end{pmatrix}$$

and Φ_ϵ^- is defined similarly using the skew-symmetry of the system. We further assume that ϵ is chosen small enough ($|\epsilon| < 1/2$) so that $\Phi_\epsilon^\pm(C) \subset C$.

When $\epsilon > 0$, ζ^+ intersects C in the upper cylindrical region C^+ and then, after passing close to the origin, winds around itself before reintersecting C a second time. When $\epsilon < 0$, ζ^+ now meets C in C^- and then takes a very different route back to C, this time winding around ζ^-.

Recall that Φ_0^+ has infinitely many fixed points in $C^\pm$. This changes dramatically when $\epsilon \neq 0$.

Proposition.   *The maps $\Phi_\epsilon^\pm$ each have only finitely many fixed points in $C^\pm$ when $\epsilon \neq 0$.*

Proof: To find fixed points of Φ_ϵ^+, we must solve

$$\theta = \frac{\sqrt{z_0}}{2}\sin\left(\theta + \log(\sqrt{z})\right)$$

$$z = \frac{\sqrt{z_0}}{2}\cos\left(\theta + \log(\sqrt{z})\right) + \epsilon$$

where $\epsilon > 0$. As in Section 16.1, we must therefore have

$$\frac{z}{4} = \theta^2 + (z - \epsilon)^2$$

so that

$$\theta = \pm\frac{1}{2}\sqrt{z - 4(z - \epsilon)^2}.$$

In particular, we must have

$$z - 4(z - \epsilon)^2 \geq 0$$

or, equivalently,

$$\frac{4(z - \epsilon)^2}{z} \leq 1.$$

This inequality holds provided z lies in the interval I_ϵ defined by

$$\frac{1}{8} + \epsilon - \frac{1}{8}\sqrt{1 + 16\epsilon} \leq z \leq \frac{1}{8} + \epsilon + \frac{1}{8}\sqrt{1 + 16\epsilon}.$$

This puts a further restriction on ϵ for Φ_ϵ^+ to have fixed points, namely, $\epsilon > -1/16$. Note that, when $\epsilon > -1/16$, we have

$$\frac{1}{8} + \epsilon - \frac{1}{8}\sqrt{1 + 16\epsilon} > 0$$

so that I_ϵ has length $\sqrt{1 + 16\epsilon}/4$ and this interval lies to the right of 0.

To determine the z-values of the fixed points, we must now solve

$$\cos\left(\pm\frac{1}{2}\sqrt{z - 4(z - \epsilon)^2} + \log(\sqrt{z})\right) = \frac{2(z - \epsilon)}{\sqrt{z}}$$

or

$$\cos^2\left(\pm\frac{1}{2}\sqrt{z - 4(z - \epsilon)^2} + \log(\sqrt{z})\right) = \frac{4(z - \epsilon)^2}{z}.$$

With a little calculus, one may check that the function

$$g(z) = \frac{4(z - \epsilon)^2}{z}$$

has a single minimum 0 at $z = \epsilon$ and two maxima equal to 1 at the endpoints of I_ϵ. Meanwhile the graph of

$$h(z) = \cos^2 \left(\pm \frac{1}{2} \sqrt{z - 4(z - \epsilon)^2} + \log(\sqrt{z}) \right)$$

oscillates between ± 1 only finitely many times in I_ϵ. Hence $h(z) = g(z)$ at only finitely many z-values in I_ϵ. These points are the fixed points for $\Phi_\epsilon^\pm$. ∎

Note that, as $\epsilon \to 0$, the interval I_ϵ tends to $[0, 1/4]$ and so the number of oscillations of h in I_ϵ increases without bound. Therefore we have

Corollary. *Given $N \in \mathbb{Z}$, there exists ϵ_N such that if $0 < \epsilon < \epsilon_N$, then Φ_ϵ^+ has at least N fixed points in C^+.* ∎

When $\epsilon > 0$, the unstable curve misses the stable surface in its first pass through C. Indeed, ζ^+ crosses C^+ at $\theta = 0$, $z = \epsilon$. This does not mean that there are no homoclinic orbits when $\epsilon \neq 0$. In fact, we have the following proposition:

Proposition. *There are infinitely many values of ϵ for which $\zeta^\pm$ are homoclinic solutions that pass twice through C.*

Proof: To show this, we need to find values of ϵ for which $\Phi_\epsilon^+(0, \epsilon)$ lies on the stable surface of the origin. Thus we must solve

$$0 = \frac{\sqrt{\epsilon}}{2} \cos \left(0 - \log(\sqrt{\epsilon}) \right) + \epsilon$$

or

$$-2\sqrt{\epsilon} = \cos \left(-\log(\sqrt{\epsilon}) \right).$$

But, as in Section 16.1, the graph of $\cos(-\log\sqrt{\epsilon})$ meets that of $-2\sqrt{\epsilon}$ infinitely often. This completes the proof. ∎

For each of the ϵ-values for which $\zeta^\pm$ is a homoclinic solution, we again have infinitely many fixed points (for $\Phi_\epsilon^\pm \circ \Phi_\epsilon^\pm$) as well as a very different structure for the attractor. Clearly, a lot is happening as ϵ changes. We invite the reader who has lasted this long with this book to go and figure out everything that is happening here. Good luck! And have fun!

16.5 Exploration: The Chua Circuit

In this exploration, we investigate a nonlinear three-dimensional system of differential equations related to an electrical circuit known as Chua's circuit. These were the first examples of circuit equations to exhibit chaotic behavior. Indeed, for certain values of the parameters these equations exhibit behavior similar to the double scroll attractor in Section 16.3. The original Chua circuit equations possess a piecewise linear nonlinearity. Here we investigate a variation of these equations in which the nonlinearity is given by a cubic function. For more details on the Chua circuit, refer to [11] and [25].

The nonlinear Chua circuit system is given by

$$x' = a(y - \phi(x))$$
$$y' = x - y + z$$
$$z' = -bz$$

where a and b are parameters and the function ϕ is given by

$$\phi(x) = \frac{1}{16}x^3 - \frac{1}{6}x.$$

Actually, the coefficients of this polynomial are usually regarded as parameters, but we will fix them for the sake of definiteness in this exploration. When $a = 10.91865\ldots$ and $b = 14$, this system appears to have a pair of symmetric homoclinic orbits as illustrated in Figure 16.13. The goal of this exploration

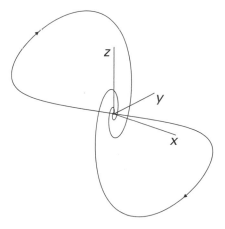

Figure 16.13   A pair of homoclinic orbits in the nonlinear Chua system at parameter values $a = 10.91865\ldots$ and $b = 14$.

is to investigate how this system evolves as the parameter a changes. As a consequence, we will also fix the parameter b at 14 and then let a vary. We caution the explorer that proving any of the chaotic or bifurcation behavior observed below is nearly impossible; virtually anything that you can do in this regard would qualify as an interesting research result.

1. As always, begin by finding the equilibrium points.
2. Determine the types of these equilibria, perhaps by using a computer algebra system.
3. This system possesses a symmetry; describe this symmetry and tell what it implies for solutions.
4. Let a vary from 6 to 14. Describe any bifurcations you observe as a varies. Be sure to choose pairs of symmetrically located initial conditions in this and other experiments in order to see the full effect of the bifurcations. Pay particular attention to solutions that begin near the origin.
5. Are there values of a for which there appears to be an attractor for this system? What appears to be happening in this case? Can you construct a model?
6. Describe the bifurcation that occurs near the following a-values:

    (a) $a = 6.58$

    (b) $a = 7.3$

    (c) $a = 8.78$

    (d) $a = 10.77$

EXERCISES

1. Prove that

$$d[(\mathbf{s}), (\mathbf{t})] = \sum_{i=-\infty}^{\infty} \frac{|s_i - t_i|}{2^{|i|}}$$

    is a distance function on Σ_2, where Σ_2 is the set of doubly infinite sequences of 0's and 1's as described in Section 16.2.
2. Prove that the shift σ is a homeomorphism of Σ_2.
3. Prove that $S \colon \Lambda \to \Sigma_2$ gives a conjugacy between σ and F.
4. Construct a dense orbit for σ.
5. Prove that periodic points are dense for σ.
6. Let $\mathbf{s}^* \in \Sigma_2$. Prove that $W^s(\mathbf{s}^*)$ consists of precisely those sequences whose entries agree with those of $\mathbf{s}^*$ to the right of some entry of $\mathbf{s}^*$.
7. Let $(\mathbf{0}) = (\ldots 00.000\ldots) \in \Sigma_2$. A sequence $\mathbf{s} \in \Sigma_2$ is called *homoclinic* to $(\mathbf{0})$ if $\mathbf{s} \in W^s(\mathbf{0}) \cap W^u(\mathbf{0})$. Describe the entries of a sequence that is

homoclinic to $(\mathbf{0})$. Prove that sequences that are homoclinic to $(\mathbf{0})$ are dense in Σ_2.

8. Let $(\mathbf{1}) = (\ldots 11.111\ldots) \in \Sigma_2$. A sequence $\mathbf{s}$ is a *heteroclinic* sequence if $\mathbf{s} \in W^s(\mathbf{0}) \cap W^u(\mathbf{1})$. Describe the entries of such a heteroclinic sequence. Prove that such sequences are dense in Σ_2.

9. Generalize the definitions of homoclinic and heteroclinic points to arbitrary periodic points for σ and reprove Exercises 7 and 8 in this case.

10. Prove that the set of homoclinic points to a given periodic point is countable.

11. Let $\Sigma_2^{\frac{1}{2}}$ denote the set of one-sided sequences of 0's and 1's. Define $\Phi: \Sigma_2^{\frac{1}{2}} \to \Sigma_2$ by

$$\Phi(s_0 s_1 s_2 \ldots) = (\ldots s_5 s_3 s_1 \cdot s_0 s_2 s_4 \ldots).$$

Prove that Φ is a homeomorphism.

12. Let X^* denote the fixed point of F in H_0 for the horseshoe map. Prove that the closure of $W^u(X^*)$ contains all points in Λ as well as points on their unstable curves.

13. Let $R: \Sigma_2 \to \Sigma_2$ be defined by

$$R(\ldots s_{-2} s_{-1} \cdot s_0 s_1 s_2 \ldots) = (\ldots s_2 s_1 s_0 \cdot s_{-1} s_{-2} \ldots).$$

Prove that $R \circ R = id$ and that $\sigma \circ R = R \circ \sigma^{-1}$. Conclude that $\sigma = U \circ R$ where U is a map that satisfies $U \circ U = id$. Maps that are their own inverses are called *involutions*. They represent very simple types of dynamical systems. Hence the shift may be decomposed into a composition of two such maps.

14. Let $\mathbf{s}$ be a sequence that is fixed by R, where R is as defined in the previous exercise. Suppose that $\sigma^n(\mathbf{s})$ is also fixed by R. Prove that $\mathbf{s}$ is a periodic point of σ of period $2n$.

15. Rework the previous exercise, assuming that $\sigma^n(\mathbf{s})$ is fixed by U, where U is given as in Exercise 13. What is the period of $\mathbf{s}$?

16. For the Lorenz system in Chapter 14, investigate numerically the bifurcation that takes place for r between 13.92 and 13.96, with $\sigma = 10$ and $b = 8/3$.

17

Existence and
Uniqueness Revisited

In this chapter we return to the material presented in Chapter 7, this time filling in all of the technical details and proofs that were omitted earlier. As a result, this chapter is more difficult than the preceding ones; it is, however, central to the rigorous study of ordinary differential equations. To comprehend thoroughly many of the proofs in this section, the reader should be familiar with such topics from real analysis as uniform continuity, uniform convergence of functions, and compact sets.

17.1 The Existence and Uniqueness
Theorem

Consider the autonomous system of differential equations

$$X' = F(X)$$

where $F: \mathbb{R}^n \to \mathbb{R}^n$. In previous chapters, we have usually assumed that F was C^∞; here we will relax this condition and assume that F is only C^1. Recall that this means that F is continuously differentiable. That is, F and its first partial derivatives exist and are continuous functions on $\mathbb{R}^n$. For the first few sections of this chapter, we will deal only with autonomous equations; later we will assume that F depends on t as well as X.

As we know, a solution of this system is a differentiable function $X: J \to \mathbb{R}^n$ defined on some interval $J \subset \mathbb{R}$ such that for all $t \in J$

$$X'(t) = F(X(t)).$$

Geometrically, $X(t)$ is a curve in $\mathbb{R}^n$ whose tangent vector $X'(t)$ equals $F(X(t))$; as in previous chapters, we think of this vector as being based at $X(t)$, so that the map $F: \mathbb{R}^n \to \mathbb{R}^n$ defines a vector field on $\mathbb{R}^n$. An *initial condition* or *initial value* for a solution $X: J \to \mathbb{R}^n$ is a specification of the form $X(t_0) = X_0$ where $t_0 \in J$ and $X_0 \in \mathbb{R}^n$. For simplicity, we usually take $t_0 = 0$.

A nonlinear differential equation may have several solutions that satisfy a given initial condition. For example, consider the first-order nonlinear differential equation

$$x' = 3x^{2/3}.$$

In Chapter 7 we saw that the identically zero function $u_0: \mathbb{R} \to \mathbb{R}$ given by $u_0(t) \equiv 0$ is a solution satisfying the initial condition $u(0) = 0$. But $u_1(t) = t^3$ is also a solution satisfying this initial condition, and, in addition, for any $\tau > 0$, the function given by

$$u_\tau(t) = \begin{cases} 0 & \text{if } t \le \tau \\ (t - \tau)^3 & \text{if } t > \tau \end{cases}$$

is also a solution satisfying the initial condition $u_\tau(0) = 0$.

Besides uniqueness, there is also the question of existence of solutions. When we dealt with linear systems, we were able to compute solutions explicitly. For nonlinear systems, this is often not possible, as we have seen. Moreover, certain initial conditions may not give rise to any solutions. For example, as we saw in Chapter 7, the differential equation

$$x' = \begin{cases} 1 & \text{if } x < 0 \\ -1 & \text{if } x \ge 0 \end{cases}$$

has no solution that satisfies $x(0) = 0$.

Thus it is clear that, to ensure existence and uniqueness of solutions, extra conditions must be imposed on the function F. The assumption that F is continuously differentiable turns out to be sufficient, as we shall see. In the first example above, F is not differentiable at the problematic point $x = 0$, while in the second example, F is not continuous at $x = 0$.

The following is the fundamental local theorem of ordinary differential equations.

The Existence and Uniqueness Theorem. *Consider the initial value problem*

$$X' = F(X), \quad X(0) = X_0$$

where $X_0 \in \mathbb{R}^n$. Suppose that $F \colon \mathbb{R}^n \to \mathbb{R}^n$ is C^1. Then there exists a unique solution of this initial value problem. More precisely, there exists $a > 0$ and a unique solution

$$X \colon (-a, a) \to \mathbb{R}^n$$

of this differential equation satisfying the initial condition

$$X(0) = X_0. \qquad ∎$$

We will prove this theorem in the next section.

17.2 Proof of Existence and Uniqueness

We need to recall some multivariable calculus. Let $F \colon \mathbb{R}^n \to \mathbb{R}^n$. In coordinates $(x_1, \ldots, x_n)$ on $\mathbb{R}^n$, we write

$$F(X) = (f_1(x_1, \ldots, x_n), \ldots, f_n(x_1, \ldots, x_n)).$$

Let DF_X be the derivative of F at the point $X \in \mathbb{R}^n$. We may view this derivative in two slightly different ways. From one point of view, DF_X is a linear map defined for each point $X \in \mathbb{R}^n$; this linear map assigns to each vector $U \in \mathbb{R}^n$ the vector

$$DF_X(U) = \lim_{h \to 0} \frac{F(X + hU) - F(X)}{h},$$

where $h \in \mathbb{R}$. Equivalently, from the matrix point of view, DF_X is the $n \times n$ Jacobian matrix

$$DF_X = \left(\frac{\partial f_i}{\partial x_j} \right)$$

where each derivative is evaluated at $(x_1, \ldots, x_n)$. Thus the derivative may be viewed as a function that associates different linear maps or matrices to each point in $\mathbb{R}^n$. That is, $DF \colon \mathbb{R}^n \to L(\mathbb{R}^n)$.

As earlier, the function F is said to be continuously differentiable, or C^1, if all of the partial derivatives of the f_j exist and are continuous. We will assume for the remainder of this chapter that F is C^1. For each $X \in \mathbb{R}^n$, we define the norm $|DF_X|$ of the Jacobian matrix DF_X by

$$|DF_X| = \sup_{|U|=1} |DF_X(U)|$$

where $U \in \mathbb{R}^n$. Note that $|DF_X|$ is not necessarily the magnitude of the largest eigenvalue of the Jacobian matrix at X.

Example. Suppose

$$DF_X = \begin{pmatrix} 2 & 0 \\ 0 & 1 \end{pmatrix}.$$

Then, indeed, $|DF_X| = 2$, and 2 is the largest eigenvalue of DF_X. However, if

$$DF_X = \begin{pmatrix} 1 & 1 \\ 0 & 1 \end{pmatrix},$$

then

$$|DF_X| = \sup_{0 \le \theta \le 2\pi} \left| \begin{pmatrix} 1 & 1 \\ 0 & 1 \end{pmatrix} \begin{pmatrix} \cos\theta \\ \sin\theta \end{pmatrix} \right|$$

$$= \sup_{0 \le \theta \le 2\pi} \sqrt{(\cos\theta + \sin\theta)^2 + \sin^2\theta}$$

$$= \sup_{0 \le \theta \le 2\pi} \sqrt{1 + 2\cos\theta\sin\theta + \sin^2\theta}$$

$$> 1$$

whereas 1 is the largest eigenvalue.   ∎

We do, however, have

$$|DF_X(V)| \le |DF_X||V|$$

for any vector $V \in \mathbb{R}^n$. Indeed, if we write $V = (V/|V|)\,|V|$, then we have

$$|DF_X(V)| = \left|DF_X\big(V/|V|\big)\right| |V| \le |DF_X||V|$$

since $V/|V|$ has magnitude 1. Moreover, the fact that $F: \mathbb{R}^n \to \mathbb{R}^n$ is C^1 implies that the function $\mathbb{R}^n \to L(\mathbb{R}^n)$, which sends $X \to DF_X$, is a continuous function.

Let $\mathcal{O} \subset \mathbb{R}^n$ be an open set. A function $F: \mathcal{O} \to \mathbb{R}^n$ is said to be *Lipschitz* on $\mathcal{O}$ if there exists a constant K such that

$$|F(Y) - F(X)| \leq K|Y - X|$$

for all $X, Y \in \mathcal{O}$. We call K a *Lipschitz constant* for F. More generally, we say that F is *locally Lipschitz* if each point in $\mathcal{O}$ has a neighborhood $\mathcal{O}'$ in $\mathcal{O}$ such that the restriction F to $\mathcal{O}'$ is Lipschitz. The Lipschitz constant of $F|\mathcal{O}'$ may vary with the neighborhoods $\mathcal{O}'$.

Another important notion is that of compactness. We say that a set $\mathcal{C} \subset \mathbb{R}^n$ is *compact* if $\mathcal{C}$ is closed and bounded. An important fact is that, if $f : \mathcal{C} \to \mathbb{R}$ is continuous and $\mathcal{C}$ is compact, then first of all f is bounded on $\mathcal{C}$ and, secondly, f actually attains its maximum on $\mathcal{C}$. See Exercise 13 at the end of this chapter.

Lemma. *Suppose that the function $F: \mathcal{O} \to \mathbb{R}^n$ is C^1. Then F is locally Lipschitz.*

Proof: Suppose that $F: \mathcal{O} \to \mathbb{R}^n$ is C^1 and let $X_0 \in \mathcal{O}$. Let $\epsilon > 0$ be so small that the closed ball $\mathcal{O}_\epsilon$ of radius ϵ about X_0 is contained in $\mathcal{O}$. Let K be an upper bound for $|DF_X|$ on $\mathcal{O}_\epsilon$; this bound exists because DF_X is continuous and $\mathcal{O}_\epsilon$ is compact. The set $\mathcal{O}_\epsilon$ is *convex*; that is, if $Y, Z \in \mathcal{O}_\epsilon$, then the straight-line segment connecting Y to Z is contained in $\mathcal{O}_\epsilon$. This straight line is given by $Y + sU \in \mathcal{O}_\epsilon$, where $U = Z - Y$ and $0 \leq s \leq 1$. Let $\psi(s) = F(Y + sU)$. Using the chain rule we find

$$\psi'(s) = DF_{Y+sU}(U).$$

Therefore

$$F(Z) - F(Y) = \psi(1) - \psi(0)$$

$$= \int_0^1 \psi'(s)\, ds$$

$$= \int_0^1 DF_{Y+sU}(U)\, ds.$$

Thus we have

$$|F(Z) - F(Y)| \leq \int_0^1 K|U|\, ds = K|Z - Y|. \qquad \blacksquare$$

The following remark is implicit in the proof of the lemma: If $\mathcal{O}$ is convex, and if $|DF_X| \leq K$ for all $X \in \mathcal{O}$, then K is a Lipschitz constant for $F \mid \mathcal{O}$.

Suppose that J is an open interval containing zero and $X: J \to \mathcal{O}$ satisfies

$$X'(t) = F(X(t))$$

with $X(0) = X_0$. Integrating, we have

$$X(t) = X_0 + \int_0^t F(X(s))\, ds.$$

This is the integral form of the differential equation $X' = F(X)$. Conversely, if $X: J \to \mathcal{O}$ satisfies this integral equation, then $X(0) = X_0$ and X satisfies $X' = F(X)$, as is seen by differentiation. Thus the integral and differential forms of this equation are equivalent as equations for $X: J \to \mathcal{O}$. To prove the existence of solutions, we will use the integral form of the differential equation.

We now proceed with the proof of existence. Here are our assumptions:

1. $\mathcal{O}_\rho$ is the closed ball of radius $\rho > 0$ centered at X_0.
2. There is a Lipschitz constant K for F on $\mathcal{O}_\rho$.
3. $|F(X)| \le M$ on $\mathcal{O}_\rho$.
4. Choose $a < \min\{\rho/M,\, 1/K\}$ and let $J = [-a, a]$.

We will first define a sequence of functions $U_0, U_1, \ldots$ from J to $\mathcal{O}_\rho$. Then we will prove that these functions converge uniformly to a function satisfying the differential equation. Later we will show that there are no other such solutions. The lemma that is used to obtain the convergence of the U_k is the following:

Lemma from Analysis. *Suppose $U_k: J \to \mathbb{R}^n$, $k = 0, 1, 2, \ldots$ is a sequence of continuous functions defined on a closed interval J that satisfy: Given $\epsilon > 0$, there is some $N > 0$ such that for every $p, q > N$*

$$\max_{t \in J} |U_p(t) - U_q(t)| < \epsilon.$$

Then there is a continuous function $U: J \to \mathbb{R}^n$ such that

$$\max_{t \in J} |U_k(t) - U(t)| \to 0 \quad \text{as } k \to \infty.$$

Moreover, for any t with $|t| \le a$,

$$\lim_{k \to \infty} \int_0^t U_k(s)\, ds = \int_0^t U(s)\, ds. \qquad \blacksquare$$

This type of convergence is called *uniform convergence* of the functions U_k. This lemma is proved in elementary analysis books and will not be proved here. See [38].

The sequence of functions U_k is defined recursively using an iteration scheme known as *Picard iteration*. We gave several illustrative examples of this iterative scheme back in Chapter 7. Let

$$U_0(t) \equiv X_0.$$

For $t \in J$ define

$$U_1(t) = X_0 + \int_0^t F(U_0(s))\, ds = X_0 + tF(X_0).$$

Since $|t| \leq a$ and $|F(X_0)| \leq M$, it follows that

$$|U_1(t) - X_0| = |t||F(X_0)| \leq aM \leq \rho$$

so that $U_1(t) \in \mathcal{O}_\rho$ for all $t \in J$. By induction, assume that $U_k(t)$ has been defined and that $|U_k(t) - X_0| \leq \rho$ for all $t \in J$. Then let

$$U_{k+1}(t) = X_0 + \int_0^t F(U_k(s))\, ds.$$

This makes sense since $U_k(s) \in \mathcal{O}_\rho$ so the integrand is defined. We show that $|U_{k+1}(t) - X_0| \leq \rho$ so that $U_{k+1}(t) \in \mathcal{O}_\rho$ for $t \in J$; this will imply that the sequence can be continued to U_{k+2}, U_{k+3}, and so on. This is shown as follows:

$$|U_{k+1}(t) - X_0| \leq \int_0^t |F(U_k(s))|\, ds$$

$$\leq \int_0^t M\, ds$$

$$\leq Ma < \rho.$$

Next, we prove that there is a constant $L \geq 0$ such that, for all $k \geq 0$,

$$\left| U_{k+1}(t) - U_k(t) \right| \leq (aK)^k L.$$

Let L be the maximum of $|U_1(t) - U_0(t)|$ over $-a \leq t \leq a$. By the above, $L \leq aM$. We have

$$|U_2(t) - U_1(t)| = \left| \int_0^t F(U_1(s)) - F(U_0(s))\, ds \right|$$

$$\leq \int_0^t K|U_1(s) - U_0(s)|\, ds$$

$$\leq aKL.$$

Assuming by induction that, for some $k \geq 2$, we have already proved

$$|U_k(t) - U_{k-1}(t)| \leq (aK)^{k-1}L$$

for $|t| \leq a$, we then have

$$|U_{k+1}(t) - U_k(t)| \leq \int_0^t |F(U_k(s)) - F(U_{k-1}(s))|\, ds$$

$$\leq K \int_0^t |U_k(s) - U_{k-1}(s)|\, ds$$

$$\leq (aK)(aK)^{k-1}L$$

$$= (aK)^k L.$$

Let $\alpha = aK$, so that $\alpha < 1$ by assumption. Given any $\epsilon > 0$, we may choose N large enough so that, for any $r > s > N$ we have

$$|U_r(t) - U_s(t)| \leq \sum_{k=N}^{\infty} |U_{k+1}(t) - U_k(t)|$$

$$\leq \sum_{k=N}^{\infty} \alpha^k L$$

$$\leq \epsilon$$

since the tail of the geometric series may be made as small as we please.

By the lemma from analysis, this shows that the sequence of functions $U_0, U_1, \ldots$ converges uniformly to a continuous function $X: J \to \mathbb{R}^n$. From the identity

$$U_{k+1}(t) = X_0 + \int_0^t F(U_k(s))\, ds,$$

we find by taking limits of both sides that

$$X(t) = X_0 + \lim_{k \to \infty} \int_0^t F(U_k(s))\, ds$$

$$= X_0 + \int_0^t \left(\lim_{k \to \infty} F(U_k(s)) \right) ds$$

$$= X_0 + \int_0^t F(X(s)) \, ds.$$

The second equality also follows from the Lemma from analysis. Therefore $X: J \to \mathcal{O}_\rho$ satisfies the integral form of the differential equation and hence is a solution of the equation itself. In particular, it follows that $X: J \to \mathcal{O}_\rho$ is C^1.

This takes care of the existence part of the theorem. Now we turn to the uniqueness part.

Suppose that $X, Y: J \to \mathcal{O}$ are two solutions of the differential equation satisfying $X(0) = Y(0) = X_0$, where, as above, J is the closed interval $[-a, a]$. We will show that $X(t) = Y(t)$ for all $t \in J$. Let

$$Q = \max_{t \in J} |X(t) - Y(t)|.$$

This maximum is attained at some point $t_1 \in J$. Then

$$Q = |X(t_1) - Y(t_1)| = \left| \int_0^{t_1} (X'(s) - Y'(s)) \, ds \right|$$

$$\leq \int_0^{t_1} |F(X(s)) - F(Y(s))| \, ds$$

$$\leq \int_0^{t_1} K|X(s) - Y(s)| \, ds$$

$$\leq aKQ.$$

Since $aK < 1$, this is impossible unless $Q = 0$. Therefore

$$X(t) \equiv Y(t).$$

This completes the proof of the theorem. ∎

To summarize this result, we have shown: Given any ball $\mathcal{O}_\rho \subset \mathcal{O}$ of radius ρ about X_0 on which

1. $|F(X)| \leq M$;
2. F has Lipschitz constant K; and
3. $0 < a < \min\{\rho/M, 1/K\}$;

there is a unique solution $X: [-a, a] \to \mathcal{O}$ of the differential equation such that $X(0) = X_0$. In particular, this result holds if F is C^1 on $\mathcal{O}$.

Some remarks are in order. First note that two solution curves of $X' = F(X)$ cannot cross if F satisfies the hypotheses of the theorem. This is an immediate consequence of uniqueness but is worth emphasizing geometrically. Suppose $X: J \to \mathcal{O}$ and $Y: J_1 \to \mathcal{O}$ are two solutions of $X' = F(X)$ for which $X(t_1) = Y(t_2)$. If $t_1 = t_2$ we are done immediately by the theorem. If $t_1 \neq t_2$, then let $Y_1(t) = Y(t_2 - t_1 + t)$. Then Y_1 is also a solution of the system. Since $Y_1(t_1) = Y(t_2) = X(t_1)$, it follows that Y_1 and X agree near t_1 by the uniqueness statement of the theorem, and hence so do $X(t)$ and $Y(t)$.

We emphasize the point that if $Y(t)$ is a solution, then so too is $Y_1(t) = Y(t + t_1)$ for any constant t_1. In particular, if a solution curve $X: J \to \mathcal{O}$ of $X' = F(X)$ satisfies $X(t_1) = X(t_1 + w)$ for some t_1 and $w > 0$, then that solution curve must in fact be a periodic solution in the sense that $X(t + w) = X(t)$ for all t.

17.3 Continuous Dependence on Initial Conditions

For the existence and uniqueness theorem to be at all interesting in any physical or even mathematical sense, the result needs to be complemented by the property that the solution $X(t)$ depends continuously on the initial condition $X(0)$. The next theorem gives a precise statement of this property.

Theorem. *Let $\mathcal{O} \subset \mathbb{R}^n$ be open and suppose $F: \mathcal{O} \to \mathbb{R}^n$ has Lipschitz constant K. Let $Y(t)$ and $Z(t)$ be solutions of $X' = F(X)$ which remain in $\mathcal{O}$ and are defined on the interval $[t_0, t_1]$. Then, for all $t \in [t_0, t_1]$, we have*

$$|Y(t) - Z(t)| \leq |Y(t_0) - Z(t_0)| \exp(K(t - t_0)). \qquad \blacksquare$$

Note that this result says that, if the solutions $Y(t)$ and $Z(t)$ start out close together, then they remain close together for t near t_0. While these solutions may separate from each other, they do so no faster than exponentially. In particular, we have this corollary:

Corollary. (Continuous Dependence on Initial Conditions) *Let $\phi(t, X)$ be the flow of the system $X' = F(X)$ where F is C^1. Then ϕ is a continuous function of X.* $\qquad \blacksquare$

The proof depends on a famous inequality that we prove first.

Gronwall's Inequality. *Let* $u: [0, \alpha] \to \mathbb{R}$ *be continuous and nonnegative. Suppose* $C \geq 0$ *and* $K \geq 0$ *are such that*

$$u(t) \leq C + \int_0^t Ku(s)\, ds$$

for all $t \in [0, \alpha]$. *Then, for all* t *in this interval,*

$$u(t) \leq Ce^{Kt}.$$

Proof: Suppose first that $C > 0$. Let

$$U(t) = C + \int_0^t Ku(s)\, ds > 0.$$

Then $u(t) \leq U(t)$. Differentiating U, we find

$$U'(t) = Ku(t).$$

Therefore,

$$\frac{U'(t)}{U(t)} = \frac{Ku(t)}{U(t)} \leq K.$$

Hence

$$\frac{d}{dt}(\log U(t)) \leq K$$

so that

$$\log U(t) \leq \log U(0) + Kt$$

by integration. Since $U(0) = C$, we have by exponentiation

$$U(t) \leq Ce^{Kt},$$

and so

$$u(t) \leq Ce^{Kt}.$$

If $C = 0$, we may apply the above argument to a sequence of positive c_i that tends to 0 as $i \to \infty$. This proves Gronwall's inequality. ∎

Proof: We turn now to the proof of the theorem. Define

$$v(t) = |Y(t) - Z(t)|.$$

Since

$$Y(t) - Z(t) = Y(t_0) - Z(t_0) + \int_{t_0}^{t} \big(F(Y(s)) - F(Z(s))\big)\, ds,$$

we have

$$v(t) \leq v(t_0) + \int_{t_0}^{t} Kv(s)\, ds.$$

Now apply Gronwall's inequality to the function $u(t) = v(t + t_0)$ to get

$$u(t) = v(t + t_0) \leq v(t_0) + \int_{t_0}^{t+t_0} Kv(s)\, ds$$

$$= v(t_0) + \int_{0}^{t} Ku(\tau)\, d\tau$$

so $v(t + t_0) \leq v(t_0)\exp(Kt)$ or $v(t) \leq v(t_0)\exp(K(t - t_0))$, which is just the conclusion of the theorem. ∎

As we have seen, differential equations that arise in applications often depend on parameters. For example, the harmonic oscillator equations depend on the parameters b (the damping constant) and k (the spring constant); circuit equations depend on the resistance, capacitance, and inductance; and so forth. The natural question is how do solutions of these equations depend on these parameters? As in the previous case, solutions depend continuously on these parameters provided that the system depends on the parameters in a continuously differentiable fashion. We can see this easily by using a special little trick. Suppose the system

$$X' = F_a(X)$$

depends on the parameter a in a C^1 fashion. Let's consider an "artificially" augmented system of differential equations given by

$$x'_1 = f_1(x_1, \ldots, x_n, a)$$

$$\vdots$$

$$x'_n = f_n(x_1, \ldots, x_n, a)$$

$$a' = 0.$$

This is now an autonomous system of $n + 1$ differential equations. While this expansion of the system may seem trivial, we may now invoke the previous result about continuous dependence of solutions on initial conditions to verify that solutions of the original system depend continuously on a as well.

Theorem. (Continuous Dependence on Parameters) *Let $X' = F_a(X)$ be a system of differential equations for which F_a is continuously differentiable in both X and a. Then the flow of this system depends continuously on a as well as X.* ∎

17.4 Extending Solutions

Suppose we have two solutions $Y(t)$, $Z(t)$ of the differential equation $X' = F(X)$ where F is C^1. Suppose also that $Y(t)$ and $Z(t)$ satisfy $Y(t_0) = Z(t_0)$ and that both solutions are defined on an interval J about t_0. Now the existence and uniqueness theorem guarantees that $Y(t) = Z(t)$ for all t in an interval about t_0 which may *a priori* be smaller than J. However, this is not the case. To see this, suppose that J^* is the largest interval on which $Y(t) = Z(t)$. If $J^* \neq J$, there is an endpoint t_1 of J^* and $t_1 \in J$. By continuity, we have $Y(t_1) = Z(t_1)$. Now the uniqueness part of the theorem guarantees that, in fact, $Y(t)$ and $Z(t)$ agree on an interval containing t_1. This contradicts the assertion that J^* is the largest interval on which the two solutions agree.

Thus we can always assume that we have a unique solution defined on a maximal time domain. There is, however, no guarantee that a solution $X(t)$ can be defined for all time. For example, the differential equation

$$x' = 1 + x^2,$$

has as solutions the functions $x(t) = \tan(t - c)$ for any constant c. Such a function cannot be extended over an interval larger than

$$c - \frac{\pi}{2} < t < c + \frac{\pi}{2}$$

since $x(t) \to \pm\infty$ as $t \to c \pm \pi/2$.

Next, we investigate what happens to a solution as the limits of its domain are approached. We state the result only for the right-hand limit; the other case is similar.

Theorem. *Let $\mathcal{O} \subset \mathbb{R}^n$ be open, and let $F\colon \mathcal{O} \to \mathbb{R}^n$ be C^1. Let $Y(t)$ be a solution of $X' = F(X)$ defined on a maximal open interval $J = (\alpha, \beta) \subset \mathbb{R}$ with $\beta < \infty$. Then, given any compact set $C \subset \mathcal{O}$, there is some $t_0 \in (\alpha, \beta)$ with $Y(t_0) \notin C$.*

This theorem says that if a solution $Y(t)$ cannot be extended to a larger time interval, then this solution leaves any compact set in $\mathcal{O}$. This implies that, as $t \to \beta$, either $Y(t)$ accumulates on the boundary of $\mathcal{O}$ or else a subsequence $|Y(t_i)|$ tends to ∞ (or both).

Proof: Suppose $Y(t) \subset \mathcal{C}$ for all $t \in (\alpha, \beta)$. Since F is continuous and $\mathcal{C}$ is compact, there exists $M > 0$ such that $|F(X)| \leq M$ for all $X \in \mathcal{C}$.

Let $\gamma \in (\alpha, \beta)$. We claim that Y extends to a continuous function $Y: [\gamma, \beta] \to \mathcal{C}$. To see this, it suffices to prove that Y is uniformly continuous on J. For $t_0 < t_1 \in J$ we have

$$|Y(t_0) - Y(t_1)| = \left| \int_{t_0}^{t_1} Y'(s) \, ds \right|$$

$$\leq \int_{t_0}^{t_1} |F(Y(s))| \, ds$$

$$\leq (t_1 - t_0)M.$$

This proves uniform continuity on J. Hence we may define

$$Y(\beta) = \lim_{t \to \beta} Y(t).$$

We next claim that the extended curve $Y: [\gamma, \beta] \to \mathbb{R}^n$ is differentiable at β and is a solution of the differential equation. We have

$$Y(\beta) = Y(\gamma) + \lim_{t \to \beta} \int_{\gamma}^{t} Y'(s) \, ds$$

$$= Y(\gamma) + \lim_{t \to \beta} \int_{\gamma}^{t} F(Y(s)) \, ds$$

$$= Y(\gamma) + \int_{\gamma}^{\beta} F(Y(s)) \, ds.$$

where we have used uniform continuity of $F(Y(s))$. Therefore

$$Y(t) = Y(\gamma) + \int_{\gamma}^{t} F(Y(s)) \, ds$$

for all t between γ and β. Hence Y is differentiable at β, and, in fact, $Y'(\beta) = F(Y(\beta))$. Therefore Y is a solution on $[\gamma, \beta]$. Since there must then be a solution on an interval $[\beta, \delta)$ for some $\delta > \beta$, we can extend Y to the interval (α, δ). Hence (α, β) could not have been a maximal domain of a solution. This completes the proof of the theorem. ∎

The following important fact follows immediately from this theorem.

Corollary. *Let C be a compact subset of the open set $\mathcal{O} \subset \mathbb{R}^n$ and let $F: \mathcal{O} \to \mathbb{R}^n$ be C^1. Let $Y_0 \in C$ and suppose that every solution curve of the form $Y: [0, \beta] \to \mathcal{O}$ with $Y(0) = Y_0$ lies entirely in C. Then there is a solution $Y: [0, \infty) \to \mathcal{O}$ satisfying $Y(0) = Y_0$, and $Y(t) \in C$ for all $t \geq 0$, so this solution is defined for all (forward) time.* ∎

Given these results, we can now give a slightly stronger theorem on the continuity of solutions in terms of initial conditions than the result discussed in Section 17.3. In that section we assumed that both solutions were defined on the same interval. In the next theorem we drop this requirement. The theorem shows that solutions starting at nearby points are defined on the same closed interval and also remain close to each other on this interval.

Theorem. *Let $F: \mathcal{O} \to \mathbb{R}^n$ be C^1. Let $Y(t)$ be a solution of $X' = F(X)$ that is defined on the closed interval $[t_0, t_1]$, with $Y(t_0) = Y_0$. There is a neighborhood $U \subset \mathbb{R}^n$ of Y_0 and a constant K such that, if $Z_0 \in U$, then there is a unique solution $Z(t)$ also defined on $[t_0, t_1]$ with $Z(t_0) = Z_0$. Moreover Z satisfies*

$$|Y(t) - Z(t)| \leq K|Y_0 - Z_0| \exp(K(t - t_0))$$

for all $t \in [t_0, t_1]$. ∎

For the proof we will need the following lemma.

Lemma. *If $F: \mathcal{O} \to \mathbb{R}^n$ is locally Lipschitz and $C \subset \mathcal{O}$ is a compact set, then $F|C$ is Lipschitz.*

Proof: Suppose not. Then for every $k > 0$, no matter how large, we can find X and Y in C with

$$|F(X) - F(Y)| > k|X - Y|.$$

In particular, we can find X_n, Y_n such that

$$|F(X_n) - F(Y_n)| \geq n|X_n - Y_n| \quad \text{for } n = 1, 2, \ldots.$$

Since C is compact, we can choose convergent subsequences of the X_n and Y_n. Relabeling, we may assume $X_n \to X^*$ and $Y_n \to Y^*$ with X^* and Y^* in C. Note that we must have $X^* = Y^*$, since, for all n,

$$|X^* - Y^*| = \lim_{n \to \infty} |X_n - Y_n| \leq n^{-1}|F(X_n) - F(Y_n)| \leq n^{-1}2M,$$

where M is the maximum value of $|F(X)|$ on C. There is a neighborhood $\mathcal{O}_0$ of X^* on which $F|\mathcal{O}_0$ has Lipschitz constant K. Also there is an n_0 such that $X_n \in \mathcal{O}_0$ if $n \geq n_0$. Therefore, for $n \geq n_0$,

$$|F(X_n) - F(Y_n)| \leq K|X_n - Y_n|,$$

which contradicts the assertion above for $n > n_0$. This proves the lemma.   ∎

Proof: The proof of the theorem now goes as follows: By compactness of $[t_0, t_1]$, there exists $\epsilon > 0$ such that $X \in \mathcal{O}$ if $|X - Y(t)| \leq \epsilon$ for some $t \in [t_0, t_1]$. The set of all such points is a compact subset C of $\mathcal{O}$. The C^1 map F is locally Lipschitz, as we saw in Section 17.2. By the lemma, it follows that $F|C$ has a Lipschitz constant K.

Let $\delta > 0$ be so small that $\delta \leq \epsilon$ and $\delta \exp(K|t_1 - t_0|) \leq \epsilon$. We claim that if $|Z_0 - Y_0| < \delta$, then there is a unique solution through Z_0 defined on all of $[t_0, t_1]$. First of all, $Z_0 \in \mathcal{O}$ since $|Z_0 - Y(t_0)| < \epsilon$, so there is a solution $Z(t)$ through Z_0 on a maximal interval $[t_0, \beta)$. We claim that $\beta > t_1$. For suppose $\beta \leq t_1$. Then, by Gronwall's inequality, for all $t \in [t_0, \beta)$, we have

$$|Z(t) - Y(t)| \leq |Z_0 - Y_0| \exp(K|t - t_0|)$$
$$\leq \delta \exp(K|t - t_0|)$$
$$\leq \epsilon.$$

Thus $Z(t)$ lies in the compact set C. By the results above, $[t_0, \beta)$ could not be a *maximal* solution domain. Therefore $Z(t)$ is defined on $[t_0, t_1]$. The uniqueness of $Z(t)$ then follows immediately. This completes the proof.   ∎

17.5 Nonautonomous Systems

We turn our attention briefly in this section to nonautonomous differential equations. Even though our main emphasis in this book has been on autonomous equations, the theory of nonautonomous (linear) equations is needed as a technical device for establishing the differentiability of autonomous flows.

Let $\mathcal{O} \subset \mathbb{R} \times \mathbb{R}^n$ be an open set, and let $F: \mathcal{O} \to \mathbb{R}^n$ be a function that is C^1 in X but perhaps only continuous in t. Let $(t_0, X_0) \in \mathcal{O}$. Consider the nonautonomous differential equation

$$X'(t) = F(t, X), \quad X(t_0) = X_0.$$

As usual, a solution of this system is a differentiable curve $X(t)$ in $\mathbb{R}^n$ defined for t in some interval J having the following properties:

1. $t_0 \in J$ and $X(t_0) = X_0$,
2. $(t, X(t)) \in \mathcal{O}$ and $X'(t) = F(t, X(t))$ for all $t \in J$.

The fundamental local theorem for nonautonomous equations is as follows:

Theorem. *Let $\mathcal{O} \subset \mathbb{R} \times \mathbb{R}^n$ be open and $F \colon \mathcal{O} \to \mathbb{R}^n$ a function that is C^1 in X and continuous in t. If $(t_0, X_0) \in \mathcal{O}$, there is an open interval J containing t and a unique solution of $X' = F(t, X)$ defined on J and satisfying $X(t_0) = X_0$.* ■

The proof is the same as that of the fundamental theorem for autonomous equations (Section 17.2), the extra variable t being inserted where appropriate. An important corollary of this result follows:

Corollary. *Let $A(t)$ be a continuous family of $n \times n$ matrices. Let $(t_0, X_0) \in J \times \mathbb{R}^n$. Then the initial value problem*

$$X' = A(t)X, \quad X(t_0) = X_0$$

has a unique solution on all of J. ■

For the proof, see Exercise 14 at the end of the chapter.

We call the function $F(t, X)$ *Lipschitz in X* if there is a constant $K \geq 0$ such that

$$|F(t, X_1) - F(t, X_2)| \leq K|X_1 - X_2|$$

for all (t, X_1) and (t, X_2) in $\mathcal{O}$. Locally Lipschitz in X is defined analogously.

As in the autonomous case, solutions of nonautonomous equations are continuous with respect to initial conditions if $F(t, X)$ is locally Lipschitz in X. We leave the precise formulation and proof of this fact to the reader.

A different kind of continuity is continuity of solutions as functions of the *data* $F(t, X)$. That is, if $F \colon \mathcal{O} \to \mathbb{R}^n$ and $G \colon \mathcal{O} \to \mathbb{R}^n$ are both C^1 in X, and $|F - G|$ is uniformly small, we expect solutions to $X' = F(t, X)$ and $Y' = G(t, Y)$, having the same initial values, to be close. This is true; in fact, we have the following more precise result:

Theorem. *Let $\mathcal{O} \subset \mathbb{R} \times \mathbb{R}^n$ be an open set containing $(0, X_0)$ and suppose that $F, G \colon \mathcal{O} \to \mathbb{R}^n$ are C^1 in X and continuous in t. Suppose also that for all $(t, X) \in \mathcal{O}$,*

$$|F(t, X) - G(t, X)| < \epsilon.$$

Let K be a Lipschitz constant in X for $F(t, X)$. If $X(t)$ and $Y(t)$ are solutions of the equations $X' = F(t, X)$ and $Y' = G(t, Y)$, respectively, on some interval J, and $X(0) = X_0 = Y(0)$, then

$$|X(t) - Y(t)| \leq \frac{\epsilon}{K}\left(\exp(K|t|) - 1\right)$$

for all $t \in J$.

Proof: For $t \in J$ we have

$$X(t) - Y(t) = \int_0^t (X'(s) - Y'(s))\, ds$$

$$= \int_0^t (F(s, X(s)) - G(s, Y(s)))\, ds.$$

Hence

$$|X(t) - Y(t)| \leq \int_0^t |F(s, X(s)) - F(s, Y(s))|\, ds$$

$$+ \int_0^t |F(s, Y(s)) - G(s, Y(s))|\, ds$$

$$\leq \int_0^t K|X(s) - Y(s)|\, ds + \int_0^t \epsilon\, ds.$$

Let $u(t) = |X(t) - Y(t)|$. Then

$$u(t) \leq K \int_0^t \left(u(s) + \frac{\epsilon}{K}\right) ds,$$

so that

$$u(t) + \frac{\epsilon}{K} \leq \frac{\epsilon}{K} + K \int_0^t \left(u(s) + \frac{\epsilon}{K}\right) ds.$$

It follows from Gronwall's inequality that

$$u(t) + \frac{\epsilon}{K} \leq \frac{\epsilon}{K} \exp\left(K|t|\right),$$

which yields the theorem. ∎

17.6 Differentiability of the Flow

Now we return to the case of an autonomous differential equation $X' = F(X)$ where F is assumed to be C^1. Our aim is to show that the flow $\phi(t, X) = \phi_t(X)$

determined by this equation is a C^1 function of the two variables, and to identify $\partial\phi/\partial X$. We know, of course, that ϕ is continuously differentiable in the variable t, so it suffices to prove differentiability in X.

Toward that end let $X(t)$ be a particular solution of the system defined for t in a closed interval J about 0. Suppose $X(0) = X_0$. For each $t \in J$ let

$$A(t) = DF_{X(t)}.$$

That is, $A(t)$ denotes the Jacobian matrix of F at the point $X(t)$. Since F is C^1, $A(t)$ is continuous. We define the nonautonomous linear equation

$$U' = A(t)U.$$

This equation is known as the *variational equation* along the solution $X(t)$. From the previous section we know that the variational equation has a solution on all of J for every initial condition $U(0) = U_0$. Also, as in the autonomous case, solutions of this system satisfy the linearity principle.

The significance of this equation is that, if U_0 is small, then the function

$$t \rightarrow X(t) + U(t)$$

is a good approximation to the solution $X(t)$ of the original autonomous equation with initial value $X(0) = X_0 + U_0$.

To make this precise, suppose that $U(t,\xi)$ is the solution of the variational equation that satisfies $U(0,\xi) = \xi$ where $\xi \in \mathbb{R}^n$. If ξ and $X_0 + \xi$ belong to $\mathcal{O}$, let $Y(t,\xi)$ be the solution of the autonomous equation $X' = F(X)$ that satisfies $Y(0) = X_0 + \xi$.

Proposition. *Let J be the closed interval containing 0 on which $X(t)$ is defined. Then*

$$\lim_{\xi \to 0} \frac{|Y(t,\xi) - X(t) - U(t,\xi)|}{|\xi|}$$

converges to 0 uniformly for $t \in J$. ∎

This means that for every $\epsilon > 0$, there exists $\delta > 0$ such that if $|\xi| \leq \delta$, then

$$|Y(t,\xi) - (X(t) + U(t,\xi))| \leq \epsilon|\xi|$$

for all $t \in J$. Thus as $\xi \to 0$, the curve $t \rightarrow X(t) + U(t,\xi)$ is a better and better approximation to $Y(t,\xi)$. In many applications $X(t) + U(t,\xi)$ is used in place of $Y(t,\xi)$; this is convenient because $U(t,\xi)$ is linear in ξ.

We will prove the proposition momentarily, but first we use this result to prove the following theorem:

Theorem. (Smoothness of Flows) *The flow* $\phi(t, X)$ *of the autonomous system* $X' = F(X)$ *is a* C^1 *function; that is,* $\partial\phi/\partial t$ *and* $\partial\phi/\partial X$ *exist and are continuous in* t *and* X.

Proof: Of course, $\partial\phi(t, X)/\partial t$ is just $F(\phi_t(X))$, which is continuous. To compute $\partial\phi/\partial X$ we have, for small ξ,

$$\phi(t, X_0 + \xi) - \phi(t, X_0) = Y(t, \xi) - X(t).$$

The proposition now implies that $\partial\phi(t, X_0)/\partial X$ is the linear map $\xi \to U(t, \xi)$. The continuity of $\partial\phi/\partial X$ is then a consequence of the continuity in initial conditions and data of solutions for the variational equation. ∎

Denoting the flow again by $\phi_t(X)$, we note that for each t the derivative $D\phi_t(X)$ of the map ϕ_t at $X \in \mathcal{O}$ is the same as $\partial\phi(t, X)/\partial X$. We call this the *space derivative* of the flow, as opposed to the *time derivative* $\partial\phi(t, X)/\partial t$.

The proof of the preceding theorem actually shows that $D\phi_t(X)$ is the solution of an initial value problem in the space of linear maps on $\mathbb{R}^n$: for each $X_0 \in \mathcal{O}$ the space derivative of the flow satisfies the differential equation

$$\frac{d}{dt}(D\phi_t(X_0)) = DF_{\phi_t(X_0)}D\phi_t(X_0),$$

with the initial condition $D\phi_0(X_0) = I$. Here we may regard X_0 as a parameter.

An important special case is that of an equilibrium solution $\bar{X}$ so that $\phi_t(\bar{X}) \equiv \bar{X}$. Putting $DF_{\bar{X}} = A$, we get the differential equation

$$\frac{d}{dt}(D\phi_t(\bar{X})) = AD\phi_t(\bar{X}),$$

with $D\phi_0(\bar{X}) = I$. The solution of this equation is

$$D\phi_t(\bar{X}) = \exp tA.$$

This means that, in a neighborhood of an equilibrium point, the flow is approximately linear.

We now prove the proposition. The integral equations satisfied by $X(t)$, $Y(t, \xi)$, and $U(t, \xi)$ are

$$X(t) = X_0 + \int_0^t F(X(s))\, ds,$$

$$Y(t,\xi) = X_0 + \xi + \int_0^t F(Y(s,\xi))\,ds,$$

$$U(t,\xi) = \xi + \int_0^t DF_{X(s)}(U(s,\xi))\,ds.$$

From these we get, for $t \geq 0$,

$$|Y(t,\xi) - X(t) - U(t,\xi)| \leq \int_0^t |F(Y(s,\xi)) - F(X(s)) - DF_{X(s)}(U(s,\xi))|\,ds.$$

The Taylor approximation of F at a point Z says

$$F(Y) = F(Z) + DF_Z(Y - Z) + R(Z, Y - Z),$$

where

$$\lim_{Y \to Z} \frac{R(Z, Y - Z)}{|Y - Z|} = 0$$

uniformly in Z for Z in a given compact set. We apply this to $Y = Y(s,\xi)$, $Z = X(s)$. From the linearity of $DF_{X(s)}$ we get

$$|Y(t,\xi) - X(t) - U(t,\xi)| \leq \int_0^t |DF_{X(s)}(Y(s,\xi) - X(s) - U(s,\xi))|\,ds$$

$$+ \int_0^t |R(X(s), Y(s,\xi) - X(s))|\,ds.$$

Denote the left side of this expression by $g(t)$ and set

$$N = \max\{|DF_{X(s)}| \mid s \in J\}.$$

Then we have

$$g(t) \leq N \int_0^t g(s)\,ds + \int_0^t |R(X(s), Y(s,\xi) - X(s))|\,ds.$$

Fix $\epsilon > 0$ and pick $\delta_0 > 0$ so small that

$$|R(X(s), Y(s,\xi) - X(s))| \leq \epsilon |Y(s,\xi) - X(s)|$$

if $|Y(s,\xi) - X(s)| \leq \delta_0$ and $s \in J$.

From Section 17.3 there are constants $K \geq 0$ and $\delta_1 > 0$ such that

$$|Y(s,\xi) - X(s)| \leq |\xi| e^{Ks} \leq \delta_0$$

if $|\xi| \leq \delta_1$ and $s \in J$.

Assume now that $|\xi|\leq\delta_1$. From the previous equation, we find, for $t\in J$,

$$g(t)\leq N\int_0^t g(s)\,ds+\int_0^t \epsilon|\xi|e^{Ks}\,ds,$$

so that

$$g(t)\leq N\int_0^t g(s)\,ds+C\epsilon|\xi|$$

for some constant C depending only on K and the length of J. Applying Gronwall's inequality we obtain

$$g(t)\leq C\epsilon e^{Nt}|\xi|$$

if $t\in J$ and $|\xi|\leq\delta_1$. (Recall that δ_1 depends on ϵ.) Since ϵ is any positive number, this shows that $g(t)/|\xi|\to 0$ uniformly in $t\in J$, which proves the proposition.

EXERCISES

1. Write out the first few terms of the Picard iteration scheme for each of the following initial value problems. Where possible, use any method to find explicit solutions. Discuss the domain of the solution.

   (a) $x'=x-2; x(0)=1$
   (b) $x'=x^{4/3}; x(0)=0$
   (c) $x'=x^{4/3}; x(0)=1$
   (d) $x'=\cos x; x(0)=0$
   (e) $x'=1/2x; x(1)=1$

2. Let A be an $n\times n$ matrix. Show that the Picard method for solving $X'=AX, X(0)=X_0$ gives the solution $\exp(tA)X_0$.

3. Derive the Taylor series for $\cos t$ by applying the Picard method to the first-order system corresponding to the second-order initial value problem

$$x''=-x; \quad x(0)=1, \quad x'(0)=0.$$

4. For each of the following functions, find a Lipschitz constant on the region indicated, or prove there is none:

   (a) $f(x)=|x|, -\infty<x<\infty$
   (b) $f(x)=x^{1/3}, -1\leq x\leq 1$

(c) $f(x)=1/x, 1 \le x \le \infty$
(d) $f(x,y)=(x+2y,-y), (x,y) \in \mathbb{R}^2$
(e) $f(x,y)=\dfrac{xy}{1+x^2+y^2}, x^2+y^2 \le 4$

5. Consider the differential equation

$$x'=x^{1/3}.$$

   How many different solutions satisfy $x(0)=0$?

6. What can be said about solutions of the differential equation $x'=x/t$?

7. Define $f: \mathbb{R} \to \mathbb{R}$ by $f(x)=1$ if $x \le 1$; $f(x)=2$ if $x>1$. What can be said about solutions of $x'=f(x)$ satisfying $x(0)=1$, where the right-hand side of the differential equation is discontinuous? What happens if you have instead $f(x)=0$ if $x>1$?

8. Let $A(t)$ be a continuous family of $n \times n$ matrices and let $P(t)$ be the matrix solution to the initial value problem $P'=A(t)P, P(0)=P_0$. Show that

$$\det P(t) = (\det P_0) \exp \left(\int_0^t \mathrm{Tr}\, A(s)\, ds \right).$$

9. Suppose F is a gradient vector field. Show that $|DF_X|$ is the magnitude of the largest eigenvalue of DF_X. (*Hint:* DF_X is a symmetric matrix.)

10. Show that there is no solution to the second-order two-point boundary value problem

$$x''=-x, \quad x(0)=0, \quad x(\pi)=1.$$

11. What happens if you replace the differential equation in the previous exercise by $x''=-kx$ with $k>0$?

12. Prove the following general fact (see also Section 17.3): If $C \ge 0$ and $u, v: [0, \beta] \to \mathbb{R}$ are continuous and nonnegative, and

$$u(t) \le C + \int_0^t u(s)v(s)\, ds$$

   for all $t \in [0, \beta]$, then $u(t) \le Ce^{V(t)}$, where

$$V(t) = \int_0^t v(s)\, ds.$$

13. Suppose $C \subset \mathbb{R}^n$ is compact and $f: C \to \mathbb{R}$ is continuous. Prove that f is bounded on C and that f attains its maximum value at some point in C.

14. Let $A(t)$ be a continuous family of $n \times n$ matrices. Let $(t_0, X_0) \in J \times \mathbb{R}^n$. Then the initial value problem

$$X' = A(t)X, \quad X(t_0) = X_0$$

has a unique solution on all of J.

15. In a lengthy essay not to exceed 50 pages, describe the behavior of all solutions of the system $X' = 0$ where $X \in \mathbb{R}^n$. Ah, yes. Another free and final gift from the Math Department.

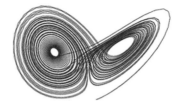

Bibliography

1. Abraham, R., and Marsden, J. *Foundations of Mechanics.* Reading, MA: Benjamin-Cummings, 1978.
2. Abraham, R., and Shaw, C. *Dynamics: The Geometry of Behavior.* Redwood City, CA: Addison-Wesley, 1992.
3. Alligood, K., Sauer, T., and Yorke, J. *Chaos: An Introduction to Dynamical Systems.* New York: Springer-Verlag, 1997.
4. Afraimovich, V. S., and Shil'nikov, L. P. Strange attractors and quasiattractors. In *Nonlinear Dynamics and Turbulence.* Boston: Pitman, (1983), 1.
5. Arnold, V. I. *Ordinary Differential Equations.* Cambridge: MIT Press, 1973.
6. Arnold, V. I. *Mathematical Methods of Classical Mechanics.* New York: Springer-Verlag, 1978.
7. Arrowsmith, D., and Place, C. *An Introduction to Dynamical Systems.* Cambridge: Cambridge University Press, 1990.
8. Banks, J. *et al.* On Devaney's definition of chaos. *Amer. Math. Monthly.* **99** (1992), 332.
9. Birman, J. S., and Williams, R. F. Knotted periodic orbits in dynamical systems I: Lorenz's equations. *Topology.* **22** (1983), 47.
10. Blanchard, P., Devaney, R. L., and Hall, G. R. *Differential Equations.* Pacific Grove, CA: Brooks-Cole, 2002.
11. Chua, L., Komuro, M., and Matsumoto, T. The double scroll family. *IEEE Trans. on Circuits and Systems.* **33** (1986), 1073.
12. Coddington, E., and Levinson, N. *Theory of Ordinary Equations.* New York: McGraw-Hill, 1955.
13. Devaney, R. L. *Introduction to Chaotic Dynamical Systems.* Boulder, CO: Westview Press, 1989.
14. Devaney, Kℓ. Math texts and digestion. *J. Obesity.* **23** (2002), 1.8.
15. Edelstein-Keshet, L. *Mathematical Models in Biology.* New York: McGraw-Hill, 1987.

16. Ermentrout, G. B., and Kopell, N. Oscillator death in systems of coupled neural oscillators. *SIAM J. Appl. Math.* **50** (1990), 125.
17. Field, R., and Burger, M., eds. *Oscillations and Traveling Waves in Chemical Systems.* New York: Wiley, 1985.
18. Fitzhugh, R. Impulses and physiological states in theoretical models of nerve membrane. *Biophys. J.* **1** (1961), 445.
19. Golubitsky, M., Josić, K., and Kaper, T. An unfolding theory approach to bursting in fast-slow systems. In *Global Theory of Dynamical Systems.* Bristol, UK: Institute of Physics, 2001, 277.
20. Guckenheimer, J., and Williams, R. F. Structural stability of Lorenz attractors. *Publ. Math. IHES.* **50** (1979), 59.
21. Guckenheimer, J., and Holmes, P. *Nonlinear Oscillations, Dynamical Systems, and Bifurcations of Vector Fields.* New York: Springer-Verlag, 1983.
22. Gutzwiller, M. The anisotropic Kepler problem in two dimensions. *J. Math. Phys.* **14** (1973), 139.
23. Hodgkin, A. L., and Huxley, A. F. A quantitative description of membrane current and its application to conduction and excitation in nerves. *J. Physiol.* **117** (1952), 500.
24. Katok, A., and Hasselblatt, B. *Introduction to the Modern Theory of Dynamical Systems.* Cambridge, UK: Cambridge University Press, 1995.
25. Khibnik, A., Roose, D., and Chua, L. On periodic orbits and homoclinic bifurcations in Chua's circuit with a smooth nonlinearity. *Int. J. Bifurcation and Chaos.* **3** (1993), 363.
26. Kraft, R. Chaos, Cantor sets, and hyperbolicity for the logistic maps. *Amer. Math Monthly.* **106** (1999), 400.
27. Lengyel, I., Rabai, G., and Epstein, I. Experimental and modeling study of oscillations in the chlorine dioxide–iodine–malonic acid reaction. *J. Amer. Chem. Soc.* **112** (1990), 9104.
28. Liapunov, A. M. *The General Problem of Stability of Motion.* London: Taylor & Francis, 1992.
29. Lorenz, E. Deterministic nonperiodic flow. *J. Atmos. Sci.* **20** (1963), 130.
30. Marsden, J. E., and McCracken, M. *The Hopf Bifurcation and Its Applications.* New York: Springer-Verlag, 1976.
31. May, R. M. *Theoretical Ecology: Principles and Applications.* Oxford: Blackwell, 1981.
32. McGehee, R. Triple collision in the collinear three body problem. *Inventiones Math.* **27** (1974), 191.
33. Moeckel, R. Chaotic dynamics near triple collision. *Arch. Rational Mech. Anal.* **107** (1989), 37.
34. Murray, J. D. *Mathematical Biology.* Berlin: Springer-Verlag, 1993.
35. Nagumo, J. S., Arimoto, S., and Yoshizawa, S. An active pulse transmission line stimulating nerve axon. *Proc. IRE.* **50** (1962), 2061.
36. Robinson, C. *Dynamical Systems: Stability, Symbolic Dynamics, and Chaos.* Boca Raton, FL: CRC Press, 1995.
37. Rössler, O. E. An equation for continuous chaos. *Phys. Lett. A* **57** (1976), 397.

38. Rudin, W. *Principles of Mathematical Analysis*. New York: McGraw-Hill, 1976.
39. Schneider, G., and Wayne, C. E. Kawahara dynamics in dispersive media. *Phys. D* **152** (2001), 384.
40. Shil'nikov, L. P. A case of the existence of a countable set of periodic motions. *Sov. Math. Dokl.* **6** (1965), 163.
41. Shil'nikov, L. P. Chua's circuit: Rigorous results and future problems. *Int. J. Bifurcation and Chaos.* **4** (1994), 489.
42. Siegel, C., and Moser, J. *Lectures on Celestial Mechanics*. Berlin: Springer-Verlag, 1971.
43. Smale, S. Diffeomorphisms with many periodic points. In *Differential and Combinatorial Topology*. Princeton, NJ: Princeton University Press, 1965, 63.
44. Sparrow, C. *The Lorenz Equations: Bifurcations, Chaos, and Strange Attractors*. New York: Springer-Verlag, 1982.
45. Strogatz, S. *Nonlinear Dynamics and Chaos*. Reading, MA: Addison-Wesley, 1994.
46. Tucker, W. The Lorenz attractor exists. *C. R. Acad. Sci. Paris Sér. I Math.* **328** (1999), 1197.
47. Winfree, A. T. The prehistory of the Belousov-Zhabotinsky reaction. *J. Chem. Educ.* **61** (1984), 661.

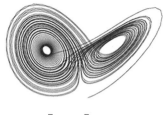

Index

Page numbers followed by "f" denote figures.

A

algebra
  equations, 26–29
  linear. *see* linear algebra
angular momentum
  conservation of, 283–284
  definition of, 283
anisotropic Kepler problem, 298–299
answers
  to all exercises, 1000.8, vii
areal velocity, 284
asymptotic stability, 175
asymptotically stable, 175
attracting fixed point, 329
attractor
  chaotic, 319–324
  description of, 310–311
  double scroll, 372–375
  Lorenz. *see* Lorenz attractor
  Rössler, 324–325
autonomous, 5, 22

B

backward asymptotic, 370
backward orbit, 320–321, 368
basic regions, 190
basin of attraction, 194, 200
basis, 90
Belousov-Zhabotinsky reaction, 231
bifurcation
  criterion, 333

definition of, 4, 8–9
discrete dynamical systems, 332–335
exchange, 334
heteroclinic, 192–194
homoclinic, 375–378
Hopf, 181–182, 270–271, 308
nonlinear systems, 176–182
period doubling, 334, 335f
pitchfork, 178–179, 179f
saddle-node, 177–178, 179f–180f, 181, 332
tangent, 332
bifurcation diagram, 8
biological applications
  competition and harvesting, 252–253
  competitive species, 246–252
  infectious diseases, 235–239
  predator/prey systems, 239–246
blowing up the singularity, 293–297
Bob. *See* Moe

C

canonical form, 49, 67, 84, 98, 111, 115
Cantor middle-thirds set, 349–352
capacitance, 260
carrying capacity, 335
Cauchy-Riemann equations, 184
center
  definition of, 44–47
  spiral, 112f
center of mass, 293
central force fields, 281–284
changing coordinates, 49–57